The Biochemistry of Plants

A COMPREHENSIVE TREATISE

Volume 9

P. K. Stumpf and E. E. Conn
EDITORS-IN-CHIEF

*Department of Biochemistry
and Biophysics
University of California
Davis, California*

THE BIOCHEMISTRY OF PLANTS

A COMPREHENSIVE TREATISE

Volume 9
Lipids: Structure and Function

P. K. Stumpf, editor

Department of Biochemistry and Biophysics
University of California
Davis, California

1987

ACADEMIC PRESS, INC.
Harcourt Brace Jovanovich, Publishers

Orlando   San Diego   New York   Austin
Boston   London   Sydney   Tokyo   Toronto

ACADEMIC PRESS, INC.
Orlando, Florida 32887

United Kingdom Edition published by
ACADEMIC PRESS INC. (LONDON) LTD.
24–28 Oval Road, London NW1 7DX

Library of Congress Cataloging in Publication Data
(Revised for vol. 9)
The Biochemistry of plants.

   Includes bibliographies and indexes.
   Contents: v. 1. The plant cell.—v. 2. Metabolism and
respiration.—[etc.]—v. 9. Lipids.
   1. Botanical chemistry—Collected works.   I. Stumpf,
P. K. (Paul Karl), Date   .   II. Conn, Eric E.
QK861.B48   581.19'2   80-13168
ISBN 0–12–675409–8 (v. 9: alk. paper)

PRINTED IN THE UNITED STATES OF AMERICA

87 88 89 90   9 8 7 6 5 4 3 2 1

Contents

List of Contributors

Numbers in parentheses indicate the pages on which the authors' contributions begin.

Roland Douce (215), Laboratoire de Physiologie Cellulaire Végétale, Département de Recherche Fondamentale, Unité Associée au CNRS no. 576, Centre d'Etudes Nucléaires et Université Scientifique, Technologique, et Médicale de Grenoble 85 X, F38041 Grenoble Cédex, France

Anthony H. C. Huang (91), Biology Department, University of South Carolina, Columbia, South Carolina 29208

Jan G. Jaworski (159), Department of Chemistry, Miami University, Oxford, Ohio 45056

Jacques Joyard (215), Laboratoire de Physiologie Cellulaire Végétale, Département de Recherche Fondamentale, Unité Associée au CNRS no. 576, Centre d'Etudes Nucléaires et Université Scientifique, Technologique, et Médicale de Grenoble 85 X, F38041 Grenoble Cédex, France

Helmut Kindl (31), Institut für Biochemie, Philipps-Universität Marburg, D-3550 Marburg, Federal Republic of Germany

Kathryn F. Kleppinger-Sparace (275), ARCO Plant Cell Research Institute, Dublin, California 94568

P. E. Kolattukudy (291), Biotechnology Center, Ohio State University, Columbus, Ohio 43210

J. Brian Mudd (275), ARCO Plant Cell Research Institute, Dublin, California 94568

N. Murata (315), Department of Regulation Biology, National Institute for Basic Biology, Okazaki 444, Japan

I. Nishida (315), Department of Regulation Biology, National Institute for Basic Biology, Okazaki 444, Japan

John B. Ohlrogge[1] (137), Northern Regional Research Center, Agricultural Research Service, U.S. Department of Agriculture, Peoria, Illinois 61604

Michael R. Pollard[2] (1), ARCO Plant Cell Research Institute, Dublin, California 94568

Allan Keith Stobart (175), Department of Botany, University of Bristol, Bristol BS8 1UG, England

P. K. Stumpf (121), Department of Biochemistry and Biophysics, University of California, Davis, California 95616

Sten Stymne (175), Department of Plant Physiology, Swedish University of Agricultural Sciences, S-750 07 Uppsala, Sweden

Brady A. Vick (53), Metabolism and Radiation Research Laboratory, Agricultural Research Service, U.S. Department of Agriculture, State University Station, Fargo, North Dakota 58105

Don C. Zimmerman (53), Metabolism and Radiation Research Laboratory, Agricultural Research Service, U.S. Department of Agriculture, State University Station, Fargo, North Dakota 58105

[1]Present address: Department of Botany and Plant Pathology, Michigan State University, East Lansing, Michigan 48824.

[2]Present address: Calgene, Inc., Davis, California 95616.

General Preface

In 1950, a new book entitled "Plant Biochemistry" was authored by James Bonner and published by Academic Press. It contained 490 pages, and much of the information described therein referred to animal or bacterial systems. This book had two subsequent editions, in 1965 and 1976.

In 1980, our eight-volume series entitled "The Biochemistry of Plants: A Comprehensive Treatise" was published by Academic Press; this multivolume, multiauthored treatise contained 4670 pages.

Since 1980, the subject of plant biochemistry has expanded into a vigorous discipline that penetrates all aspects of agricultural research. Recently a large number of research-oriented companies have been formed to explore and exploit the discipline of plant biochemistry, and older established chemical companies have also become heavily involved in plant-oriented research. With this in mind, Academic Press and the editors-in-chief of the treatise felt it imperative to update these volumes. Rather than have each chapter completely rewritten, it was decided to employ the approach used so successfully by the editors of *Methods in Enzymology,* in which contributors are invited to update those areas of research that are most rapidly expanding. In this way, the 1980 treatise constitutes a set of eight volumes with much background information, while the new volumes both update subjects that are rapidly developing and discuss some wholly new areas. The editors-in-chief have therefore invited the editors of the 1980 volumes to proceed on the basis of this concept. As a result, new volumes are forthcoming on lipids; general metabolism, including respiration; carbohydrates; amino acids; molecular biology; and photosynthesis. Additional volumes will be added as the need arises.

Once again we thank our editorial colleagues for accepting the important task of selecting authors to update chapters for their volumes and bringing their volumes promptly to completion. And once again we thank Mrs. Billie Gabriel and Academic Press for their assistance in this project.

P. K. Stumpf
E. E. Conn

Preface to Volume 9

Since the publication in 1980 of "Lipids: Structure and Function," Volume 4 of "The Biochemistry of Plants," so much progress has been made that the original volume no longer serves as the source of current information on plant lipid biochemistry. As a result, this new volume is designed to update Volume 4; 12 chapters have been written to cover the impressive progress made since 1980 in each of the 12 areas of plant lipid research.

The development of recent physical techniques employed in lipid research is covered in Chapter 1, and the increased knowledge on ß-oxidation systems in higher plants and the oxidative modifications of unsaturated fatty acids are discussed in Chapters 2 and 3, respectively. An entire chapter is devoted to lipases (Chapter 4). The resolution of the enzymology of the plant fatty acid synthetases is covered in Chapter 5, and the remarkable progress in understanding the function of acyl carrier protein is discussed in Chapter 6. New information on desaturases is reviewed in Chapter 7. The biochemistry of the complex lipids such as triacylglycerols, galactolipids, and sulfolipids is covered in Chapters 8, 9, and 10, respectively. Chapter 11 brings together the new information about surface waxes and their participation in the defense mechanisms of plants, and finally Chapter 12 reviews the new developments in the lipids of the blue-green algae.

A number of chapters in Volume 4 were not revised and included in this new volume. These will be updated in future volumes.

Once again I extend my warm appreciation to all the contributors of this volume who kept to their deadlines and submitted chapters of high caliber. And once again I thank Mrs. Billie Gabriel for her usual excellent secretarial assistance.

P. K. Stumpf

Analysis and Structure Determination of Acyl Lipids

1

MICHAEL R. POLLARD

I. INTRODUCTION

It is a statement of the obvious that the development of plant lipid biochemistry depends on our knowledge of the structures of the lipid molecules and on the methods available to separate and quantitate them. It is the purpose of this chapter to foster an awareness of the chemists' tools to perform the tasks of structure determination and quantitative analysis. Historically, organic chemists isolated and identified the major components in fats and oils using methods such as distillation, fractional crystallization, and chemical degradations (Gunstone, 1978). Fats were assayed by such indices as saponification, iodine, and hydroxyl numbers (Rossell, 1986). The early pioneers did an excellent job, but only abundantly occurring natural products were accessible to these time-consuming methods. With the development of chromatography and spectroscopy it has become possible to detect and characterize more complex molecules in nature in ever-smaller amounts, and to

The Biochemistry of Plants, Vol. 9

1

increase the speed, sensitivity, and resolution of quantitative assays, thus increasing our knowledge of biochemical processes. This chapter discusses the analysis and structure determination of lipids, especially acyl lipids, with emphasis on the more recent advances. Where possible, examples are chosen from the plant literature. However, the purpose of this chapter is not to review all classes of plant lipids but to show the approaches and limitations of various methods to problems of a chemical nature.

There could be many reasons for initiating a lipid analysis and structure determination. Perhaps a taxonomic study is being undertaken, or a plant oil exhibits interesting bioactive properties, or a membrane fraction needs characterization, or a tracer study has revealed interesting intermediates. Whatever the reason, the first step will almost certainly involve extraction with organic solvents. The lipid mixture will then be separated into its individual components, probably by a combination of chromatographic methods. And finally, the compounds will need identification. Physicochemical methods will be used, especially mass spectroscopy (MS) and nuclear magnetic resonance (NMR) spectroscopy. With the lipid structure defined, the scientist can further develop methods of quantitative assay. It should be noted that "separation" and "identification" are interdependent processes. The chromatographic behavior of a molecule is a function of its molecular structure. Thus elution behavior will give clues to the structure of the molecule. Conversely, the definition of the structure or a partial structure of a molecule may suggest improved purification or assay techniques. At the end of the chapter a worked example of a simple isolation and structure determination recently performed in our laboratory will be described.

II. EXTRACTION

Extraction is the first step in purification. It is almost always accomplished by the use of organic solvents. Common methods include extraction with chloroform–methanol (Folch *et al.*, 1957; Bligh and Dyer, 1959) or hexane–isopropanol (Hara and Radin, 1978) followed by addition of a salt solution to result in partition between an aqueous and an organic phase. These may be regarded as general methods for the extraction of both neutral and polar lipids, but there are a variety of more specialized extraction methods. Steam distillation is used to produce the volatile essential oils, in which mono- and sesquiterpenes dominate, while continuous extraction with hot solvents is accomplished by soxhlet extraction. A very simple extraction method is the short-duration dip in organic solvents for epicuticular lipids. In some cases a separation of lipid classes can be accomplished by partition during an extraction procedure. A case in point is the separation of neutral and polar lipids from very polar lipids and acyl-CoAs from acyl-ACPs in the procedure devised by Mancha *et al.* (1975). Sometimes "extraction" can involve a chemical treatment. The depolymerization of cutin and suberin layers is required before the monomers are soluble in organic solvents (Kolattukudy, 1980).

There are several points to consider when designing a new extraction or utilizing an existing one. These include problems of incomplete extraction, the release of lipolytic enzymes, and oxidative and other chemical stabilities of the extracted compounds. Taking this list in order, an extraction method should always be standardized for exhaustive extraction of the lipids and, where possible, with regrinding of tissue. Losses in partition methods may include highly polar lipids which partition into the aqueous phase or to insoluble material at the interface. In this respect cereal grains require forcing conditions of extraction of certain lipids. Sphingolipids (Fujino and Ohnishi, 1976) and lysophospholipids (Fishwick and Wright, 1977) require direct extraction with water-saturated n-butanol. Also, acyl-CoAs are notorious for sticking to interfaces or binding to proteinaceous materials at the interface, so care in quantitation in tracer experiments is essential. Homogenization of plant tissues will release lipolytic enzymes. Even if the tissue is being homogenized in organic solvents, problems can arise. In particular, phospholipase D is active in organic solvents, and can give rise to phosphatidic acid. In the presence of alcohols a transphosphatidylase reaction also occurs, giving rise to phosphatidyl alcohols. A short heat treatment of the tissue is usually used to inactivate the phospholipase D and other lipolytic enzymes prior to homogenization of the tissue in organic solvent (Colborne and Laidman, 1975; Phillips and Privett, 1979). A greater problem can exist if the researcher wishes to isolate a particular membrane fraction from a tissue and then perform a membrane analysis on it. Moreau (1984) has documented the extreme variability of phospholipase activity in a range of plant tissues. In potato tubers, which have been most extensively characterized, there is extensive phospholipase and galactosidase activity (Galliard, 1970). However, even within the potato there are large varietal differences (Moreau, 1985). A final problem when considering extractions is that of chemical stability. Oxidation can be a problem, particularly with polyunsaturated fatty acids (Holman, 1967). Precautions to minimize oxidation include the use of peroxide-free solvents, the addition of antioxidants such as ethoxyquin or butylated hydroxytoluene (although the natural antioxidants in the extract will often be sufficient), shielding from strong light in order to prevent photooxidation and photoisomerization, and the use of nitrogen or argon to evaporate and degas solvents. Stability to acidic and basic conditions is also a prime consideration. At the end of a process of extraction, purification, and identification the researcher should always ask whether any compounds isolated could be artifacts arising from incorrect extraction and handling techniques.

III. CHROMATOGRAPHIC METHODS OF SEPARATION

Chromatography is the term used to describe the separation of components in a mixture based on sample partitioning between a mobile (gas or liquid) phase and a stationary (liquid or solid) phase. Its discovery is usually credited to Tswett (1906), who first described the separation of leaf pigments by passage of petroleum ether down a calcium carbonate column. The theoretical principles of chromatography were first described by Martin and Synge (1941). Since then, the practice of chromatography has expanded tremendously, to take a central position in separation sciences with techniques like thin-layer chromatography (TLC), column chromatography, gas–liquid chromatography (GLC), and most recently, high-performance liquid chromatography (HPLC). A description of modern chromatographic practices, including simple theoretical treatments, is given by Williard *et al.* (1981) and by Poole and Schuette (1984). In this section we will confine discussion to fatty acids and acyl lipids, but the principles are general. Separation and analysis can therefore be classified as the definition of lipid classes, of the acyl composition, and of the molecular species of lipids. A comprehensive compendium of the chromatography of lipids has been edited by Mangold (1984). Books by Christie (1982a) and Kates (1972) are excellent laboratory guides to the earlier, and still highly relevant techniques of lipid analysis.

A.   Gas–Liquid Chromatography

In GLC the mobile phase is an inert gas while the stationary phase is usually a liquid coating, either a capillary wall or an inert column packing. With a nonpolar stationary phase, separation is based largely on the difference in boiling points, while with increasingly polar stationary phases interactions between polarized (and polarizable) groups on the solute molecules and polar groups on the stationary phase become increasingly important. Thus more polar solute molecules are retained longer on the column. Detection and quantitation of the components eluting off the GC column can be by a variety of devices. These include thermal conductivity and electron-capture detectors, which are nondestructive, and flame-photometric, flame ionization, and alkali flame ionization (N,P) detectors. GC-MS, which will be considered in Section IV,A, can be used as a universal detector (total ion current monitoring) or as a tuneable, specific detector (selective-ion monitoring). The flame ionization detector is the workhorse detector for lipid analysis, as it is a nonspecific carbon detector. Its sensitivity can be 100 pg/sec, which means that 1–10 ng of sample is detectable. Generally, however, sample loadings of 0.01–0.1 μg (capillary column GC) or 1–10 μg (packed-column GC) per component are used. The sample is injected in a small volume of organic solvent. For a homologous series of compounds the flame ionization detector response is often proportional to percentage carbon, so that over a limited range within the series peak areas in the chromatogram approximate to masses. However, for precise quantitation and for comparisons between different classes of lipid molecules, calibration of the flame ionization detector response is required. The major developments in GC in the last decade have been in the automation of data processing, the widespread use of capillary columns, new stationary phases, and new derivatization methods. The latter three will be considered in relation to the analysis of fatty acids and lipids. The basic principles of GC have been described by McNair and Bonelli (1969).

With the commercial availability of fused-silica capillary columns, which are easier to handle than glass capillary columns, there has been a surge in the use of capillary GC. Reliable injection systems and stationary phases chemically bonded to the column wall are improvements also worth noting. The driving force behind the use of capillary GC is the much higher resolution obtainable. A 30-m capillary (or WCOT, wall-coated open tubular) column with an internal diameter of 0.25 mm and a stationary-phase thickness of 0.25 μm will have an efficiency of about 90,000 theoretical plates, compared with about 5000 theoretical plates for a 2 m × 2 mm packed column. The difference can be appreciated by comparing peak widths in Fig. 1A and B. Obviously components unresolved on a packed column can be resolved on a capillary column with a similar stationary phase.

A large number of stationary phases are available. The general polarity of these is defined by McReynolds constants (McReynolds, 1970). Earlier stationary phases such as Apiezon greases and polyesters are being replaced by polysiloxane and carborane phases which are capable of use at higher temperatures. Molecules of increasing polarity will be increasingly retained by stationary phases of increasing polarity. This is illustrated in Fig. 1 using fatty acid phenacyl esters derived from *Cuphea lutea* seed oil. The order of retention of the C_{18} fatty acids on the apolar SE-30 column is $18:2 < 18:1 < 18:0$, while on the highly polar SP-2330 column it is reversed because the polarizable double bonds interact with cyano groups of the SP-2330 phase. Often, a preliminary identification of a compound can be made by GLC on the basis of its retention time compared to standards. This identification will have a stronger basis if the retention time is identical with the authenticated standard on several GC columns with stationary phases of widely differing polarity. There is a large amount of information in the literature on the GC behavior of lipids, and particularly of fatty acids. Retention behavior of fatty acid esters is often described by equivalent chain length (ECL) values. Ackman (1969) and Jamieson (1970) review their use and the early literature on the GC behavior of fatty acid esters. Corrections to the basic assumption of linear proportionality between carbon number and the logarithm of retention time can be required (Nelson, 1974). Positional isomers elute at different retention times, facilitating their identification. This topic has been systematically explored in the laboratory of Gunstone (1976). Geometric isomers can also separated by GLC. For example, cis- and trans-unsaturated fatty acid esters can be separated directly, even with packed GC columns, using very high polarity stationary phases (Heckers *et al.*, 1977). Providing molecules are stable to elevated temperatures, GC analysis of higher MW neutral molecules is possible. Triacylglycerols can be determined on short packed columns. For example, Phillips *et al.* (1971) used a 60-cm OV-1 (apolar) column temperature programmed to 390°C to quantitate meadowfoam seed triacylglycerols, up to $C_{69}H_{122}O_6$ (MW = 1046). Triacylglycerols have also been separated on polar columns by degree of unsaturation (Takagi and Itabashi, 1977). Mono- and digalactosyl diacylglycerols can also be analyzed intact after derivatization (Williams *et al.*, 1975; Myher *et al.*, 1978). Capillary GC of high-MW neutral lipids has also been reported, such as for the determination of tocopherols and

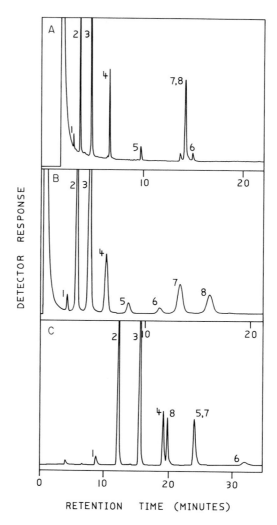

Fig. 1. Analysis of fatty acid esters by three chromatographic systems. Phenacyl esters of fatty acids from *Cuphea lutea* seeds are analyzed by (A) capillary GLC, using a fused-silica column (30 m × 0.25 mm i.d.) wall-coated with 0.25-μm SE-30 (nonpolar) stationary phase, by (B) packed-column GLC (2 m × 2 mm i.d.) containing 10% SP-2330 (highly polar) stationary phase on Supelcoport 100/120, and by (C) reversed-phase HPLC, using a 20% octadecyl column (25 cm × 4 mm, 5-μm particle size), and eluting with a linear gradient of 80% acetonitrile, 20% water to 100% acetonitrile over 10 min, followed by by isocratic elution with 100% acetonitrile. The capillary and packed-column GLC analyses used isothermal conditions of 250 and 270°C, respectively, with flame ionization detection. HPLC detection was by UV absorption at 254 nm. The individual fatty acid esters are (1) caprylate, (2) caprate, (3) laurate, (4) myristate, (5) palmitate, (6) stearate, (7) oleate, and (8) linoleate.

sterols (Slover *et al.*, 1983). A full analysis of the free fatty acid, mono-, di-, and triacylglycerol content of processed vegetable oils used a shortened capillary column and silylation (D'Alonzo *et al.*, 1982).

Derivatization methods, and particularly those that reduce polarity, are an important aspect of GC technique (Kates, 1972; Knapp, 1979; Blau and King, 1978; Christie, 1982a). Commonly encountered derivatizations are (1) strong acid, BF_3, or NaOMe-catalyzed transmethylations of acyl lipids to give methyl esters, (2) base-catalyzed hydrolysis of acyl lipids to give fatty acids, (3) esterification of carboxylic acids, and (4) silylation, acetylation, or methylation of hydroxyl groups. The use of quaternary ammonium salts for transmethylation has been described (McCreary *et al.*, 1978; Metcalfe and Wang, 1981), and a rapid NaOMe procedure has been described by Christie (1982b). To accomplish the transesterification of very lipophilic molecules, such as jojoba wax, more specialized conditions have been employed (Miwa, 1971). A cautionary note is required here, in that under certain conditions fatty acids with unusual functional groups may undergo reactions other than transesterification. For example, ring opening will occur for cyclopropenyl-, cyclopropyl- and 1,2-epoxy fatty acids when transesterification is accomplished in strong acids. Derivatization of free fatty acids for GC is most easily accomplished by the use of diazomethane, to produce methyl esters. There are many other methods in the literature.

The handling of shorter chain fatty acids and esters can create problems. Short-chain esters may be lost due to their volatility, especially when solutions have to be concentrated, while very short-chain fatty acids can partition appreciably into the aqueous phase in extractions. To increase the size of the alkyl group, as in butyl esters prepared by transesterification in butanol or by reaction of fatty acids with dimethylformamide dibutyl acetals (Thenot and Horning, 1972), is one answer. A second approach is reaction GC, as described by Johnson (1976) or by Richards *et al.* (1975). A third approach is direct GC analysis of free fatty acids. A fourth approach, which we have used to demonstrate malonyl decarboxylation to acetate and for short chain fatty acid analysis in *Cuphea* seeds, is to concentrate acids as their quaternary ammonium salts and then prepare phenacyl esters.

Another technique that the reader should be aware of is the use of halogenated alcohols or phenols for fatty acid derivatization. These derivatives, such as pentafluorobenzyl esters (Pempkowiak, 1983), give greatly enhanced sensitivity of detection when an electron-capture detector is used. And finally, Barron and Mooney (1968) have described a reductive technique for the GC determination of the acyl portion of acyl thioesters, but this method was later shown to have insufficient specificity (Nichols and Safford, 1973).

GLC is not only a tool for mass analysis. It can be adapted to become a preparative method. Milligram amounts of highly purified compounds can be trapped from packed columns if a stream splitter is placed before the flame ionization detector, or if a thermal conductivity detector is used instead of a flame ionization detector. This method of purification is likely to become less attractive with the advent of HPLC. A second modification is the addition of a radioactivity detector (Roberts, 1978). Radio-GC, as it is known, has been a standard technique of the lipid biochemist for many years. Gross *et al.* (1980) have described a system for simultaneous flame ionization and ^{14}C detection for capillary columns.

An interesting development in chromatography, included here for completeness, is supercritical fluid chromatography (SFC). It employs a supercritical gas such as carbon dioxide, pentane, or nitrous oxide as the mobile phase. In many ways it can be regarded as a hybrid between GLC and HPLC. GC capillary columns are used for separations, and detection is by flame ionization or fluorescence detectors, or by interfacing with a mass spectrometer. In GC elution is accomplished with a temperature gradient, while in SFC a pressure gradient is employed. In particular, SFC may be useful for high-resolution analysis of mixtures of MW up to 2000 that are too unstable to run by conventional GC. The subject has recently been reviewed by Fjeldsted and Lee (1984).

A review of recent developments in GC by Karasek *et al.* (1984) contains reference to such topics as headspace analysis, multidimensional GC, backflushing, GC-FT-IR and other detection systems, the use of multiple detectors, and theoretical and instrumental aspects. Recent reviews on packed and capillary column GC of lipids have been written by Hammond (1986) and Ackman (1986), respectively.

B. Thin-Layer and Liquid Chromatography

Liquid chromatography, as the name implies, is separation based on the passage of a mobile liquid phase over either a solid (for adsorption) or liquid (for partition) stationary phase. This section will cover thin-layer chromatography (TLC); column chromatography; reverse-phase, ion-exchange, and silver ion chromatography; high-performance liquid chromatography (HPLC); and droplet countercurrent chromatography (DCC). Liquid chromatography is particularly suited for separations by lipid class, with reverse-phase and silver ion methods used for separations of molecular species of lipids.

TLC is described by Stahl (1969) and by Skipski and Barclay (1969), with recent advances covered by Sherma and Fried (1984). The most commonly used stationary phase for lipid separations is silica gel. It is a relatively quick, simple, and inexpensive method which is still very widely used for neutral (Fig. 2) and polar lipid class separations. Unlike a column, a TLC plate has the advantage that the entire sample can be seen. Detection of mass can take several forms. Common, general, destructive techniques are charring with sulfuric acid or phosphomolybdic acid. A wide range of functional group-specific spray reagents are available (Christie, 1982a; Kates, 1972; Mangold, 1984; Skipski and Barclay, 1969). Common nondestructive techniques include staining with iodine, or the use of UV-sensitive reagents such as rhodamine or 2,7-dichlorofluorescein. Iodine is strongly adsorbed by unsaturated compounds, but quantitative recovery of compounds from the TLC plate after iodine staining cannot always be expected. Quantitation of mass can be achieved by densitometry (Bitman and Wood, 1981; Baron and Coburn, 1984). Another method is the use of TLC–flame ionization

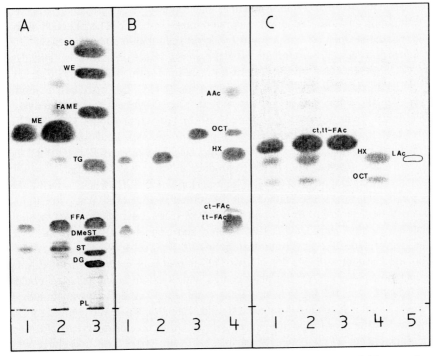

Fig. 2. TLC separations of fragrance compounds from ambrette seeds. A lipid extract from ambrette seeds was examined by silicic acid TLC using a two-thirds development with 70:30:1 (v/v/v) hexane–diethyl–ether:acetic acid followed by a full development with 95:5 (v/v) hexane–diethyl ether (A). The standards run in lane 3A are squalene (SQ), wax esters (WE), fatty acid methyl esters (FAME), triacylglycerols (TG), free fatty acids (FFA), 4,4-dimethylsterols (DMeST), sitosterol (ST), 1,2-diacylglycerols (DG), and phospholipids (PL). The monoester (ME) fraction from the ambrette extract, which was the fragrance fraction, was further characterized by silver nitrate–silicic acid TLC (B) and by reversed-phase TLC (C). The individual components are saturated alkyl acetates (AAc), 16-hexadecenolide (HX), 18-octadecenolide (OCT), 2-*cis*,6-*trans*-farnesyl acetate (ct-FAc) and 2-*trans*,6-*trans*-farnesyl acetate (tt-FAc). Silver ion TLC was achieved by spraying silicic acid TLC plates with a 10% aqueous silver nitrate solution, air-drying, and then activating the plates at 120°C for 30 min, followed by multiple development of the plates in 95:5 (v/v) toluene–ethyl acetate. Reversed-phase TLC used octadecyl coated plates developed halfway then fully with 99:1 (v/v) methanol–water. Detection in all cases was by spraying with ethanolic phosphomolybdic acid and charring at 120°C. In the case of lauryl acetate (LAc) used on the reverse phase plates, charring does not occur, but a lighter spot than background is observed.

detection (TLC-FID). The Iatroscan is a system in which TLC is carried out on silica gel fused to a quartz support rod and then the rod slowly passed through a flame ionization detector (Ackman, 1981). However, as pointed out by Crane *et al.* (1983), with complex relationships between detector response and such variables as sample loading, lipid class, and flame ionization detector scan rate, it is not possible to use the TLC-FID system for accurate mass quantitation without extensive standardization. Radioactivity can be monitored by radio-TLC scanners, which operate using an open counting chamber under a positive gas pressure. However, if maximum resolution is required, autoradiography is better. The radioactive bands can then be scraped off and assayed by liquid scintillation counting.

There is a copious literature on solvent systems for the separation of lipid classes by TLC. Some simple solvent systems for plant lipids using silica gel TLC are given by Wilson and Rinne (1974). One-dimensional TLC can also utilize multiple developments with several solvent systems, as described for the analysis of chloroplast lipids (Sparace and Mudd, 1982). Two-dimensional (2D) TLC will obviously increase the potential resolution of a complex lipid mixture (Christie, 1982a; Skipski and Barclay, 1969), while Duck-Chong and Baker (1983) have utilized plastic-backed TLC plates to produce a "three-way" system.

A recent development in TLC is high-performance TLC, where a well-defined 5- μm particle size is used rather than the normal average of 20 μm (with a greater range of particle sizes). Resolution is improved from about 500 to 5000 theoretical plates (Fenimore and Davis, 1981; Poole and Schuette, 1984). Representative high-performance thin-layer chromatograms for lipid separations are described by Yao and Rastetter (1985). Preparative TLC is routinely used for small-scale lipid preparations. A 20 × 20 cm plate coated with a 1-mm silica gel layer could be loaded with up to 100 mg of neutral lipid sample for a "crude" separation, although the maximum is much less for a neutral lipid separation where the components have similar R_f values, or for polar lipid separations. After nondestructive localization of the components on the plate, the desired bands are scraped off and eluted with a more polar solvent to recover the lipid. Recent modifications of preparative TLC include the use of wedge-shaped coatings (Felton, 1982) and centrifugal TLC (Stahl and Muller, 1982).

A powerful method of TLC separation first described by Morris (1966) is the inclusion of silver nitrate in silica gel layers. Because double bonds complex with silver ions, a separation based on the degree and the stereochemistry of unsaturation is possible. This has been extensively exploited for fatty acid ester separation, and also for intact lipids. Examples in the plant literature include the separation of meadowfoam triacylglycerols (Pollard and Stumpf, 1980) and monogalactosyl diacylglycerols (Williams et al., 1976). A problem with preparative silver nitrate TLC is the failure to recover unsaturated lipids quantitatively from TLC plates (Chen et al., 1976). Another powerful method developed early in the history of TLC was reversed-phase TLC. Plates were coated with viscous hydrocarbon oils. Today the use of chemically bonded, cross-linked, and capped reverse-phase plates, which are commercially available, has made reverse-phase TLC much easier and more reproducible. The octadecyl phase has become dominant, but many others, such as octyl, ethyl, phenyl, and cyanopropyl are available. Detection of mass on bonded reversed-phase TLC plates is more problematic than for conventional adsorption TLC, but it appears that phosphomolybdic acid and iodine represent the best destructive and nondestructive methods, respectively (Sherma and Bennett, 1983; Domnas et al., 1983). Experience in our laboratory would suggest that these methods are suitable only for molecules containing unsaturation (Fig. 2C). For fatty acid analysis the use of phenacyl esters is helpful (Gattavecchia *et al.*, 1983). We have routinely used fatty phenacyl esters on C_{18}-coated plates impregnated with zinc silicate phosphors, where under 254-nm UV light a loading of 5 μg can readily be detected as a dark spot against a bright- green background. On a reverse-phase TLC plate there are critical pairs of fatty acids, most notably palmitate and oleate. These can be separated on one TLC plate using a 2D method. Jee and Ritchie (1984) describe the use of multiple-zone TLC plates, with a C_{18} phase in one dimension and silver nitrate–silica gel in the other. A variation on this theme utilizes "home made" multiple-zone TLC plates (Svetashov and Zhukova, 1985). In this case the use of decane as the reversed phase avoids detection problems, as it can be evaporated off prior to detection. The use of cartridges for quick separations and sample concentrations is increasing. Rogers *et al.* (1984) have used an octadecyl cartridge to concentrate organics from aquatic plants, while Juaneda and Rocquelin (1985) used a silica gel cartridge for a rapid separation of neutral and polar lipids.

A description of HPLC follows on naturally from TLC. Johnson and Stevenson (1978) describe the basic principles. The theory of chromatography predicts that the decreased particle size of the stationary phase will improve resolution. A typical 25 cm × 4 mm HPLC column containing a 5-μm packing will have a resolution equivalent to about 25,000 theoretical plates. Another major difference with TLC is that HPLC is a closed system. The small particle size requires high pressures and hence a specialized pumping system to maintain an adequate flow of mobile phase through the column. A third difference is the continuous detection system. Early work on the HPLC of lipids often used refractive index detectors, but these are not very sensitive and require careful calibration and thermostatting. UV detection is the dominant form of detection in HPLC, but for most lipids the chromophore will be either the ester carbonyl or an isolated double bond, so absorption will occur at very short wavelengths, around 200 to 210 nm, and the extinction coefficient will be low. These limitations for most acyl lipids have hindered the development of HPLC analysis for plant lipids, but despite these disadvantages the method can be used with low-UV cut off solvents (190 nm). Derivatization with a strong UV chromophore or a fluorescent label is to be preferred. Fatty acids have been analyzed by a variety of such derivatives, including phenacyl esters (Borch, 1975; Jordi, 1978) (Fig. 1), naphthacyl esters (Wood and Lee, 1983), pentafluorobenzyl esters (Netting and Duffield, 1984), and isopropylidene hydrazides (Agrawal and Schulte, 1983). Such reagents are not developed for phospholipid and glycolipid derivatizations. Recent developments in HPLC detectors, along with other developments such as multidimensional liquid chromatography and microbore columns are given in the review of Majors *et al.* (1984). LC-MS interfaces are discussed in Section IV,A. The development of flame ionization detectors for lipid analysis by HPLC is under way (Phillips *et al.*, 1982, 1984), while commercial radioisotope flow monitors are available.

HPLC is suitable for separation of a very wide range of compounds. Unlike GLC, HPLC has no limitations of MW or thermal stability, and is suitable for nonpolar and highly polar lipids alike. A plant lipid extract will most likely be separated into lipid classes first, followed by separation of individual lipid classes into true molecular species, or into subgroups, such as by number of double bonds. Silica columns are still the basic method for lipid class separations. Most reports concern nonplant lipids (Chen and Kou, 1982; Alam *et al.*, 1982), but separations of plant phospho- and galactolipids are beginning to appear (Marion *et al.*, 1984; Demandre *et al.*, 1985). A key use for HPLC is likely to be in separations of molecular species. This has been achieved for seed oil triacylglycerols using silver nitrate-impregnated silica (Smith *et al.*, 1980), and by reverse phase. Plattner (1981) reviews the latter topic and describes the use of equivalent carbon numbers (ECNs) and the problem of overlapping molecular species. The use of both silver ion and reverse-phase fractionation techniques to resolve and analyze cottonseed triacylglycerols has been reported (Bezard and Ouedraogo, 1980). However, structural assignments will require more information than can be gleaned from retention data alone. The separation of molecular species of plant polar lipids, either intact or after specific digestion, to leave the 1,2-diacylglycerol moiety intact, is an area where increased emphasis can be expected. Kesselmeier and Heinz (1985) have reported C_6 reverse-phase HPLC separation of chloroplast lipids (mono- and digalactosyl-, and sulfoquinovosyl-diacylglycerols, and phosphatidylglycerols). They point out critical pairs, and state that a reverse-phase separation of a complex lipid mixture is unlikely. A lipid class separation is needed first. Furthermore, they show that for UV detection at 200 nm for chloroplast lipids the detector response is a function of the number of double bonds. The use of *p*-anisoyl derivatives of 1,2-diacylglycerols derived from these lipids by phospholipase treatment (for phospholipids) or a periodic acid oxidation–1,1-dimethylhydrazine treatment for galactolipids (Heinze *et al.*, 1984), with UV monitoring at 250 nm, bypasses this problem. Demandre *et al.* (1985) have used silica absorption HPLC followed by C_{18} reverse-phase HPLC to fractionate galacto- and phospholipids into molecular species. A study by Crawford *et al.* (1980) showed the separation of soybean phosphatidylcholines by reverse-phase HPLC. Particularly interesting was their use of UV detection at 206 nm to monitor total phosphatidylcholine species, and at 234 nm to measure oxidized species by their conjugated diene chromophore. HPLC separation of acyl-CoAs has been achieved by C_{18} stationary phase with counter ion in the mobile phase (Baker and Schoolery, 1981).

Although the use of silica gel and octadecyl reverse phases have been dominant in liquid chromatography, other stationary phases have been used, including florisil (Radin, 1969), and anion exchangers such as di- and triethylaminoethyl celluloses (Rouser *et al.*, 1969). Alumina, being a highly basic chromatographic medium, is likely to cause hydrolysis or acyl rearrangement, and is not often used. Gel permeation chromatography has not unexpectedly had minimal use in lipid isolation, but it has been used as a method to separate acyl-ACPs from acyl-CoAs and free fatty acids in *in vitro* assays (Slabas *et al.*, 1982). The potential of HPLC for separation of enantiomers, either by use of chiral stationary phases or by inclusion of chiral ion-pairing agents in the mobile phase, has been realized by organic chemists, but apparently has yet to find application to plant lipids. And finally, the reader should be aware of the usefulness of droplet countercurrent chromatography as a method of separation, particularly of polar compounds (Hostettmann *et al.*, 1984).

IV. PHYSICAL AND CHEMICAL METHODS OF STRUCTURE DETERMINATION

For the chemist the two most powerful tools in determination of structure will most likely be NMR and MS. With the basic spectral data assembled [NMR, MS, UV, and IR (infrared)], the chemist will need to decide on what further physical and chemical methods are required, if any, for further determination of structure, and particularly stereochemistry. Once a structure has been determined, unequivocal proof is often considered to be either X-ray structure determination or chemical synthesis, both topics outside the scope of this review. The subject of organic spectroscopy is covered in many texts (Silverstein *et al.*, 1981; Dyke *et al.*, 1971; Gordon and Ford, 1972). The subject of fatty acid spectroscopy and chemistry has been reviewed by Gunstone (1967, 1975).

A. Mass Spectroscopy

Mass spectroscopy (MS) is in essence the analysis of the mass/charge ratios of ions produced from the sample molecules. In its simplest description it involves the processes of sample introduction, sample ionization, and mass analysis. A mass spectrum is a plot of the intensity of ions produced from a parent molecule, ions that will generally have a single charge (Fig. 3). The molecular ion will give the nominal MW of the compound, while a high-resolution mass determination will give a molecular formula. The fragmentation pattern will produce a fingerprint for the molecule and provide clues to its structure. MS has been extensively applied to the structure determination and analysis of lipids, and particularly of fatty acids. One of the great advantages of the technique is its high sensitivity. A mass spectrum can be obtained from microgram or nanogram quantities of sample. A comprehensive text on the subject, with a detailed description of many classes of organic compounds, is given by Waller (1972) and Waller and Dermer (1980).

Today the lipid chemist will have a large choice in the type of MS experiment he can perform. Sample introduction can be by direct-inlet probe, or by GC-MS or LC-MS interfaces. Sample ionization can occur in the gas phase by electron impact (EI) or chemical-ionization (CI) techniques, or by direct ionization from the condensed phase, often termed desorption ionization (Busch and Cooks, 1982). The more common techniques of desorption ionization include secondary-ion mass spectroscopy (SIMS), field desorption (FD), desorption–chemical ionization (DCI), and fast atom bombardment (FAB). These desorption ionization modes have more applicability to higher MW and polar molecules. In terms of mass analysis, instruments can be of the quadrupole or magnetic sector type. The magnetic sector

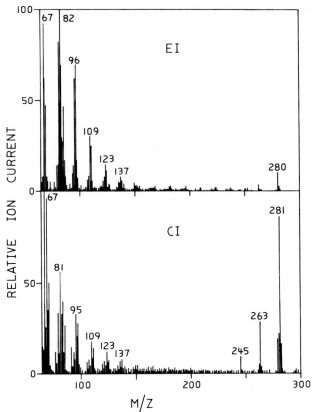

Fig. 3. Direct inlet probe mass spectra of 18-octadec-9-enolide run under electron impact (EI, 70 eV) and chemical-ionization (CI, isobutane as reagent gas) conditions. The upper, EI spectrum shows the low-intensity molecular-ion peak (M $^+$, m/e = 280) typical of the technique, whereas the lower, CI spectrum shows the typical high-intensity adduct molecular-ion peak [(M + H) $^+$, m/e = 281]. Successive loss of water molecules (m/e = 263.245) is also evident. Because of the simple structure of the molecule there are no diagnostic fragments other than the clusters separated by 14 amu typical of alkyl chain fragmentation.

instruments can have the capability for high-resolution MS, and, with the development of laminated magnets they are now capable of the rapid scan rates needed in capillary GC-MS. Recent technology has pushed the mass range of moderninstruments beyond 10^4 amu. Fourier transform–mass spectroscopy (FT-MS) is still at the development stage (Gross and Rempel, 1984). If these choices are not enough, + ve and − ve ion spectra can be selected, along with a variety of specialized techniques such as tandem mass spectroscopy (MS-MS) and selective-ion monitoring (mass chromatography). A detailed review of recent developments in MS is given by Burlingame *et al.* (1984).

The early work on the analysis of acyl lipids using MS is dominated by fatty acid analysis. McCloskey (1969, 1970) has reviewed the subject of fatty acid EI-MS.

Much emphasis has been placed on the use of MS to locate the position of a functional group along the hydrocarbon chain. Considering just the location of double bonds, a large number of derivatives have been tried. These include O-trimethylsilyl, O-boronate, O-isopropylidene, methoxy, and methoxyhalogeno derivatives. A comparison of the methods based on vicinal diol intermediates led Dommes *et al.* (1976) to suggest that O-trimethylsilyl ethers were the best for the analysis of polyunsaturated fatty acids. A recent method, using methoxybromo derivatives seems suitable for the analysis of conjugated and nonconjugated polyunsaturated fatty acids (Shantha and Kaimal, 1984). A second approach to functional group localization is the use of an appropriate head group on the otherwise underivatized molecule. Acyl pyrrolidides appear to be particularly useful in this respect (Andersson, 1978). An area where MS has been extensively used in structure determination is in the analysis of oxygenated fatty acids from cutin depolymerization (Holloway, 1982). The EI-MS method is limited by its gas-phase ionization, requiring molecules of reasonable volatility. However, it has been applied to neutral lipids quite successfully (Natale, 1977), and even to more polar molecules if they can be modified to increase their volatility. One of the weaknesses of EI-MS is that it often gives weak or absent molecular ion peaks (Fig. 3). Nevertheless, it remains a very important method, partly because the largest existing data bases are for EI (70 eV)-MS, facilitating identification by fingerprinting the fragmentation patterns.

The early 1970s saw the emergence of chemical ionization as the first of the "soft-ionization" techniques. Unlike EI, where extensive fragmentation of the molecule occurs after direct impact with high-energy electrons, CI spectra will have dominant molecular-ion peaks, with limited fragmentation (Fig. 3). Ionization is by an intermediate reagent gas, which can produce different molecular ion peaks depending on the reagent gas (e.g., MH^+ for isobutane, or MNH_4^+ for ammonia). Nitrous oxide is useful for producing $-$ve ion CI-MS in cases where the molecule can stabilize a $-$ve charge, such as in p-nitrophenyl esters. A description of CI-MS is given by Harrison (1983). CI-MS has also been extensively applied to the analysis of fatty acids and lipids (Murata, 1977; Agriga *et al.*, 1977; Games, 1978; Plattner and Spencer, 1983).

The latter part of the 1970s saw the introduction of desorption ionization methods. FD was the first of these, whereas FAB is probably now the most popular (Rinehart, 1982). Desorption ionization involves ionization of molecules in a condensed phase with subsequent direct desorption of ions into the gas phase. Unlike EI or CI modes, there is no neutral gas-phase intermediate. Less fragmentation occurs with desorption ionization modes, and there is the possibility of multiple charged species and of ion adducts. A crucial difference with gas-phase ionization is that neutral, volatile molecules are not required. In fact, ionic derivatives are favored. Thus, desorption ionization methods are ideally suited for higher MW, more polar compounds, although which specific desorption technique is best suited for which particular polar molecule is still an art. An interesting suggestion is the direct analysis of TLC spots by FAB-MS (Chang *et al.*, 1984). Examples of the use of desorption ionization methods on lipids include the ammonia

DCI-MS of triacylglycerols (Merritt *et al.*, 1982), of sucrose esters from tobacco leaves (Einolf and Chan, 1984), and of phosphatidylcholines (Crawford and Plattner, 1983), while Fujino *et al.* (1985) have reported the FD-MS of mono-, di-, and triglycosylceramides from wheat grain. These latter spectra show the typical ion adduct molecular-ion peaks (MH^+, MNa^+, and MK^+) common in desorption spectra.

Very important aspects of MS are the on-line techniques, particularly GC-MS and LC-MS. GC-MS is a mature technology: packed columns are usually interfaced through a jet separator, while capillary columns can be inserted directly into the ionization source of the spectrometer. GC-MS has been used extensively in the analysis of plant lipids and fatty acids. Acyl composition of lipids can be quantitated by GC-MS. An example of this is the determination of molecular species of diacylglycerols as their *t*-butyldimethylsilyl derivatives by EI-GC-MS (Myher *et al.*, 1978). Molecular weight was determined by a strong M-57 peak, while acyl composition is determined by M − RCOO fragments. A similar analysis has been applied to diacylglycerol species derived from chloroplast lipids (Siebertz *et al.*, 1979). There are reports of fragmentation patterns for stereospecific isomers of acyl lipids. Odham and Stenhagen (1972) show that the deacyloxymethylene fragment for EI-MS of triacylglycerols is stronger for the α-position, while Ohashi (1984) observed different intensities of deacyl and deacyloxy fragments for different pairs of phosphatidylcholine isomers by SIMS-MS. Kuksis *et al.* (1984, 1985) have reviewed their data on the mass spectra of 1,2-diacylglycerols. The rate of loss of fatty acids from the primary and secondary positions can vary, giving different intensity deacyloxy fragments $(M − RCOO)^+$ In some cases inspection of the 1,2-diacylglycerol mass spectra should indicate the major isomer. However, a general, quantitative method based on relative intensities of fragment ions to perform stereospecific analyses on natural samples of acylglycerolipids, without recourse to enzymatic digestion, seems unlikely. A variation of GC-MS is selective-ion monitoring, sometimes known as mass chromatography or mass fragmentography. The technique has been demonstrated in the analysis of molecular species of plant cerebrosides and ceramides (Ohnishi *et al.*, 1983; Fujino *et al.*, 1985). In a specific example, the mass spectrometer was used to monitor ions diagnostic of glucosyl, 4,8-sphingadiene, 4-hydroxy-8-sphinganine, 2-hydroxy-palmitate, 2-hydroxydocosanate, and 2-hydroxytetracosanate, from GC of tri-methylsilyl derivatives of the monoglucosylceramide fraction from spinach leaves (Fig. 4). Of the three major species detected by total-ion current, the coincidence of monitored ions showed that the molecular species were 1-*O*-glucosyl-*N*-2′-hydroxypalmityl-4,8-sphingadiene, 1-*O*-glucosyl-*N*-2′-hydroxydocosanoyl-4-hydroxy-8-sphinganine, and 1-*O*-glucosyl-*N*-2′-hydroxytetracosanoyl-4-hydroxy-8-sphinganine.

LC-MS interfaces are a more recent development, and there has yet to develop a consensus as to which type of interface is best, moving-belt or thermospray (Vestal, 1984; Desiderio *et al.*, 1984). Privett and Erdahl (1985) have utilized a moving-belt interface which incorporates a reactor heater to degrade thermally acyl lipids

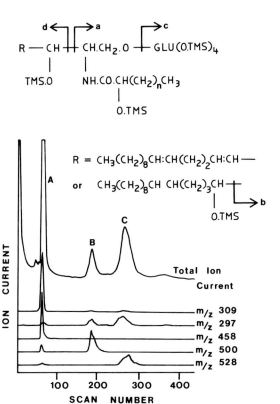

Fig. 4. Selective-ion monitoring GC-MS of monoglucosylceramides from spinach leaves (with permission from Ohnishi *et al.*, 1983). GC of the per-trimethylsilylated (TMS) monoglucosyl-ceramides, as monitored by the total-ion current from the mass spectrometer, shows three major components. In order to identify directly the molecular species, ions characteristic of parts of the structure were monitored. These were $m/z = 309$ $(M - a)^+$ and $m/z = 297$ $(M - b)^+$, characteristic of the 4,8-sphingadiene and 4-hydroxy-8-sphinganine moieties, respectively; and $m/z = 458$ $(M - c - d + 73)^-$, and $m/z = 500,528$ $(M - c - d + 31)^+$, which are characteristic of 2-hydroxy palmitate, docosanoate, and tetracosanoate respectively. The co-incidence of $m/z = 309$ with $m/z = 458$, and of $m/z = 297$ with $m/z = 500$ and 528, shows the principal molecular species to be 1-O-glucosyl-N-2'-hydroxypalmityl-4,8-sphingadiene, 1-O-glucosyl-N-2'-hydroxydocosanoyl-4-hydroxy-8-sphinganine, and 1-O-glucosyl-N-2'-hydroxytetracosanoyl-4-hydroxy-8-sphinganine.

deposited on the belt. The fatty acids split off are analyzed by CI-MS as $RCOOH_2^+$ and RCO^+ ions. This system for monitoring the acyl composition from an LC effluent is not commercially available. Another LC-MS interface system is direct liquid injection. Either microbore HPLC or normal HPLC with a stream splitter can be used to introduce a solvent flow directly into the ionization source of the mass spectrometer at a low enough rate for the vacuum pumps to handle. Kuksis *et al.* (1984, 1985) have used this technique, and review the use of GC-MS and LC-MS interfaces for the analysis of acyl lipids. Particular emphasis is placed on reversed-phase HPLC separations of triacylglycerols and the *t*-butyldimethylsilyl ethers of 1,2-diacylglycerols, followed by direct-inlet CI–MS. Examination of the MH^+ and $(MH - RCOOH)^+$ ions gives the molecular association of acyl groups within isologs. The method is suitable for quantitative analysis only following careful calibration. A potentially useful development for polar lipid analysis is LC-MS-FAB (Stroh *et al.*, 1985).

Tandem mass spectroscopy (MS-MS) is another recent development that may have application to the analysis of lipid mixtures (McLafferty, 1981). In this technique an ion is selected by the primary mass analyzer, representing the molecule of interest within the mixture. The selected ion is then dissociated by collisional activation and the fragments analyzed by the second mass analyzer. MS-MS can therefore be operated to observe secondary ions produced from a selected primary ion, but it can also be run in a mode that selects primary ions that yield a common secondary ion or a constant neutral loss ion in order to characterize a class of molecules with a constant structural determinant. MS-MS is most useful with soft-ionization techniques such as CI, DCI, or FAB. Its great advantage is that it is a rapid analytical technique for detecting specific components in a complex mixture while bypassing difficult chromatographic or other separations. Its use has been described for the identification of molecular species of jojoba wax (Spencer and Plattner, 1984) and soybean phosphatidylcholine (Plattner and Crawford, 1983).

Although the emphasis has been placed on structure determination, it is important to realize that MS has considerable potential for biochemical studies. The use of stable isotopes in metabolic labeling studies has been frequent, and in fact historically preceded the use of radioisotopes. In his review of the classical investigations on the stereochemistry of fatty acid desaturation, Morris (1970) shows the use of deuterium labeling to advantage. With the advent of "benchtop," dedicated, quadrupole GC-MS systems, the possibility of using stable isotopes for routine biochemical assays, without tying up a major instrument, seems possible. In this light, the use of $[2-^{13}C]$malonyl-CoA to assay fatty acid synthesis with measurement of the individual fatty acids seems entirely reasonable (Ohashi *et al.*, 1985).

B. Nuclear Magnetic Resonance (NMR)

The NMR phenomenon arises when a nucleus with a nuclear-spin quantum number $I \geq 1/2$ is placed in an external magnetic field. The nuclear-spin energy states become non-degenerate. Transitions between these states then occur with the absorption or emission of a quantum of radiation in the radio-frequency region of the electromagnetic spectrum. The energy of the transition is directly proportional to the strength of the external magnetic field, and since the local chemical environment will make its own small contribution to the magnetic field experienced by the nucleus, it is possible to make chemical inferences from the transition frequencies of different nuclei in the molecule. The chemical shift of a nucleus in a molecule is a measure of this transition formulated to be independent of the applied magnetic field. Further structural information is obtained from spin–spin coupling between adjacent magnetic nuclei. The theory of NMR is quite complex, and the reader lacking a basic knowledge should consult one of the many standard texts which describe both the basics of the NMR experiment and spectral interpretation (Dyke *et al.*, 1971; Gordon and Ford, 1972; Abraham and Loftus, 1978; Becker, 1980; Silverstein *et al.*, 1981; Sohar, 1983). For lipids, the most commonly studied nuclei will be ^{1}H and ^{13}C. ^{31}P is sometimes useful, more so in biological than in chemical studies, while ^{2}H, ^{19}F (and ^{13}C) are useful in stable-isotope labeling studies. In addition to the routine use of NMR by the lipid chemist, NMR has also become an increasingly useful tool in biochemistry. Its ability to probe the dynamics of biological molecules, and particularly of membranes, has been extensively exploited (Jardetzky and Roberts, 1981; Shulman, 1979). The use of high-resolution NMR in the study of the chemistry and biochemistry of lipids has been reviewed by Pollard (1986). A separate topic is the use of wide-line NMR in measurement of bulk phases, such as the oil and water content in oilseeds (Waddington, 1986).

Like other areas of spectroscopy, NMR spectroscopy has benefited from a continued improvement in instrumentation. The three major advances have been the use of superconducting magnets, the FT-NMR experiment, and computer-controlled instruments. The high-field, superconducting magnets (200–500 MHz) have allowed better resolution of overlapping resonances, particularly for ^{1}H spectra, and improved sensitivity. The FT-NMR spectrometer gives considerably greater sensitivity over the older, coherent-wave NMR spectrometers, and, when coupled with computer-controlled pulse sequence generation, data storage, and manipulation, has allowed exciting new NMR experiments (Shoolery, 1984; Jelinski, 1984). From the biochemists' view, the greatly increased sensitivity will make many more compounds, isolated only in small amounts, amenable to detailed structure determination by NMR. Workable ^{1}H and natural abundance ^{13}C spectra of acyl lipids should be obtainable on samples as small as 50 μg and 1 mg, respectively, in several hours of acquisition time, or less. These are relatively large amounts compared to MS, UV, and FT-IR spectroscopy, but NMR is an inherently insensitive technique. Aspects of sample preparation are covered in the comprehensive text of Martin *et al.* (1981).

Figure 5 shows 300-MHz ^{1}H and 75-MHz ^{13}C NMR spectra for 18-octadec-*cis*-9-enolide, the worked example. Some general points need to be made. Usually the ^{1}H spectrum of lipids has a narrow chemical shift range (0–10 ppm.) and exhibits ^{1}H—^{1}H coupling. An integral is given for the ^{1}H spectrum, as signal strength reflects the number of protons producing that signal. An example of quantitative ^{1}H NMR is the assay of the cyclopropenoid content of vegetable oils down to the 1% level by following the ring methylene signal versus the terminal methyl resonance (Pawloski *et al.*, 1972). The ^{13}C spectrum has a wider chemical-shift range (0–200 ppm) and, since it is obtained in a ^{1}H noise-decoupled mode, each resonance is represented as a single line. Thus resonances that overlap in a ^{1}H spectrum may well be resolved in a ^{13}C spectrum, allowing greater definition of structure. For normal ^{13}C spectra the integral is not quantitative over all ^{13}C atoms. However, with

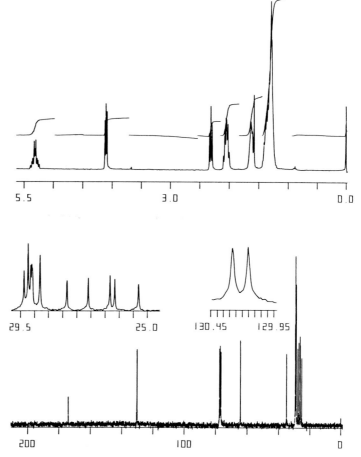

Fig. 5. ¹H- and ¹³C NMR spectra of 18-octadec-9-enolide obtained at 300 and 75.4 MHz, respectively. Chemical shifts are quoted as parts per million (ppm) downfield from tetramethylsilane (0 ppm). The ¹H-NMR spectrum shows the following resonances, defined by the carbons to which the protons are bonded: 5.31 ppm (olefinic, C-9 and C-10), 4.1 ppm (C-18), 2.31 ppm (C-2), 2.04 ppm (allylic, C-8 and C-11), 1.6 ppm (C-3, C-17), and 1.3 ppm ("bulk methylenes"). Conspicuously lacking is the triplet at 0.9 ppm characterizing the terminal methyl of the alkyl chain. The integral corresponds to the number of protons per resonance. The ¹³C-NMR spectrum is assigned as follows: 174.17 ppm (C-1), 130.35 and 130.22 ppm (olefinic, C-9 and C-10, shown in inset), 64.34 ppm (C-18), 35.02 ppm (C-2), and 25.38–29.45 ppm (the remaining methylenes, expanded in inset).

gated decoupling and long pulse delays, a quantitative integral can be obtained, allowing quantitation by ¹³C NMR, as demonstrated for palm oil (Ng and Ng, 1983). In some cases it may be possible to obtain quantitative data while acquiring the spectrum under normal conditions, if resonances have similar spin-lattice relaxation times and similar nuclear Overhauser effects (Pfeffer *et al.*, 1977).

Most classes of lipid have been examined by NMR. The review by Pollard (1986) lists many of these studies and some pertinent chemical-shift data, and describes the approaches available for structure elucidation. Fatty acids, and particularly unsaturated fatty acids, have been extensively examined by NMR (Frost and Gunstone, 1975; Khan and Scheinmann, 1977). ^{13}C NMR can be especially powerful at unraveling the finer points of fatty acid structure, such as position of functional groups and stereochemistry. This is partly because the acyl chain is very amenable to a simple, empirical, parameterized approach to calculate chemical shifts. To interpret NMR spectra fully, ^{1}H and ^{13}C resonances need to be assigned to specific atoms in the molecule. Although complete assignment is an objective in itself, often a partial assignment will be sufficient for structure definition. To begin, the spectroscopist needs to be aware of the purity of the sample. Hydrogen and carbon atoms in the molecule can be counted and assignments made on the basis of literature tabulations and empirical relationships for the chemical-shift and coupling-constant data extracted from the spectra. In interpretation it is often necessary to make use of other chromatographic, chemical, and/or physical data on the isolated compound(s). After these deliberations the basic ^{1}H and ^{13}C spectra may be insufficient and further NMR experiments required.

Further NMR experiments can take a variety of forms, depending on need. Molecules can be specifically labeled with ^{1}H or ^{13}C, either chemically or biosynthetically, to assist in assignments. A simple but useful approach is to run spectra in CD_3OD. This will not only distinguish exchangeable protons, but is helpful in that isotope substituent effects on neighboring carbon atoms can also aid assignments (Gaddamidi et al., 1985). Derivatization and observation of changes in chemical shifts, as in acetylation, is another common approach. Lanthanide-induced shift reagents [tri-β-diketo complexes of Eu(III) or Pr(III)] have been a tool in structure determination, particularly in the days before ^{13}C and high field ^{1}H-NMR capabilities were routinely available. The shift reagents are used to pull apart overlapping resonances, particularly in ^{1}H spectra. Examples include examination of hydroxyfatty acids (Wineberg and Swern, 1974), and the positional analysis of triacylglycerols (Wedmid and Litchfield, 1976), a possibility first noted by Pfeffer and Rothbart (1972). Though eclipsed by other NMR methods today, a relatively new application is the use of chiral lanthanide-induced shift reagents in stereochemistry (Rinaldi, 1982; Sohar, 1983). Chiral-shift reagents can be used to determine configuration and to quantitate enantiomers in racemic mixtures. Chiral-shift reagents have been applied to the stereospecific analyses of model triacylglycerols, with some success (Bus et al., 1976; Bus and Lok, 1978), but it is questionable whether this is the most suitable technique for this thorny problem.

Finally, the NMR spectroscopist today has an impressive arsenal of specialized experiments for structure determination (Shoolery, 1984; Jelinski, 1984; Pollard, 1986). Space does not permit a description of these, and their application to plant acyl lipids has been very limited. However, they have been used extensively in natural products chemistry, and the reader should be aware of their potential. One class of experiments are the spectral-editing techniques based on spin–echo J modulation. These are methods for determining the multiplicity of protonated carbons which have superseded the "off-resonance" method. A second class of experiment is the 2D-NMR experiment, in which the resultant spectra have two frequency axes and one intensity axis. 2D-NMR experiments allow the examination of all possible ^{1}H-decoupling experiments, to ^{1}H and to ^{13}C resonances, in two elegant experiments. Or, it is possible to correlate ^{1}H and ^{13}C resonances, or resolve overlapping spin–multiplet systems. The 2D-NMR spectra of biotin produced by Ikura and Hikichi (1982) represent a comprehensive example. An exciting 2D method is the double-quantum coherence experiment that relies on natural abundance ^{13}C—^{13}C bonded pairs. In this experiment the carbon–carbon connectivity of a molecule can be mapped out, as in the case of fusarin C (Gaddamidi *et al.*, 1985). A caveat to the 2D-NMR experiment concerns sensitivity. The experiments require much larger sample sizes and instrument times than for simple spectra.

In the application of NMR to the biology of lipids the study of membrane dynamics has already been mentioned. NMR has also found use in the study of lipid–protein interactions (Gally *et al.*, 1978; Parks *et al.*, 1983), and in the nondestructive quantitation of the lipids of oilseeds (Schaefer and Stejskal, 1974, 1975; Rutar *et al.*, 1977). In the latter aspect the new technique of cross-polarization–magic angle spinning, which has been developed for the NMR of solids, may offer improved resolution for whole-seed analysis (Haw and Maciel, 1983). The use of stable isotopes for tracer experiments with NMR analysis is possible, but the sensitivity of the technique is low when compared to mass spectral analysis or the use of radioisotopes. Ashworth *et al.* (1985) describe [^{13}C]acetate labeling experiments performed on corn suspension cells, a study that traditionally would be accomplished with [^{14}C]acetate. An interesting study was performed by Schaefer *et al.* (1975) showing the power of NMR not merely to quantitate products in a tracer experiment, but, through precise definition of the positions of the label, to understand carbon metabolism better. Soybean plants were exposed to [^{13}C]carbon dioxide and the carbohydrate and lipid pools examined by ^{13}C NMR. From a consideration of ^{13}C—^{13}C pairing the authors were able to estimate the extent of recycling of carbon, from sucrose to lipid, through the pentose-phosphate cycle. And finally, the reader should be cognizant of the possibility of performing NMR experiments on living plant tissue or cells (Roberts, 1984).

C. Other Spectroscopic Methods

The natural products chemist will rely heavily on MS and NMR, but other spectroscopic techniques are available to help in structure determination, and can sometimes be used as quantitative assays. For the lipid chemist, IR spectroscopy will probably be the most useful of these, as it allows definition of some common functional groups. FT-IR has greatly enhanced the sensitivity of the technique. UV spectroscopy will be useful less frequently. The texts of Dyke *et al.* (1971), Gordon and Ford (1972), and Silverstein *et al.* (1981) contain general information, while Gunstone (1967, 1975) has data more pertinent to acyl lipids.

D. Degradative Methods

A variety of chemical and enzymatic methods for the degradation of acyl lipids are available. Most of these are well-established methods. They can be classified as (1) the analysis of the constituent moieties of a complex lipid, (2) the degradation of the acyl chain, and (3) stereospecific analysis. Apart from their use in structure determination, a primary use of these methods is to locate the label from tracer experiments.

Analysis of the constituent parts of a complex lipid has already been covered to a large part in Sections III,A and B, particularly in regard to the acyl content and composition. Quantitative assay for the constituent moieties of complex lipids (fatty acids, glycerol, phosphorus, carbohydrates, nitrogenous bases, etc.) are given in the standard texts of lipid methodology (Kates, 1972; Christie, 1982a). Degradative methods to locate label in the hydrocarbon chain include double-bond cleavage, followed by radio-GLC analysis of the fragments. The method of choice is probably ozonolysis, and the subject has been reviewed by Ackman *et al.* (1981). Saturated acids (and alcohols), including those derived from unsaturated acids by hydrogenation, can be subjected to chemical α-oxidation to produce a mixture of chain-shortened fatty acids. Specific activity measurements of the components in this mixture, derived by successive decarboxylations, will give the distribution of label. The most updated procedure is that of Netting (1978). A highly controlled decarboxylation (i.e., one carbon atom is removed at a time in a series of reactions that can be recycled to produce successively chain-shortened fatty acids) is best achieved by the procedure of Dauben *et al.* (1953). An example of the use of all three types of hydrocarbon chain degradation is to be found in the location of the label in long-chain fatty acids after incubating meadowfoam seeds with [1-^{14}C]acetate (Pollard and Stumpf, 1980). The use of stable-isotope labeling, and in this case, specifically of [1-^{13}C]acetate labeling and ^{13}C NMR, would have been a more elegant way of performing the experiment, bypassing the traditional methods of degradative analysis. Unfortunately, the level of acetate incorporation in this tissue was too low to perform the experiment effectively. The possibility of using stable isotopes rather than the more common radioisotopes should always be considered in such experiments.

A third category of degradative analysis is the cleavage of lipids by lipases. This is necessary in order to assess the positional distribution of fatty acids in glycerolipids. The use of phospholipases (C and D) and galactosidases to cleave head groups, of phospholipase A_2 to cleave secondary acyl linkages in phospholipids, and of pancreatic lipases to cleave primary acyl linkages, is well documented (Christie, 1982a). The stereospecific analysis of natural triacylglycerols represents a particularly challenging problem, which has been reviewed by Bockerhoff (1971), Smith (1972), and Christie (1982a, 1986). An additional

approach propounded by Kuksis *et al.* (1985) is the preferential hydrolysis of
sn-1,2-diacylglycerophospholipids over the *sn*-2,3 species (both derived from the
triacylglycerols), catalyzed by phospholipase C, and analysis of the intact
diacylglycerols by HPLC-CI-MS.

V. WORKED EXAMPLE

A recent project in our laboratory has involved the analysis of fragrance
compounds in ambrette (*Hibiscus abelmoschus*) seeds (Nee *et al.*, 1986). During
this analysis a novel macrocyclic musk was identified in addition to the well-known
macrocyclic lactone musk ambrettolide (16-hexadec-*cis*-7-enolide, or oxacyclo-
heptadec-*cis*-8-en-2-one). The isolation and characterization of this new musk,
18-octadec-*cis*-9-enolide, or oxacyclononadec-*cis*-10-en-2-one, represents a simple
example of modern methods of chromatography and structure determination, and is
one we will now follow through.

Extraction of ambrette seeds with organic solvents gave an extract which, when
examined by adsorption TLC (Fig. 2A), contained a major fraction with a polarity
between that of methyl oleate and triolein. This fraction was the fragrance fraction
of the seeds, could be purified by silicic acid column chromatography or preparative
TLC, and is called the "monoester fraction" subsequently. IR spectroscopy showed
the presence of ester carbonyl stretch (at 1740 cm^{-1}), while UV spectroscopy
revealed the absence of any strong chromophore. The monoester fraction was
amenable to analysis by GLC, and retention time and GC-MS suggested that the
late-running compound was a homolog of ambrettolide (5%), as well as giving
identifications on the alkyl acetates (6%), the farnesyl acetates (70%), and
ambrettolide (14%). In order to effect a separation of the suspected C_{18} musk, the
monoester fraction was first subjected to silver nitrate chromatography (Fig. 2B). A
monoester fraction enriched in monoenes was obtained, purified away from the
saturated alkyl acetates and the sesquiterpene acetates. Interestingly, separation of
hexadecenolide and octadecenolide occurred by silver nitrate TLC alone. Linear
compounds would not be expected to be resolved. With column scale silver
nitrate–silicic acid chromatography the resolution of the two macrocyclic musks was
lost, and a further stage of the purification by preparative reverse-phase (C_{18}) HPLC
was used. The corresponding TLC plate is shown in Fig. 2C. A pure fraction of the
suspected musk was obtained, as analyzed by GLC, which was then used for
physical and chemical characterization studies. Mass spectra are shown in Fig. 3,
and ^{1}H and ^{13}C-NMR spectra in Fig. 5. The compound, being a simple structure,
gave no diagnostic fragments in its EI and CI mass spectra, but a nominal MW of
280 was obtained. A high-resolution mass measurement to define the molecular
formula was not considered necessary, as this was adequately defined by NMR. The
^{1}H- and ^{13}C-NMR spectra showed resonances characteristic of an isolated double
bond and of a single ester group linked to long alkyl and acyl hydrocarbon chains.
From the ^{13}C-NMR spectrum 15 distinct resonances were observed, but as three of

these in the "methylene envelope" region were of sufficient magnitude to contain 2 carbons, the compound could be tentatively shown to contain 18 carbon atoms. The ^1H integral fitted octadecenolide as the structure of the molecule, while the molecular formula $C_{18}H_{32}O_2$ gave a nominal MW of 280, coincident with that obtained by MS. The observation that there must be three degrees of unsaturation in the molecule (derived from the molecular formula) yet there were only two assignable unsaturated bonds (the olefinic and the carbonyl π-bonds) meant that the molecule must contain a ring system. Another very diagnostic feature of the NMR spectra was the absence of terminal methyl resonances, which would indicate that the molecule must be a macrocycle. The absence of methyl groups was best demonstrated by the use of the APT spectral-editing sequence to examine the multiplicities of the ^{13}C resonances.

So far the MS and NMR data have defined the structure down to 18-octadecenolide. The remaining topic for a full structure determination is the stereochemistry and position of the double bond. There are no chiral centers in the molecule to complicate matters. The stereochemistry of the double bond was easily deduced by reexamining the IR and ^{13}C-NMR spectra. The IR spectrum showed no absorbance at 970 cm^{-1} characteristic of a trans double bond, while the allylic ^{13}C resonances were indicative of a cis rather than trans double bond (27.3 versus 32.7 ppm, respectively). The question of double-bond position could not be inferred from the ^{13}C-NMR spectrum. Extensive ^{13}C data exist to evaluate double-bond position in olefinic fatty acids, where the acyl chain has freedom of motion (Batchelor *et al.*, 1974; Bus *et al.*, 1977; Gunstone *et al.*, 1977). However, a macrocycle will have limited conformations, so that the application of the tabulated data will have a degree of uncertainty. Double-bond position in the macrocycle was therefore determined by ring opening of the 18-octadecenolide by transmethylation to produce methyl 18-hydroxyoctadecenoate, and then ozonolysis and GLC analysis of the aldehydic fragments produced. An alternative method might have been trimethylsilylation and mass spectral analysis, as has been conducted for cutin monomers (Holloway, 1982). Thus the novel musk was characterized as a direct derivative of oleic acid.

ACKNOWLEDGMENTS

I wish to thank ARCO Chemical Co. for allowing me the time and facilities to write this article, Jim Bourell for running the NMR spectra, David Hirano for running the mass spectra, and to my colleagues for critical comments on the manuscript.

REFERENCES

Abraham, R. J., and Loftus, P. (1978). "Proton and Carbon-13 NMR Spectroscopy." Heyden, London.
Ackman, R. G. (1969). *In* "Methods in Enzymology" (J. M. Lowenstein, ed.), Vol. 14, pp. 329–381. Academic Press, New York.

Ackman, R. G. (1981). *In* "Methods in Enzymology" (J. M. Lowenstein, ed.), Vol. 72, pp. 205–252. Academic Press, New York.

Ackman, R. G. (1986). *In* "Analysis of Fats and Oils" (R. J. Hamilton and J. B. Rossell, eds.), pp. 137–206. Elsevier, Amsterdam.

Ackman, R. G., Sebedio, J.-L., and Ratnayake, W. N. (1981). *In* "Methods in Enzymology " (J. M. Lowenstein, ed.), Vol. 72, pp. 253–276. Academic Press, New York.

Agrawal, V. P., and Schulte, E. (1983). *Anal. Biochem.* **131,** 356–359.

Agriga, T., Araki, E., and Murata, T. (1977). *Anal. Biochem.* **83,** 474–483.

Alam, I., Smith, J. B., Silver, M. J., and Ahern, D. (1982). *J. Chromatogr.* **234,** 218–221.

Andersson, B. A. (1978). *Prog. Chem. Fats Other Lipids* **16,** 279–308.

Ashworth, D. J., Adams, D. O., Giang, B. Y., Cheng, M. T., and Lee, R. Y. (1985). *Anal. Chem.* **57,** 710–715.

Baker, F. C., and Schoolery, D. A. (1981). *In* "Methods in Enzymology" (J. M. Lowenstein, ed.), Vol. 72, pp. 41–52. Academic Press, New York.

Baron, C. B., and Coburn, R. F. (1984). *J. Liq. Chromatogr.* **7,** 2793–2801.

Barron, E. J., and Mooney, L. A. (1968). *Anal. Chem.* **40,** 1742–1744.

Batchelor, J. G., Cushley, R. J., and Prestegard, J. H. (1974). *J. Org. Chem.* **39,** 1698–1705.

Becker, E. D. (1980). "High Resolution NMR: Theory and Chemical Applications." Academic Press, New York.

Bezard, J. A., and Ouedraogo, M. A. (1980). *J. Chromatogr.* **196,** 279–293.

Bitman, J., and Wood, D. L. (1981). *J. Liq. Chromatogr.* **4,** 1023–1034.

Blau, K., and King, G. S. (1978). "Handbook of Derivatives for Chromatography." Heyden, London.

Bligh, E. G., and Dyer, W. J. (1959). *Can. J. Biochem. Physiol.* **37,** 911–917.

Bockerhoff, H. (1971). *Lipids* **6,** 942–956.

Borch, R. F. (1975). *Anal. Chem.* **47,** 2437–2439.

Burlingame, A. L., Whitney, J. O., and Russell, D. H. (1984). *Anal. Chem.* **56,** 417R-467R.

Bus, J., and Lok, C. M. (1978). *Chem. Phys. Lipids* **21,** 253–260.

Bus, J., Lok, C. M., and Groenewegen, A. (1976). *Chem. Phys. Lipids* **16,** 123–132.

Bus, J., Sies, L., and Lie Ken Jie, M. S. F. (1977). *Chem. Phys. Lipids* **18,** 130–144.

Busch, K. L., and Cooks, R. G. (1982). *Science* **218,** 247–254.

Chang, T. T., Lay, J. O., and Francel, R. J. (1984). *Anal. Chem.* **56,** 109–111.

Chen, S. L., Stein, R. A., and Mead, J. F. (1976). *Chem. Phys. Lipids* **16,** 161–166.

Chen, S. S.-H., and Kou, A. Y. (1982). *J. Chromatogr.* **227,** 25–31.

Christie, W. W. (1982a). "Lipid Analysis." Pergamon, Oxford.

Christie, W. W. (1982b). *J. Lipid Res.* **23,** 1072–1075.

Christie, W. W. (1986). *In* "Analysis of Fats and Oils" (R. J. Hamilton and J. B. Rossell, eds.), pp 313–339. Elsevier, Amsterdam.

Colborne, A. J., and Laidman, D. L. (1975). *Phytochemistry* **14,** 2639–2645.

Crane, R. T., Goheen, S. C., Larkin, E. C., and Rao, G. A. (1983). *Lipids* **18,** 74–80.

Crawford, C. G., and Plattner, R. D. (1983). *J. Lipid Res.* **24,** 456–460.

Crawford, C. G., Plattner, R. D., Sessna, D. J., and Rackis, J. J. (1980). *Lipids* **15,** 91–94.

D'Alonzo, R. P., Kozarek, W. J., and Wade, R. L. (1982). *J. Am. Oil Chem. Soc.* **59,** 292–295.

Dauben, W. G., Hoerger, E., and Petersen, J. W. (1953). *J. Am. Chem. Soc.* **75,** 2347–2351.

Demandre, C., Tremolieres, A., Justin, A.-M., and Mazliak, P. (1985). *Phytochemistry* **24,** 481–485.

Desiderio, D. M., Fridland, G. H., and Stout, C. B. (1984). *J. Liq. Chromatogr.* **7,** 317–351.

Dommes, V., Wirtz-Peitz, F., and Kunau, W. H. (1976). *J. Chromatogr. Sci.* **14,** 360–366.

Domnas, A. J., Warner, S. A., and Johnson, S. L. (1983). *Lipids* **18,** 87–89.

Duck-Chong, C. G., and Baker, G. (1983). *Lipids* **18,** 387–389.

Dyke, S. F., Floyd, A. J., Sainabury, M., and Theobald, R. S. (1971). "Organic Spectroscopy: An Introduction." Penguin, London.

Einolf, W. N., and Chan, W. C. (1984). *J. Agric. Food Chem.* **32,** 785–789.

Felton, H. R. (1982). *Chem. Abstr.* **97,** 155650p.

Fenimore, D. C., and Davis, C. M. (1981). *Anal. Chem.* **53,** 252A-262A.

Fishwick, M. J., and Wright, A. J. (1977). *Phytochemistry.* **16,** 1507–1710.

Fjeldsted, J. C., and Lee, M. L. (1984). *Anal. Chem.* **56,** 619A-628A.

Folch, J., Lees, M., and Sloane-Stanley, G. H. (1957). *J. Biol. Chem.* **226,** 497–509.

Frost, D. J., and Gunstone, F. D. (1975). *Chem. Phys. Lipids* **15,** 53–85.

Fujino, Y., and Ohnishi, M. (1976). *Chem. Phys. Lipids* **17,** 275–289.

Fujino, Y., Ohnishi, M., and Ito, S. (1985). *Lipids* **20,** 337–342.

Gaddamidi, V., Bjeldanes, L. F., and Shoolery, J. N. (1985). *J. Agric. Food Chem.* **33,** 652–654.

Galliard, T. (1970). *Phytochemistry* **9,** 1725–1734.

Gally, H. U., Spencer, A. K., Armitage, 1.M., Prestegard, J. H., and Cronan, J. E. (1978). *Biochemistry* **17,** 5377–5382.

Games, D. E. (1978). *Chem. Phys. Lipids* **21,** 389–400.

Gattavecchia, E., Tonelli, D., and Bertocchi, G. (1983). *J. Chromatogr.* **260,** 517–521.

Gordon, A. J., and Ford, R. A. (1972). "The Chemist's Companion: A Handbook of Practical Data, Techniques and References." Wiley (Interscience), New York.

Gross, D., Gutekunst H., Blaser, A., and Hambock H. (1980). *J. Chromatogr.* **198,** 389–396.

Gross, M. L., and Rempel, D. L. (1984). *Science* **226,** 261–268.

Gunstone, F. D. (1967). "An Introduction to the Chemistry and Biochemistry of Fatty Acids and Their Glycerides." Chapman & Hall, London.

Gunstone, F. D. (1975). *In* "Recent Advances in the Chemistry and Biochemistry of Plant Lipids" (T. Galliard and E. I. Mercer, eds.), pp 21–42. Academic Press, New York.

Gunstone, F. D. (1976) *Chem Ind. (London)* 243–251.

Gunstone, F. D. (1978). *Trends Biol. Sci.* **3,** N54.

Gunstone, F. D., Pollard, M. R., Scrimgeour, C. M., and Vedanayagam, H. S. (1977). *Chem. Phys. Lipids* **18,** 115–129.

Hammond, E. W. (1986) *In* "Analysis of Fats and Oils" (R. J. Hamilton and J. B. Rossell, eds.), pp. 113–135. Elsevier, Amsterdam.

Hara, A., and Radin, N. S. (1978). *Anal. Biochem.* **90,** 420–426.

Harrison, A. G. (1983). "Chemical Ionisation Mass Spectroscopy." CRC Press, Boca Raton, Florida.

Haw, J. F., and Maciel, G. E. (1983). *Anal. Chem.* **55,** 1262–1267.

Heckers, H., Dittmar, K., Melcher, H. O., and Kalinowski, H. O. (1977). *J. Chromatogr.* **135,** 93–107.

Heinze, F. J., Linscheid, M., and Heinz, E. (1984). *Anal. Biochem.* **139,** 126–133.

Holloway, P. J. (1982). *In* "The Plant Cuticle" (D. F. Cutler, K. L. Alvin, and C. E. Price, eds.), pp. 45–85. Academic Press, London.

Holman, R. T. (1967). *Prog. Chem. Fats Other Lipids* **9,** 3–12.

Hostettmann, K., Appolonia, C., Domon, B., and Hostettmann, M. (1984). *J. Liq. Chromatogr.* **7,** 231–242.

Ikura, M., and Hikichi, K. (1982). *Org. Magn. Reson.* **20,** 266–273.

Jamieson, G. R. (1970) *In* "Topics in Lipid Chemistry" (F. D. Gunstone, ed.) Vol. 1, pp. 107–159. Logos Press, London.

Jardetzky, O., and Roberts G. C. K. (1981). "NMR in Molecular Biology." Academic Press, New York.

Jee, M. H., and Ritchie, A. S. (1984). *J. Chromatogr.* **299,** 460–464.

Jelinski, L. W. (1984). *Chem. Eng. News* **62,** 26–47.

Johnson, C. B. (1976). *Anal. Biochem.* **71,** 594–596.

Johnson, E. L., and Stevenson, R. (1978) "Basic Liquid Chromatography." Varian, Palo Alto, California.

Jordi, H. C. (1978). *J. Liq. Chromatogr.* **1,** 215–230.

Juaneda, P., and Rocquelin, G. (1985). *Lipids* **20,** 40–41.

Karasek, F. W., Onuska, F. I., Yang, F. J., and Clement, R. E. (1984). *Anal. Chem.* **56,** 174R-199R.

Kates, M. (1972). *In* "Laboratory Techniques in Biochemistry and Molecular Biology. Techniques of Lipidology: Isolation, Analysis and Identification of Lipids" (T. S. Work and E. Work, eds.), pp. 269–610. Elsevier, Amsterdam.

Kesselmeier, J., and Heinz, E. (1985). *Anal. Biochem.* **144,** 319–328.

Khan, G. R., and Scheinmann, F. (1977). *Prog. Chem. Fats Other Lipids* **15**, 343–367.

Knapp, D. R. (1979). "Handbook of Analytical Derivatization Reactions." Wiley, New York.

Kolattukudy, P. E. (1980). *In* "The Biochemistry of Plants: a Comprehensive Treatise; Vol. 4; Lipids: Structure and Function" (P. K. Stumpf, ed.), pp. 571–645. Academic Press, New York.

Kuksis, A., Myher, J. J., and Marai, L. (1984). *J. Am. Oil Chem. Soc.* **61**, 1582–1589.

Kuksis, A., Myher, J. J., and Marai, L. (1985). *J. Am. Oil Chem. Soc.* **62**, 762–767.

McCloskey, J. A. (1969). *In* "Methods in Enzymology" (J. M. Lowenstein, ed.), Vol. 14, pp. 382–450. Academic Press, New York.

McCloskey, J. A. (1970). *In* "Topics in Lipid Chemistry" (F. D. Gunstone, ed.), pp 369–440. Wiley, New York.

McCrear, D. K., Kossa, W. C., Ramachandran, S., and Kurtz, R. R. (1978). *J. Chromatogr. Sci.* **16**, 329–331.

McLafferty, F. W. (1981). *Science* **214**, 280–287.

McNair, H. M., and Bonelli, E. J. (1969). "Basic Gas Chromatography. 6th Ed." Varian, Palo Alto, California.

McReynolds, W. O. (1970). *J. Chromatogr. Sci.* **8**, 685–691.

Majors, R. E., Barth, H. G., and Lochmuller, C. H. (1984). *Anal. Chem.* **56**, 300R–349R.

Mancha, M., Stokes, G. B., and Stumpf, P. K. (1975). *Anal. Biochem.* **68**, 600–608.

Mangold, H. K., ed. (1984). "CRC Handbook of Chromatography: Lipids," Vols. 1 and 2. CRC Press, Boca Raton, Florida.

Marion, D., Gandemer, G., and Douillard, R. (1984). *In* "Structure, Function and Metabolism of Plant Lipids" (P.-A. Siegenthaler and W. Eichenberger, eds.), pp. 139–143. Elsevier, Amsterdam.

Martin, A. J. P., and Synge, R. L. M. (1941). *Biochem. J.* **35**, 1358–1368.

Martin, M. L., Delpuech, J. J., and Martin, G. J. (1981). "Practical NMR Spectroscopy." Heyden, London.

Merritt, C., Vajdi, M., Kayser, S. G., Halliday, J. W., and Bazinet, M. L. (1982). *J. Am. Oil Chem. Soc.* **59**, 422–432.

Metcalfe, L. D., and Wang, C. N. (1981). *J. Chromatogr. Sci.* **19**, 530–535.

Miwa, T. K. (1971). *J. Am. Oil Chem. Soc.* **48**, 259–264.

Moreau, R. A. (1984). *Plant Physiol.* **75**, Suppl., Abstr. 439.

Moreau, R. A. (1985). *Phytochemistry* **24**, 411–414.

Morris, L. J. (1966). *J. Lipid Res.* **7**, 717–732.

Morris, L. J. (1970). *Biochem. J.* **118**, 681–693.

Murata, T. (1977). *Anal. Chem.* **49**, 2209–2213.

Myher, J. J., Kuksis, A., Marai, L., and Yeung, S. K. F. (1978). *Anal. Chem.* **50**, 557–561.

Natale, N. (1977). *Lipids* **12**, 847–856.

Nee, T., Cartt, S., and Pollard, M. R. (1986). *Phytochemistry* **25**, 2157–2161.

Nelson, G. J. (1974). *Lipids* **9**, 254–263.

Netting, A. G. (1978). *Anal. Biochem.* **86**, 580–588.

Netting, A. G., and Duffield, A. M. (1984). *J. Chromatogr.* **336**, 115–123.

Ng, S., and Ng, W. L. (1983). *J. Am. Oil Chem. Soc.* **60**, 1266–1268.

Nichols, B. W., and Safford, R. (1973). *Chem. Phys. Lipids* **11**, 222–227.

Odham, G., and Stenhagen, E. (1972). *In* "Biochemical Applications of Mass Spectroscopy" (G. R. Waller, ed.), pp. 229–249. Wiley (Interscience), New York.

Ohashi, K., Otsuka, H., and Seyama, Y. (1985). *J. Biochem. (Tokyo)* **97**, 867–875.

Ohashi, Y. (1984). *Biomed. Mass Spectrom.* **11**, 383–385.

Ohnishi, M., Ito, S, and Fujino, Y. (1983). *Biochim. Biophys. Acta* **752**, 416–422.

Parks, J. S., Cistola, D. P., Small, D. M., and Hamilton, J. A. (1983). *J. Biol. Chem.* **258**, 9262–9269.

Pawloski, N. E., Nixon, J. E., and Sinnhuber, J. E. (1972). *J. Am. Oil Chem. Soc.* **49**, 387–392.

Pempkowiak. J. (1983). *J. Chromatogr.* **258**, 93–102.

Pfeffer, P. E., and Rothbart, H. L. (1972). *Tetrahedron. Lett.* pp. 2533–2536.

Pfeffer, P. E., Sampugna, J., Schwartz, D. P., and Shoolery, J. N. (1977). *Lipids* **12**, 869–871.

Phillips, B. E., Smith, C. R., and Tallent, W. H. (1971). *Lipids* **6**, 93–99.

Phillips, F. C., and Privett, O. S. (1979). *Lipids* **14**, 949–952.

Phillips, F. C., Erdahl, W. L., and Privett, O. S. (1982). *Lipids* **17**, 992–997.

Phillips, F. C., Erdahl, W. L., Schmit, J. A., and Privett, O. S. (1984). *Lipids* **19**, 880–887.

Plattner, R. D. (1981). *In* "Methods in Enzymology" (J. M. Lowenstein, ed.), Vol. 72, pp. 21–34. Academic Press, New York

Plattner, R. D., and Crawford, C. G. (1983). *Abstr. Ann. Conf. Mass Spectrom. Allied Top.*, **31st** pp. 761–762.

Plattner, R. D., and Spencer, G. F. (1983). *Lipids* **18**, 68–73.

Pollard, M. R. (1986). *In* "Analysis of Fats and Oils" (R. J. Hamilton and J. B. Rossell, eds.), pp. 401–434. Elsevier, Amsterdam.

Pollard, M. R., and Stumpf, P. K. (1980). *Plant Physiol.* **66**, 649–655.

Poole, C. F., and Schuette, S. A. (1984). "Contemporary Practice of Chromatography." Elsevier, Amsterdam.

Privett, O. S., and Erdahl, W. L. (1985). *J. Am. Oil Chem. Soc.* **62**, 786–792.

Radin, N. S. (1969). *In* "Methods in Enzymology" (J. M. Lowenstein, ed.), Vol. 14, pp. 268–272. Academic Press, New York.

Richards, R. G., Mendenhall, C. L., and Macgee, J. (1975). *J. Lipid Res.* **16**, 395–397.

Rinaldi, P. L. (1982). *Prog. NMR Spectrosc.* **15**, 291–352.

Rinehart, K. L. (1982). *Science* **218**, 254–260.

Roberts, J. K. M. (1984). *Annu. Rev. Plant Physiol.* **35**, 375–386.

Roberts, T. R. (1978). "Radiochromatography: The Chromatography and Electrophoresis of Radiolabelled Compounds." Elsevier, Amsterdam.

Rogers, V. A., van Aller, R. T., Pessoney, G. F., Watkins, E. J., and Leggett, H. G. (1984). *Lipids* **19**, 304–306.

Rossell, J. B. (1986) *In* "Analysis of Fats and Oils" (R. J. Hamilton and J. B. Rossell, eds.), pp. 1–90. Elsevier, Amsterdam.

Rouser, G., Kritchevsky, G., Yamamoto, A., Simon, G., Galli, C., and Bauman, A. J. (1969). *In* "Methods in Enzymology" (J. M. Lowenstein, ed.), Vol. 14, pp. 272–317. Academic Press, New York.

Rutar, V., Burgar, M., and Blinc, R. (1977). *J. Magn. Reson.* **27**, 83–90.

Schaefer, J., and Stejskal, E. O. (1974). *J. Am. Oil Chem. Soc.* **51**, 210–213.

Schaefer, J., and Stejskal, E. O. (1975). *J. Am. Oil Chem. Soc.* **52**, 366–369.

Schaefer, J., Stejskal, E. O., and Beard, C. F. (1975). *Plant Physiol.* **55**, 1048–1053.

Shantha, N. C., and Kaimal, T. N. B. (1984). *Lipids* **19**, 971–974.

Sherma, J., and Bennett, S. (1983). *J. Liq. Chromatogr.* **6**, 1193–1211.

Sherma, J., and Fried, B. (1984). *Anal. Chem.* **56**, 48R-68R.

Shoolery, J. N. (1984). *J. Nat. Prod.* **47**, 226–259.

Shulman, R. (1979). "Biological Applications of Magnetic Resonance." Academic Press, New York.

Siebertz, H. P., Heinz, E., Linscheid, M., Joyard, J., and Douce, R. (1979). *Eur. J. Biochem.* **101**, 429–438.

Silverstein, R. M., Bassler, G. C., and Merrill, T. C. (1981). "Spectrophotometric Identification of Organic Compounds." Wiley, New York.

Skipski, V. P., and Barclay, M. (1969). *In* "Methods in Enzymology" (J. M. Lowenstein, ed.), Vol. 14, pp. 530–598. Academic Press, New York.

Slabas, A., Roberts, P., and Ormesher, J. (1982). *In* "Biochemistry and Metabolism of Plant Lipids" (J. F. G. M. Wintermans and P. J. C. Kuiper, eds.), pp. 251–256. Elvesier, Amsterdam.

Slover, H. T., Thompson, R. H., and Merola, G. V. (1983). *J. Am. Oil Chem. Soc.* **60**, 1524–1528.

Smith, C. R. (1972). *In* "Topics in Lipid Chemistry, Volume 3" (F. D. Gunstone, ed.), pp. 89–124. Logos Press, London.

Smith, E. C., Jones, A. D., and Hammond, E. W. (1980). *J. Chromatogr.* **188**, 205–212.

Sohar, P. (1983). "Nuclear Magnetic Resonance Spectroscopy," Vols. I, II, and III. CRC Press, Boca Raton, Florida.

Sparace, S. A., and Mudd, J. B. (1982). *Plant Physiol.* **70,** 1260–1264.

Spencer, G.F, and Plattner, R. D. (1984). *J. Am. Oil Chem. Soc.* **61,** 90–94.

Stahl, E. (1969). "Thin Layer Chromatography: A Laboratory Handbook. 2nd Ed." Springer-Verlag, Berlin and New York.

Stahl, E., and Muller, J. (1982). *Chromatographia* **15,** 493–497.

Stroh, J. G., Cook, J. C., Milberg, R. M., Brayton, L., Kihara, T., Huang, Z., Rinehart, K. L., and Lewis, L. A. S. (1985). *Anal. Chem.* **57,** 989–991.

Svetashov, V. I., and Zhukova, N. V. (1985). *J. Chromatogr.* **330,** 396–399.

Takagi, T., and Itabashi, Y. (1977). *Lipids* **12,** 1062–1068.

Thenot, J.-P., and Horning, E. C. (1972). *Anal. Lett.* **5,** 519–529.

Tswett, M. (1906). *Ber. Dtsch. Bot. Ges.* **24,** 316–323.

Vestal, M. L. (1984). *Science* **226,** 275–281.

Waddington, R. (1986). *In* "Analysis of Fats and Oils" (R. J. Hamilton and J. B. Rossell, eds.), pp. 341–399. Elsevier, Amsterdam.

Waller, G. R. (1972). "Biochemical Applications of Mass Spectroscopy." Wiley (Interscience), New York.

Waller, G. R., and Dermer, O. C. (1980). "Biochemical Applications of Mass Spectroscopy. 1st Supplement" Wiley (Interscience), New York.

Wedmid, Y., and Litchfield, C. (1976). *Lipids* **11,** 189–193.

Williams, J. P., Watson, G. R., Khan, M., Leung, S., Kuksis, A., Stachnyk, O., and Myher, J. J. (1975). *Anal. Biochem.* **66,** 110–122.

Williams, J. P., Watson, G. R., and Leung, S. P. K. (1976). *Plant Physiol.* **57,** 179–184.

Williard, H. H., Merritt, L. L., Dean, J. A., and Settle, F. A. (1981). "Instrumental Methods of Analysis. 6th Ed." Van Nostrand, New York.

Wilson, R. F., and Rinne, R. W. (1974). *Plant Physiol.* **57,** 179–184.

Wineberg, J. P., and Swern, D. (1974). *J. Am. Oil Chem. Soc.* **51,** 528–533.

Wood, R., and Lee, T. (1983). *J. Chromatogr.* **254,** 237–246.

Yao, J. K., and Rastetter, G. M. (1985). *Anal. Biochem.* **150,** 111–116.

β-Oxidation of Fatty Acids by Specific Organelles

2

HELMUT KINDL

I. INTRODUCTION

Fatty acids are contained in the structures of storage lipids and structural membrane lipids. Their degradation proceeds mainly by oxidation in the β-position to the carboxyl group and sequential removal of C units. These core reactions of fatty acid β-oxidation are flanked by additional processes. They allow unsaturated fatty acids to be fed into the pathway and include reactions effectually removing

The Biochemistry of Plants, Vol. 9

the product of the reaction chain, acetyl-CoA. Several special features — that is, irreversible steps as means to keep the concentration of products at very low levels, channeling of intermediates by microcompartments, and the organization of the whole sequence within a single organelle — make the β-oxidation system in plants a highly efficient structural entity.

A β-oxidation apparatus arranged similar to that in higher plants has been found in several fungi. Bacteria achieve another kind of compartmentation by forming a multienzyme complex. In contrast to these strategies, the β-oxidation system of animal cells always seems to be distributed among two organelles, namely mitochondria and peroxisomes.

II. COMPARTMENTS OF FATTY ACID DEGRADATION

Various compartments are active in fatty acid degradation under different metabolic situations. In plants, these compartments always seem to be one or another form of microbodies. The role played by mitochondria bears upon the recycling of cosubstrates but not on the β-oxidation and chain shortening itself. This is in contrast to the situation in liver and probably all other animal cells where the role of peroxisomes compared to mitochondria is more or less complementary.

A. Glyoxysomes

Glyoxysomes represent a compartment designed for the degradation of fatty acids. The product of the pathway is removed *in situ* and its concentration is kept at a very low level by coupling β-oxidation with another reaction sequence with unidirectional flow. In this respect, the coupling of β-oxidation with citrate cycle and respiration in mitochondria resembles the coupling of β-oxidation with glyoxylate cycle in glyoxysomes (Cooper and Beevers, 1969). This kind of assembly is optimal, in the former case for complete oxidation of the carbon skeleton, in the latter case for the conversion of fatty acids into building blocks suitable for gluconeogenesis and thus carbohydrate synthesis.

If β-oxidation, with or without combination with glyoxylate cycle, is viewed as compartmentalized channeling, the pool sizes and recycling processes for the cosubstrates involved cannot be overlooked. β-Oxidation requires CoA and ATP for fatty acid activation. This process may take place at the external face of the glyoxysomal membrane. CoA is also required in the thiolase reaction but is recovered in the malate synthase and citrate synthase reaction, thus giving rise to a balance of the CoA pool within the organelle. This is not the case for NADH, which is formed during both β-oxidation and the glyoxylate cycle. Three possibilities have been considered as to how the reoxidation of NADH could take place: (1) direct oxidation of NADH by means of a rudimentary electron transport chain (Hicks and Donaldson, 1982) in the glyoxysome membrane, (2) transport of NADH through the glyoxysomal membrane to a cytosolic site of oxidation, or (3) transfer of redox equivalents by a complex shuttle mechanism. At present it is not known whether the transport of low-molecular-weight material through the peroxisome membrane requires a specific carrier system. It might be that organellar proteins, by their binding capacity, control the enhanced pool of substrates measured within the organelle. If so, a rather unspecific pore protein would be sufficient to account for the permeability observed. Within glyoxysomes, an intraorganellar organization becomes evident, in that a few enzymes (i.e., malate synthase, citrate synthase, malate dehydrogenase, and the multifunctional protein) are more tightly associated with the membrane than others. These enzymes, also characterized by alkaline pI, are considered as peripheral membrane proteins (Bieglmayer *et al.*, 1973; Huang and Beevers, 1973). The product of the glyoxysomal fatty acid degradation is succinate, which can be further converted within mitochondria. Hence, lipid bodies, glyoxysomes, and mitochondria must form an intracellular superstructure for triglyceride utilization.

Since their discovery in 1967 (Breidenbach and Beevers, 1967), glyoxysomes have been detected in a number of storage tissues of seeds, including endosperm of *Ricinus communis*, cotyledons of *Cucumis sativus*, *Helianthus annuus*, aleuron tissue of Gramineae, and also in germinating conidia (Armentrout and Maxwell, 1981), acetate-grown (Wanner and Theimer, 1982), ethylamine-grown (Zwart *et al.*, 1983), and alkane-grown fungi (Tanaka *et al.*, 1982), and heterotrophically grown algae.

B. Leaf Peroxisomes

The main function of peroxisomes in green leaves relates to photorespiration by which these specialized leaf peroxisomes form glycine and metabolize serine. Leaf peroxisomes also contain the enzymes of fatty acid β-oxidation, although it is not yet clear what the sources of fatty acids are. Constant turnover of structural membrane lipids may yield long-chain fatty acids, and degradation of proteins and various lipid structures may furnish branched-chain fatty acids.

Irrespective of the poorly understood physiological function of β-oxidation in microbodies of green leaves, the lack of the glyoxylate cycle brings up an additional question, as to how the acetyl-CoA produced by β-oxidation is transferred into other compartments.

C. Other Specialized Peroxisomes

Acetyl-CoA can be regarded as the central metabolite of peroxisomes. It can be assumed that most peroxisomes described house the enzyme apparatus for fatty acid β-oxidation and/or acetyl-CoA metabolism. This holds true for yeast peroxisomes, and glycosomes, a well-described peroxisome-related organelle of unicellular protozoa (*Trypanosoma*) (Visser *et al.*, 1981). Acetyl-CoA can be formed and metabolized into malate and aspartate in yeasts growing on ethylamine as the sole carbon source (Veenhuis *et al.*, 1983). Specialized microbodies are capable of oxidizing ethylamine to acetaldehyde and using the malate synthase reaction to feed acetyl-CoA into anabolic pathways. In cells of *Hansenula polymorpha* previously grown on ethylamine and subsequently transferred to glucose medium, a selective degradation of the microbodies by lytic compartments has been observed. If cells were grown on C compounds (e.g., methanol), the peroxisomes form methanal and remove it by conversion into dihydroxyacetone.

D.   Unspecialized Peroxisomes

The term unspecialized peroxisome has been coined for microbody forms which do not participate in a major metabolic pathway. Their protein components are, therefore, not present in the cell at a marked level. It is not yet clear whether they participate in the fatty acid metabolism, albeit at low rates, or whether they solely constitute the minimal peroxisomal entity. Such a proform of an organelle is needed by the cell if induction leads to pronounced formation of specialized microbodies or if, in dividing cells, mother cells have to transmit the inheritance of peroxisomal membranes to daughter cells.

Microbodies from tubers, roots, and hypocotyls are regarded as belonging to this group. They may also be considered as properoxisomes in terms of microbody proliferation. They may serve a function as prestage in the biogenesis of specialized peroxisomes, for example, when etiolated leaves are illuminated. But there is not sufficient evidence to generalize that every specialized peroxisome has to derive from an unspecialized form.

E.   Possible Role of Mitochondria

Acyl-CoA dehydrogenase, which catalyzes the first oxidative step of the mitochondrial β-oxidation sequence, has not yet been demonstrated for plant mitochondria. For plants, it is at issue whether a fatty acid β-oxidation system exists in mitochondria (Macey, 1983; Macey and Stumpf, 1983; Wood *et al.*, 1986).

In analogy to the role of mitochondria in animal cells (Bremer and Osmundsen, 1984) (e.g., adipocytes or hepatocytes), originally the fatty acid β-oxidation system was thought to be located in plant mitochondria, with the exception of seed germination where the conversion of fat into carbohydrate represents a special metabolic situation. Later, it became increasingly evident that, in the presence of glyoxysomes, mitochondria do not play any direct role in β-oxidation. It may turn out that in any plant tissue investigated, β-oxidation is solely or mainly located in peroxisomes.

The enzymes indicative of microbody β-oxidation were observed in green tissue. In etiolated leaves, green leaves, hypocotyls and tubers, acyl-CoA oxidase and the multifunctional protein (enoyl-CoA hydratase/3-hydroxyacyl-CoA dehydrogenase), both are unique for microbody measurement of the synthesis of these proteins (Gerhardt, 1984; H. Kindl, unpublished observations).

In contrast to reports (Thomas and Wood, 1986; Wood et al., 1984; Burgess and Thomas, 1986) on carnitine-dependent oxidation of palmitoyl-CoA, Gerhardt (1984), employing improved organelle separation and extensive enzymological investigations, ruled out the existence in plants of an acyl-CoA dehydrogenase. In cases where a careful separation of microbodies and mitochondria was carried out and the type of enzymes determined in detail, the microbody type of β-oxidation was found. Irrespective of the localization of fatty acid degradation, the physiological function of β-oxidation in cells like hypocotyls, tubers, or green leaves is not clear at all. Minor amounts of triglycerides, structural lipids, fatty acids other than the one originating from fat, or carboxylic acids derived from amino acid catabolism may be the physiological substrates. Mitochondria may be involved if the acetyl-CoA produced in peroxisomes is not used for anabolic processes but rather for respiration.

III. MECHANISMS OF FATTY ACID DEGRADATION

It is a salient feature of plant metabolism that anabolic processes prevail over catabolic metabolism. Accordingly, degradation of fat represents a conversion of fat into carbohydrates rather than a complete oxidation to CO_2. From this point of view we have to consider β-oxidation in context with the steps that are linked with it. For reviews covering the relevant literature before 1978 see Vol. 4 of this series, Chapter 4 (Beevers, 1980). Valuable data can be found in the reviews of Beevers (1979, 1982), Tolbert (1971, 1981, 1982), Kindl (1982b, 1984), Trelease (1984), and in the books of Gerhardt (1978) and Huang et al. (1983).

A. β-Oxidation Pathway

β-Oxidation encompasses activation of the carboxyl group, oxidative steps in the carbon side chain, and cleavage of the C—C bond. Fatty acid, CoA, ATP, O_2, NAD^+ are the substrates yielding acetyl-CoA, NADH, diphosphate, and AMP as products (Fig. 1). Thioesters of fatty acids are formed by virtue of a simultaneous cleavage of ATP to AMP and diphosphate. In this respect, the activation of long-chain fatty acids is analogous to the activation of cinnamic acids, or amino acids in formation of transport antibiotica. A FAD-dependent oxidation leads to a trans-β-unsaturated derivative, and a subsequent trans-addition of water to L-3-hydroxyacyl-CoA. Following NAD^+-dependent oxidation, the 3-ketoacyl-CoA is cleaved by a thiolytic step with an additional molecule of CoA. On the basis of turnover numbers and activities present in organelles, reactions catalyzed by acyl-CoA oxidase and thiolase are probably the rate-limiting steps. The two steps between them are catalyzed by a single protein containing the activities of a

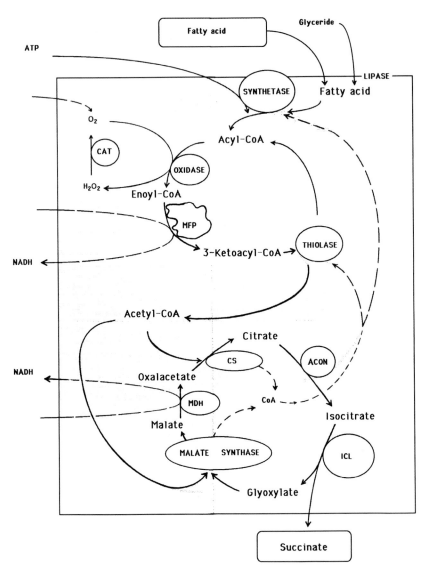

Fig. 1. β-Oxidation in combination with glyoxylate cycle localized in glyoxysome. Fatty acids are converted into the CoA esters and then subjected to dehydrogenation, catalyzed by acyl-CoA oxidase (designated as oxidase). Enoyl-CoA is converted into 3-ketoacyl-CoA by cleavage of water and dehydrogenation, both reactions catalyzed by a multifunctional protein (MFP). Acetyl-CoA is split from the β-keto ester by reacting with the thiol CoA leading to an acyl-CoA having a carbon chain that is two carbon atoms shorter (thiolase reaction). Acetyl-CoA is fed into malate synthase and citrate synthase (CS) reaction, and the glyoxylate cycle is completed by the function of aconitase (ACON), isocitrate lyase (ICL), and malate dehydrogenase (MDH).

hydratase and a 3-hydroxyacyl-CoA dehydrogenase, whose catalysis is character-ized by both efficient channeling and very high turnover numbers. In steady-state flow, it is decisive that a sufficient supply of educts is provided and that the removal of the product takes place by an efficient system. The latter is guaranteed much better by a virtually irreversible reaction than by transport processes. In this context, the combination of β-oxidation with the glyoxylate cycle (Fig. 1) is optimal. Both acetyl-CoA-consuming processes, citrate synthesis and malate formation from acetyl-CoA and glyoxylate, proceed with a high negative free energy.

The chain-length specificity of β-oxidation seems to be somewhat different if various peroxisomes are surveyed (Alexson and Cannon, 1984; Gerhardt, 1984, 1985). Although a thorough investigation is lacking in the case of acyl-CoA synthase, and although its contribution to the selectivity can only be estimated, it is highly likely that differences in the specificity of acyl-CoA oxidase parallel the overall rate of β-oxidation of fatty acids with various chain lengths. Probably, also within a given tissue the selectivity toward fatty acid derivatives with certain chain length may vary with developmental stages.

The β-oxidation apparatus involved in the degradation of unbranched fatty alcohols (Moreau and Huang, 1979) is also in oxidation of branched-chain fatty acids or chain shortening of cholestanoic acids.

B.   Sequences Preceding β-Oxidation

Fatty acid degradation begins, in most cases, with the release of fatty acid from a structural lipid or reserve triglyceride (Huang *et al.*, 1976; Huang, this volume, Chapter 4). Uncertainties exist as to the subcellular site of the lipase-catalyzed step. For example, the hydrolytic reaction may take place at different compartments when various species are compared. The cellular organization of lipolysis in fat-rich seeds falls into at least two categories. One group has lipase at the glyoxysomal membrane, the enzyme exhibiting an optimal activity at slightly alkaline pH values. In the other group, the lipase activity was attributed to a not yet characterized light membrane fraction. Endosperm of *Ricinus communis* does not fall into either of the two groups in that it has a lipase highly active at acidic pH value and confined to the lipid body membrane.

Since most plant lipids are rich in unsaturated fatty acids and these have cis double bonds, additional reactions can be envisaged to feed these compounds into the standard β-oxidation system. There is a principal difference whether 3-*cis*-enoyl-CoA or 4-*cis*-enoyl-CoA is formed during the course of chain shortening. While the conversion of 3-*cis*-enoyl-CoA into 2-*trans*-enoyl-CoA requires only one additional step, an isomerase reaction (Fig. 2), the degradation of fatty acids with cis double bonds in even-numbered positions is characterized by a reductive step. Acyl-CoAs with a 4-*cis* double bond may be oxidized by acyl-CoA oxidase to 2-*trans*-4-*cis*-enoyl-CoA, which is then reduced to 3-*trans*-enoyl-CoA by a NADPH-requiring reductase.

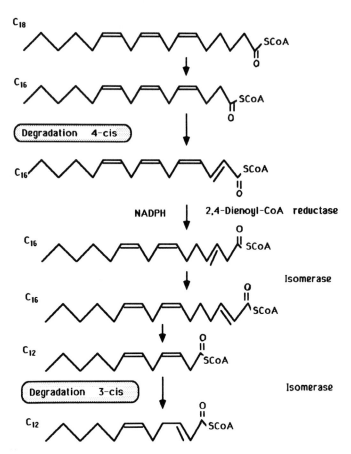

Fig. 2. Modification of fatty acid degradation pathway if 4-*cis*- or 3-*cis*-unsaturated derivatives appear during β-oxidation.

For the degradation of hydroxylated fatty acids such as ricinoleic acid (12-D-hydroxyoctadec-*cis*-9-enoic acid), a sequence has been postulated but not sufficiently verified by enzymatic studies. Hutton and Stumpf (1971) proposed a pathway including a 2-keto acid and a subsequent α-oxidation. Mechanistically, a D-4-hydroxyacyl-CoA can also be converted by dehydratation to a 3-*cis*-enoyl-CoA, which is then further converted to a 2-*trans*-enoyl-CoA.

C.   Channeling by Linking β-Oxidation with the Glyoxylate Cycle

By combination of the two pathways, succinate is the product of the degradation sequence (Fig. 1). The glyoxylate cycle is driven by the thermodynamics of two acetyl-CoA-consuming steps. All other reactions of the glyoxylate cycle are *in vitro* reversible and may also function, under certain physiological situations, in the reverse direction. Glyoxysomes probably have a balanced CoA pool as a part of their metabolic compartmentation. The glyoxylate cycle may also introduce amino acids into the carbon metabolism, that is, by glutamate:oxaloacetate transaminase reaction and by reversion of the isocitrate dehydrogenase reaction.

D. Related Pathways

If fat degradation continues in illuminated cotyledons, the peroxisomes house, in addition to β-oxidation and glyoxylate cycle, enzymes for glycollate oxidation and serine metabolism. Fat may then be converted into glycine and serine. Evidence infavor of this hypothesis was presented with isolated microbodies from greening cucumber cotyledons and used to support the view of continuous transition within the different forms of peroxisomes.

Under particular circumstances, the physiological function of peroxisomal β-oxidation could serve to generate H_2O_2 for peroxidative reactions. Studies with a lignin-destroying fungus (*Phanerochaete chrysosporium*, basidiomycete) indicated that the extracellular H_2O_2, required as a substrate for a heme-containing oxidase, originates from intracellular fat degradation. Some fungi, such as *Hansenula polymorpha*, use the principle, already shown with acetyl-CoA in glyoxysomes, that a key compound formed in peroxisomes is immediately consumed in the same organelle. Thus, formaldehyde derived in peroxisome by an oxidase-dependent reaction from methanol, is used, within the peroxisome, for dihydroxyacetone synthesis (Goodman, 1985). Among metabolic sequences which have been considered by analogy to the function of other peroxisomes but not established for plant microbodies are plasmalogen biosynthesis (Hajra and Bishop, 1982; Ballas *et al.*, 1984) and chain shortening of tetracyclic steroids.

IV. BIOCHEMICAL CHARACTERIZATION OF THE COMPONENTS

Most enzymes functioning in β-oxidation and the glyoxylate cycle have been purified and characterized as proteins and studied as to their catalytic properties. The molecular properties (Table I)—size and protein structure—were found to be similar when the respective enzymes were compared from different plant sources. Immunological cross-reactivity among analogous enzymes was also a general feature. The kinetic parameters, however, vary considerably between enzymes from different plants.

A. Enzymes of β-Oxidation

1. Acyl-CoA Synthetase

Long-chain fatty acyl-CoA's are amphiphilic compounds and form, as do detergents, molecular solutions only at low concentrations. As the concentration is

TABLE I
Properties of Enzymes Involved in β-Oxidation and Glyoxylate Cycle

Enzyme	Cofactors or prosthetic groups	Size M_r	Size Subunit M_r	Specific activity of purified enzyme ($\mu mol/min/mg$)	Relative activity present in glyoxysomes (related to thiolase)
Acyl-CoA oxidase	FAD	160,000	72,000	27	30
Multifunctional protein		75,000	75,000		
Enoyl-CoA hydratase activity	—	75,000	75,000	550	800
3-Hydroxyacyl-CoA dehydrogenase	(NAD^+)			13	20
Thiolase	—	90,000	45,000	4	1
Dienoyl-CoA reductase	(NADPH)	?	?	?	4
Malate synthase	—	520,000	63,000	25	80
Citrate synthase	—	200,000 (100,000)	49,000	108	80
Malate dehydrogenase	(NAD^+)	70,000 (210,000)	34,000	4,000	4,000
Aconitase	Fe—S	?	83,000(?)	—	—
Isocitrate lyase	—	255,000	64,000	10	30
Catalase	Heme	225,000	54,000	30,000	3,000

increased, the critical micelle concentration (CMC) is reached and association into micelles begins. Depending on the environment, the CMC is reached at concentrations of 10–50 μM. The formation of long-chain acyl-CoA's is catalyzed by a membrane-bound CoA ligase which uses ATP as energy source, cleaving it to AMP and diphosphate.

Acyl-CoA synthetases (E.C. 6.2.1.3) of different intracellular sites were investigated in liver cells (Hashimoto, 1982). Unlike most other microbody enzymes, the synthetase from peroxisomes was immunologically not distiguishable from the mitochondrial and the microsomal form.

2. Acyl-CoA Oxidase

In contrast to the animal mitochondrial acyl-CoA dehydrogenase, which uses an electron transfer flavoprotein as electron acceptor generating an anionic semiquinone, the plant acyl-CoA oxidase transfers electrons to FAD to form $FADH_2$, which is then autoxidized by molecular oxygen to generate H_2O_2 and FAD. The two enzymes differ in the chemical reactivity of their product charge–transfer complexes.

Gerhardt (1983, 1985) has shown that acyl-CoA oxidase from leaf peroxisomes preferentially reacts on CoA esters of tetradecanoic acid, butyryl-CoA being converted only at one-third of the rate. The purified enzyme from glyoxysomes (Kirsch and Kindl, 1986) revealed an even higher preference for long-chain derivatives compared to butyryl-CoA. The K_m value for short-chain acyl-CoA is much higher than for 18:0 or 18:1 acyl derivatives. In contrast to that, the undifferentiated peroxisomes of mung bean hypocotyls contain an acyl-CoA oxidase with two optima of chain-length specificity, at C-14 *and* C-4 (Gerhardt, 1985).

3. Multifunctional Protein

This protein, containing at least the two activities of enoyl-CoA hydratase (E.C. 4.2.1.17) and 3-hydroxyacyl-CoA dehydrogenase (E.C. 1.1.1.35), is a unique marker of microbodies throughout eukaryotic cells. The protein from cucumber cotyledons was purified 600-fold by chromatography on CM-cellulose and blue dextran Sepharose (Frevert and Kindl, 1980a). It shares with several other glyoxysomal enzymes the property of an extreme alkaline isoelectric point. Unlike other peroxisome enzymes, the bifunctional protein neither contains a prosthetic group nor does it oligomerize. The protein with at least two active centers is a monomer with M_r 75,000. A similar protein, including also the activity of 3-hydroxyacyl-CoA epimerase and an isomerase converting 3-*trans*-enoyl-CoA into 2-*trans*-enoyl-CoA, was described for the β-oxidation complex of *Escherichia coli* (Yang and Schulz, 1983). In *Candida tropicalis* (Moreno de la Garza *et al.*, 1985), the multifunctional protein possesses enoyl-CoA hydratase, 3-hydroxyacyl-CoA dehydrogenase, and 3-hydroxyacyl-CoA epimerase activity. Compared to the plant enzyme which is a monomer with M_r 75,000, the fungal protein shows a slightly larger subunit M_r and the property of oligomerization (2 × 102,000 for *Candida* enzyme, 4 × 93,000 for the enzyme from *Neurospora crassa*; W. H. Kunau, personal communication; Kionka and Kunau, 1985).

Within the group of plants examined, antibodies raised against the glyoxysomal multifunctional protein from cucumber cotyledons showed cross-reactivity toward the respective proteins from green leaves, etiolated leaves, and hypocotyls. Enoyl-CoA hydratase activity was determined in peroxisomes isolated from bundle sheath as well as from mesophyll cells (Ohnishi *et al.*, 1985).

4. Thiolase (Acyl-CoA:Acetyl-CoA C-Acyltransferase, E.C. 2.3.1.16)

The glyoxysomal enzyme is, like most other thiolases obtained from different mammalian tissues or bacteria, a homodimer with a subunit M_r of 90,000. Its purification was greatly facilitated because of its rather alkaline p*I*. By a combination of ion-exchange chromatography and affinity chromatography on blue dextran Sepharose, a homogeneous enzyme with a rather low specific activity of 65 mkat/kg was prepared (Frevert and Kindl, 1980b). Also the total activity of thiolase in most fat-degrading plant tissues was found to be significantly lower than that of the preceding enzymes (Table I). This suggests that the thiolase reaction is rate limiting and may cause an accumulation of acetoacetyl-CoA, which in turn is an effective inhibitor of the crotonase reaction.

B.   Other Enzymes Related to Fatty Acid Degradation

1.   Dienoyl-CoA Reductase

Following the demonstration of 2,4-dienoyl-CoA reductase in both mitochondria and peroxisomes of rat liver (Dommes et al., 1981), the NADPH-dependent enzyme was also detected in fungi and plants. Studies with extracts from oleate-grown cells of C. tropicalis containing fatty acid β-oxidation revealed that 4-cis-decenoyl-CoA was only further degraded when the 2,4-dienoyl-CoA reductase step was not blocked by lack of NADPH. The partially purified enzyme converted 2-trans,4-cis-decadienoyl-CoA into 3-trans-decenoyl-CoA (Fig. 2). For further degradation, an isomerase leading to 2-trans-enoyl-CoA is required. Similar properties were found for the enzyme from E. coli (Dommes et al., 1982) and C. tropicalis (Dommes et al., 1983).

2.   Malate Synthase

In glyoxysomes, malate synthase (EC 4.1.3.2) is the dominant organellar protein (Table I). It behaves as a peripheral membrane protein, and can be solubilized as an octameric enzyme in the presence of 100 mM Mg^{2+}. Once solubilized, the solution of the octamer is stable at 2 mM Mg^{2+}. In the absence of cations, malate synthase aggregates to approximately 100 S forms (Kruse and Kindl, 1983a).

A similar form of malate synthase was observed in fat-degrading cells prior to the peak of fat mobilization. By various centrifugation techniques it was demonstrated that the aggregated form of malate synthase was not part of the endoplasmic reticulum (ER) (Köller and Kindl, 1980). Gentle immunological techniques, too, were employed to separate the malate synthase from the ER.

3.   Citrate Synthase

Citrate synthase activity (E.C. 4.1.3.7) is located in glyoxysomes, and mitochondria as well. The two enzymes have been purified from fat-mobilizing cotyledons (Köller and Kindl, 1977; Kindl, 1987) and castor bean endosperm (Zehler et al., 1984). The two isoenzymes can be distinguished by several biochemical means, including the differences in pI, and by immunological methods. The enzyme activities per plant and the rates of de novo synthesis during seed germination are considerably higher for the glyoxysomal enzyme than for the mitochondrial enzyme. Both isoenzymes are subject to partial proteolysis during purification. The subunit M_r of citrate synthases prepared quickly and by taking special precautions against proteolysis are higher by 2000 than when purified by standard procedures.

4.   Isocitrate Lyase

Isocitrate lyase (E.C. 4.1.3.1) catalyzes the reversible cleavage of isocitrate to glyoxylate and succinate. Besides preparations from fat-mobilizing plant tissues

(Khan *et al.*, 1977; Frevert and Kindl, 1978), the enzyme has been isolated from bacteria, a free-living soil nematode (Patel and McFadden, 1977), fern (Gemmrich, 1981), fungi (Johanson *et al.*, 1974), and protozoa (Blum, 1982). The active center possesses a histidine which can be modified by diethylpyrocarbonate (Jameel *et al.*, 1985). Bromopyruvate selectively blocks the other functional group in the active center, a cysteine residue. Itaconate epoxide was found to be an irreversible active site-directed inhibitor reacting with a carboxylate group of the protein.

5.   Aconitase

This is the only enzyme in microbodies that has not been purified and investigated extensively. Aconitases (E.C. 4.2.1.3) from various sources were found to be easily deactivated (Ramsay and Singer, 1984). The very sensitive Fe–S cluster can be reactivated under anaerobic conditions with an excess of Fe^{2+} and thiol.

6.   Malate Dehydrogenase

Up to five isoenzymes of malate dehydrogenase (E.C. 1.1.1.37) have been detected in a single cell type. The peroxisomal form is distinguished by an alkaline p*I* (Walk and Hock, 1976). Separation of the isoenzymes can elegantly be performed by affinity chromatography on 5′-AMP–Sepharose. The mature glyoxysomal enzyme was differentiated from the other forms by immunological methods.

7.   Catalase

Catalase (E.C. 1.11.1.6) has been known for a long time as a marker of peroxisomes. Although H_2O_2 is a substrate which can be acted on by catalase with high turnover number, catalase might be viewed as a "buffer" for keeping H_2O_2 concentration at low levels. Fungal and plant genetics (Hamilton *et al.*, 1982; Chandlee and Scandalios, 1984) have revealed that more than one gene encoding the catalase protein can be present. In few cases, two catalase forms distinguishable by their subunit M_r have been detected in peroxisomes (Schiefer *et al.*, 1976; Yamaguchi *et al.*, 1984; Gerdes and Kindl, 1986). As has been observed for the liver enzyme, plant catalase is very sensitive to proteolysis. Consequently, the highly active enzyme prepared from homogenates may not be the original form in peroxisomes.

8.   Glutamate:Oxaloacetate Transaminase

This aminotransferase (E.C. 2.6.1.1), together with malate dehydrogenase, has been proposed (Mettler and Beevers, 1980) to form a shuttle system which transfers from glyoxysomes to mitochondria the reduction equivalents derived from β-oxidation and glyoxylate cycle. For this reason, a set of these two enzymes has to be located in microbodies and mitochondria. In glyoxysomes, such a transaminase with high activity was indeed found. In cucumber cotyledons, two of six isoenzymes were attributed to glyoxysomes (Liu and Huang, 1977). The plastidic, cytosolic, and glyoxysome forms of the enzyme were expressed at different stages of development.

In leaf peroxisomes and some of the unspecialized peroxisomes, both malate dehydrogenase and glutamate:oxalacetate transaminase may be absent, thus the question of a shuttle mechanism for the NADH originating in fatty acid β-oxidation remains unsolved. Other NADH-consuming reactions may take over the task, such as hydroxypyruvate reductase in microbodies of green leaves.

V. CONTROL OF FATTY ACID DEGRADATION

Fatty acid β-oxidation is regulated by the number, the volume, and the activity of organelles associated with this system. The control of fatty acid degradation is primarily a control of gene expression, which in turn is decisive for the quality and quantity of enzymes packaged into the specific microbodies. Besides that, there must be fine tuning, probably based on intermediates dammed up and feedback caused by substrates of the rate-limiting steps.

A. Biosynthesis of the Specific Organelle

Although the fatty acid β-oxidation apparatus constitutes part of a peroxisome, and although even properoxisomes in rapidly dividing cells show assembly of newly synthesized enzyme such as the multifunctional protein discussed earlier, most studies on biosynthesis make direct use of the observation that during certain induction periods the synthesis of peroxisomal precursors and of organelle enzymes is increased drastically. Both during stages of formation of specialized forms of peroxisomes and during the shift of one specialized form by another specialized peroxisome, the sequence of pools as well as the properties of precursor peptides, compared to the mature enzyme, can be investigated without the handicap of using extremely sensitive techniques. Investigations on organelle biosynthesis rely on both the survey and orientation by experiments *in vivo* under more or less physiological conditions *and* on experiments *in vitro* which permit analysis of single steps in more detail.

1. *Studies* in Vivo *to Survey Intracellular Pools*

If the protein constituents of an organelle are not synthesized within the organelle, the question arises as to where synthesis takes place and where intermediary protein pools exist en route to their final destination. For glyoxysomal proteins, it has been established that quantitatively significant pools of precursors occur, namely in the cytosol. In a few cases, the precursors in the early pool can be differentiated from the mature form in the organelles by differences in size of the subunit (Gietl and Hock, 1984; Furuta *et al.*, 1982; Fujiki *et al.*, 1985). In all other cases the size of subunits are not distinguishable, but differentiation is feasible by the observation that in the cytosol the precursor peptide is monomeric, does not oligomerize, and lacks a prosthetic group (Kruse and Kindl, 1983b), while mature enzymes in the glyoxysomes are mostly oligomers, some of them having acquired prosthetic groups (Lazarow and DeDuve, 1973a,b; Fujiki and Lazarow, 1985; Kirsch and Kindl, 1986).

The monomeric precursor form of an organelle precursor should be accumulated in the cytosol when cells are induced and be amenable to further purification (Kindl and Kruse, 1983). If this is not feasible, kinetic studies must be applied to analyze the appearance of the precursor. Even without quantitative separation of cell compartments, the labeling kinetics can reveal that with short-time labeling a monomeric form is synthesized at first. The use of plant protoplasts or yeast spheroblasts, or the application of suitable conditions as slightly acidic pH value (Goodman et al., 1984) can facilitate the preparation of cytosol not being heavily contaminated with organellar matrix.

The observation that, during onset of induction at the early stage of glyoxysome growth, precursors of glyoxysomal enzymes are accumulated in the cytosol and then aggregated led to a suggestion that malate synthase and citrate synthase were ER-bound forms, and this in turn led to the proposal that glyoxysome enzymes, and glyoxysomes, were synthesized via ER (Beevers, 1979, 1982). Electron-microscopic investigations on possible continuities between the ER membrane and the peroxisomal membrane added much to the controversy on the role of the ER in peroxisome ontogeneity. The two seemingly incompatible viewpoints, namely, synthesis on free polysomes or bound polysomes, could be resolved by kinetic analyses and by unraveling the hydrophobic properties of the precursors. The most convincing evidences for the role of cytosolic pools in the biosynthesis of peroxisomes came from experiments showing that the labeling of cytosolic pools precedes the labeling of the peroxisomes or glyoxysomes. In the case of one of the glyoxysomal membrane proteins (i.e., malate synthase), an exact precursor–product relationship has been obtained, implying an unequivocal sequence of pools: cytosolic monomer immediately prior to glyoxysomal forms (Kindl et al., 1980; Kindl, 1982a; Kruse and Kindl, 1982; Kindl and Kruse, 1983).

2. *Studies Demonstrating Steps* in Vitro

mRNA was prepared from bound and free polysomes and employed for its capacity to govern the synthesis of subunits of peroxisomal enzymes. In several cases, it could be established that free polysomes are the sites for the synthesis of microbody proteins (Goldman and Blobel, 1978; Roa and Blobel, 1983; Rachubinski et al., 1984; Fujiki et al., 1984). Developmental changes in mRNA activity during germination were observed (Becker et al., 1978). Experiments including ER vesicles as possible acceptors of growing peptide chains ruled out ER as being involved in a cotranslational or posttranslational transport (Roberts and Lord, 1982).

Posttranslational import into glyoxysomes and correct intraorganellar assembly of the oligomeric enzyme was first demonstrated with malate synthase (Kruse et al., 1981). Import followed by partial proteolytic processing was observed in the case of malate dehydrogenase (Gietl and Hock, 1984). By using a heterologous system, the investigators demonstrated that ionophors and uncouplers did not interfere with the processing of malate dehydrogenase at the glyoxysomal membrane (Gietl *et al.*, 1986).

Figure 3 summarizes the present concepts of steps in peroxisome biosynthesis. Specific recognition sites on the organelle membrane (by virtue of a complex protein receptor and/or glycolipid) seem to be a requirement; cleavage of a signal peptide not. Recent reports on the DNA sequences of genes coding for methanol oxidase and its flanking sequences (Ledeboer *et al.*, 1985) and of peroxisomal β-oxidation enzymes in liver (Osumi *et al.*, 1985) do not indicate that any N-terminal signals had to be cleaved off. By nuclease S_1 mapping, the initiation site of transcription was found to be 100 bp upstream of the translation initiation site. Coding DNA sequences compared with the enzyme's N-terminus, detected by sequence analysis at the protein level, proved that, during the course of the enzymes assembly in the peroxisome, only methionine is removed from the N-terminus posttranslationally. A similar situation was found for another peroxisomal protein, that is, urate oxidase (Nguyen *et al.*, 1985).

B.   Control of Gene Expression

Induction phenomena convert properoxisomes containing a minimum equivalent of peroxisomal entity into specialized peroxisomes which constitute a compatively large compartment and participate in a metabolism quantitatively relevant for a cell in the particular metabolic situation. Induction can, however, also allow a

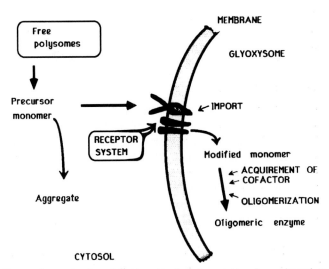

Fig. 3.   Survey of steps implicated in biosynthesis and assembly of peroxisomal or glyoxysomal enzymes. The receptor system indicated at the organelle membrane is hypothetical.

specialized form of a peroxisome to be altered into another. In the latter cases, the question arises of how these transitions take place. During the lifetime of a particular cell, more than one transition can be observed.

As outlined for cucumber cotyledons (Figure 4), the cells are formed at a very early stage of seed ripening, in a rapidly dividing tissue of the embryo. Later, during late stages of seed ripening and then during seed germination and greening as well, cells do not divide but enlarge, and change their organellar equipment. Figure 4 shows that at different stages new and unique peroxisomal parameters appear and disappear. In this particular instance, at least three forms of peroxisomes and the respective transitions (designated at T in the figure) can be studied. All peroxisome forms share the property of containing enzymes of fatty acid degradation. Similar observations were made with maturing castor bean seeds (Hutton and Stumpf, 1969) and cottonseeds (for review see Trelease, 1984).

1. Conversion of Properoxisomes into Specialized Peroxisomes

It is proposed that every somatic cell contains at least one peroxisome. This peroxisome is designated as a properoxisome. It contains the minimum set of proteins, that is, the enzymes of β-oxidation and receptors; it must divide by fission when the cell divides. A parental cell or a somatic cell transmits a properoxisome

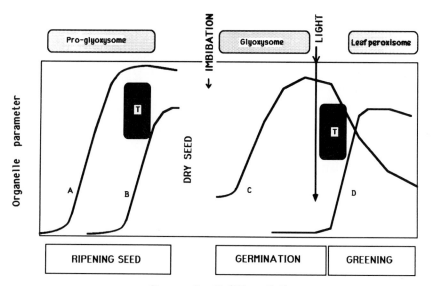

Fig. 4. Appearance of peroxisome forms at different stages of cell differentiation. The organelle parameters of fat-storing and fat-degrading cells of cotyledons of epigeously germinating seedling are shown, examplified by cucumber cotyledons. The peroxisomal parameters are as follows: A; catalase, enzymes of β-oxidation; B; malate synthase; C; all enzymes of β-oxidation and glyoxylate cycle; D; glycollate oxidase, β-hydroxypyruvate dehydrogenase, serine : glyoxylate transaminase. A different scale prior to the stage of dry seed and after imbibation is applied.

to daughter cells in order to maintain the entity of peroxisomal membrane. Accordingly, also a membrane is considered as a material inherited from one cell to another.

Cells of ripening seeds at early stages should possess properoxisomes. They are converted into small but mature glyoxysomes at a later stage of seed ripening (Trelease, 1984). The small glyoxysomes increase in size when dry seeds were imbibed and germinate.

Properoxisomes or unspecialized peroxisomes are found in hypocotyls, tubers, and etiolated leaves. The conversion of properoxisomes into specialized leaf peroxisomes was studied by using etiolated leaves of *Lens culinaris* and by following the appearance of glycollate oxidase-containing microbodies upon illumination. Both forms of peroxisomes are characterized by the enzymes of β-oxidation; these enzymes are continuously synthesized *before* and *after* illumination. The marker enzymes of the specialized microbody form (e.g., glycollate oxidase), however, are synthesized and assembled within the organelle in large amounts only *after* induction (Gerdes *et al.*, 1982; Gerdes and Kindl, 1986).

2. Transition Forms during Exchange of Specialized Peroxisomes by a New Form of Specialized Peroxisome

In particular cases, cells active in fatty acid β-oxidation and the glyoxylate cycle can be induced, by illumination, to synthesize enzymes of photorespiration, localized in peroxisomes. It is important to know whether two independent populations of microbodies are present at this stage of cell development or whether fatty acid degradation and glycollate oxidation take place in the same organelle. Electron-microscopic investigations led Trelease *et al.* (1971) to suggest a model of interconversion from one population of microbodies into another. Electron-microscopic investigations of fat-degrading cells capable of greening when illuminated showed microbodies appressed to both a lipid body and a chloroplast (Schopfer *et al.*, 1976; Bergfeld *et al.*, 1978). This kind of assembly appeared only at the stage of greening, and the number of microbodies arranged in this way was highest at the height of the transient period. This led to the proposal of an intermediate form, the glyoxyperoxisome.

Biochemical studies proved that the *de novo* synthesis of glyoxylate cycle enzymes continued at this early stage of greening (Köller and Kindl, 1978; Betsche and Gerhardt, 1978). Furthermore, when the pool size of newly synthesized peroxisomal and glyoxysomal protein precursor in the cytosol was determined at the transient period the amount of precursors destined for glyoxysomes remained virtually constant while the amount of precursors to be transferred into peroxisomes dramatically increased (Kindl, 1982a). However, the precursor pool of the microbody proteins that had been found to occur in both glyoxysomes and leaf peroxisomes also remained constant during the beginning of the transient period. This evidence rules out the possibility that the latter proteins were used for the assembly of an independent new organelle (two-population theory). Definitive answers to the controversy of one or two populations at the transition states were provided by immunohistochemical investigations and biochemical channeling experiments. Applying gold particles of different sizes and monospecific antibodies, electron-microscopic studies provided evidence for a simultaneous occurrence of leaf peroxisomal enzymes and glyoxysomal enzymes in a single microbody (Titus and Becker, 1985). Experiments using an entirely different approach—biochemical means to demonstrate a channeling and intersystem crossing between two pathways—proved that products of fatty acid β-oxidation, including intermediates of the glyoxylate cycle, can be fed into the path glycollate → glycine and vice versa without leaving the particular microbody (Kindl, 1982b).

Both—the studies bearing on the details during the transition of peroxisome forms and the concept that the cytosolic precursor pool largely determines the kind of peroxisome being formed—are of general interest in that they propose the occurrence of peroxisome forms with varying capacities of fat degradation.

Distinguishable forms of microbodies can also be detected in liver cells, if different means of inducing peroxisomal constituents are applied. Treatment with clofibrate causes the increase of peroxisomal proteins, the enzymes of fatty acid β-oxidation and catalase being induced at very different rates. Stages of transition characterized by intensive synthesis of peroxisomal constituents have been found during seed ripening (Choinski and Trelease, 1978; Frevert *et al.*, 1980; Miernyk and Trelease, 1981a–c). To extend our knowledge of peroxisome function in other than fat-rich seeds it would be of advantage to have mutants available lacking peroxisomal functions or distinguished by errors in peroxisome assembly. Mutants with relevant implications in peroxisomal metabolism have been detected in humans. Several clinically well-described syndromes (Zellweger syndrome, adrenoleukodystrophy, refsum diseases) have in common the accumulation of very long-chain fatty acids and the lack of plasmalogen (Arias *et al.*, 1985).

3. Other Changes in Gene Expression

While the perception of light and the subsequent signal chain via phytochrome has been unraveled in several aspects, other induction processes are much less understood. First insights into a possible role of hormones in inducing glyoxysomal activities have been gained (Longo *et al.*, 1979; Gonzales and Delsol, 1981; Martin *et al.*, 1984; Dommes and Northcote, 1985). Biotic signals, too, are capable of turning on the transcription of nuclear genes coding for peroxisomal enzymes. Biotic signals during the course of establishing a symbiosis between root cells of leguminosae and bacteria (Newcomb *et al.*, 1985) greatly enhance the synthesis of urate oxidase, a so-called noduline (Hanks and Tolbert, 1982; Bergmann *et al.*, 1983).

C. Metabolic Regulations

As fatty acid degradation produces, during the course of gluconeogenesis, more NADH (and indirectly ATP) than can be used for anabolic reactions, energy supply in these tissues is not a limiting factor. The competition for NAD^+ by various dehydrogenases links β-oxidation with other processes in the cell. The question of how much the various NADH-consuming reactions account for the formation of NAD$^+$ required in the peroxisomes under varying metabolic situations needs detailed studies in the future.

REFERENCES

Alexson, S. E. H., and Cannon, B. (1984). *Biochim. Biophys. Acta* **796,** 1–10.
Arias, J. A., Moser, A. B., and Goldfischer, S. L. (1985). *J. Cell Biol.* **100,** 1789–1792.
Armentrout, V. N., and Maxwell, D. P. (1981). *Exp. Mycol.* **5,** 295–309.
Ballas, L. M., Lazarow, P. B., and Beli, R. M. (1984). *Biochim. Biophys. Acta* **795,** 297–300.
Becker, W. M., Leaver, C. J., Weir, E. M., and Riezman, H. (1978). *Plant Physiol.* **62,** 542–549.
Beevers, H. (1979). *Annu. Rev. Plant Physiol.* **30,** 159–193.
Beevers, H. (1980). *In* "The Biology of Plants" (P. K. Stumpf and E. E. Conn, eds.), Vol. 4, pp. 117–130, Academic Press, New York.
Beevers, H. (1982). *Ann. N.Y. Acad. Sci.* **386,** 243–253.
Bergfeld, R., Hong, Y. N., Kuhne, T., and Schopfer, P. (1978). *Planta* **143,** 297–307.
Bergmann, H., Preddie, E., and Verma, D. P. S. (1983). *EMBO J.* **2,** 2333–2339.
Betsche, T., and Gerhardt, B. (1978). *Plant Physiol.* **62,** 590–597.
Bieglmayer, C., Graf, J., and Ruis, H. (1973). *Eur. J. Biochem.* **37,** 379–389.
Blum, J. J. (1982). *Ann. N.Y. Acad. Sci.* **386,** 217–227.
Breidenbach, R. W., and Beevers, H. (1967). *Biochem. Biophys. Res. Commun.* **27,** 462–469.
Bremer, J., and Osmundsen, H. (1984). *In* "Fatty Acid Metabolism and its Regulation" (S. Numa, ed.), pp. 113–147. Elsevier, Amsterdam.
Burgess, N., and Thomas, D. R. (1986). *Planta* **67,** 58–65.
Chandlee, J. M., and Scandalios, J. G. (1984). *Proc. Natl. Acad. Sci. U.S.A.* **81** 4903–4907.
Choinski, J. S., Jr., and Trelease, R. N. (1978). *Plant Physiol.* **62,** 141–145.
Cooper, T. G., and Beevers, H. (1969). *J. Biol. Chem.* **244,** 3514–3520.
Dommes, J., and Northcote, D. H. (1985). *Planta* **166,** 550–556.
Dommes, P., Dommes, V., and Kunau, W.-H. (1983). *J. Biol. Chem.* **258,** 10846–10852.
Dommes, V., Baumgart, C., and Kunau, W.-H. (1981). *J. Biol. Chem.* **256,** 8259–8262.
Dommes, V., Luster, W., Cvetanovic, M., and Kunau, W.-H. (1982). *Eur. J. Biochem.* **125,** 335–341.
Frevert, J., and Kindl, H. (1978). *Eur. J. Biochem.* **92,** 35–43.
Frevert, J., and Kindl, H. (1980a). *Eur. J. Biochem.* **107,** 79–86.
Frevert, J., and Kindl, H. (1980b). *Hoppe-Seyler's Z. Physiol. Chem.* **361,** 537–542.
Frevert, J., Köller, W., and Kindl, H. (1980). *Hoppe-Seyler's Z. Physiol. Chem.* **361,** 1557–1565.
Fujiki, Y., and Lazarow, P. B. (1985). *J. Biol. Chem.* **260,** 5603–5609.
Fujiki, Y., Rachubinski, R. A., and Lazarow, P. B. (1984). *Proc. Natl. Acad. Sci. U.S.A.* **81,** 7127–7131.
Fujiki, Y., Rachubinski, R. A., Mortensen, R. M., and Lazarow, P. B. (1985). *Biochem. J.* **226,** 697–704.
Furuta, S., Hashimoto, T., Miura, S., Mori, M., and Tatibana, M. (1982). *Biochem. Biophys. Res. Commun.* **105,** 639–646.
Gemmrich, A. R. (1981). *Z. Pflanzenphysiol.* **102,** 69–80.
Gerdes, H. H., and Kindl, H. (1986). *Planta* **167,** 166–174.
Gerdes, H. H., Behrends, W., and Kindl, H. (1982). *Planta* **156,** 572–578.
Gerhardt, B. (1978). "Microbodies/Peroxisomen pflanzlicher Zellen." Springer-Verlag, Berlin and New York.
Gerhardt, B. (1983). *Planta* **159,** 238–246.
Gerhardt, B. (1984) *In* "Structure, Function and Metabolism of Plant Lipids" (P.-A. Siegenthaler and W. Eichenberger, eds.), pp. 189–192, Elsevier, Amsterdam.

Gerhardt, B. (1985). *Phytochemistry* **24,** 351–352.

Gietl, C., and Hock, B. (1984). *Planta* **162,** 261–267.

Gietl, C., Lottspeich, F., and Hock, B. (1985). *Planta* **169,** 555–558.

Goldman, B. M., and Blobel, G. (1978). *Proc. Natl. Acad. Sci. U.S.A.* **75,** 5066–6070.

Gonzales, E., and Delsol, M. A. (1981). *Plant Physiol.* **67,** 550–554.

Goodman, J. M. (1985). *J. Biol. Chem.* **260,** 7108–7113.

Goodman, J. M., Scott, C. W., Donahue P. N., and Atherton, J. P. (1984). *J. Biol. Chem.* **259,** 8485–8493.

Hajra, A. K., and Bishop, J. E. (1982). *Ann. N.Y. Acad. Sci.* **386,** 170–182.

Hamilton, B., Hofbauer, R., and Ruis, H. (1982). *Proc. Natl. Acad. Sci. U.S.A.* **79,** 7609–7613.

Hanks, J., and Tolbert, N. E. (1982). *Ann. N.Y. Acad. Sci.* **386,** 420–421.

Harder, W. (1983). *Adv. Microb. Physiol.* **24,** 1–40.

Hashimoto, T. (1982). *Ann. N.Y. Acad. Sci.* **386,** 5–12.

Hicks, D. B., and Donaldson, R. P. (1982). *Arch. Biochem. Biophys.* **215,** 280–288.

Huang, A. H. C., and Beevers, H. (1973). *J. Cell Biol.* **58,** 379–389.

Huang, A. H. C., Liu, K. D. F., and Youle, R. J. (1976). *Plant Physiol.* **58,** 110–113.

Huang, A. H. C., Trelease, H. N., and Moore, T. S., Jr. (1983). "Plant Peroxisomes." Academic Press, New York.

Hutton, D., and Stumpf, P. K. (1969). *Plant Physiol.* **44,** 508–516.

Hutton, D., and Stumpf, P. K. (1971). *Arch. Biochem. Biophys.* **142,** 48–60.

Jameel, S., El-Gul, T., and McFadden, B. A. (1985). *Arch. Biochem. Biophys.* **236,** 72–81.

Johanson, H. A., Hill, J. M., and McFadden, B. A. (1974). *Biochim. Biophys. Acta* **364,** 327–340.

Khan, F. R., Salemuddin, M., Siddiqi, M., and McFadden, B. A. (1977). *Arch. Biochem. Biophys.* **183,** 15–23.

Kindl, H. (1982a). *Ann. N.Y. Acad. Sci.* **386,** 314–328.

Kindl, H. (1982b). *Int. Rev. Cytol.* **80,** 193–229.

Kindl, H. (1984). *In* "Comprehensive Biochemistry: Fatty Acid Metabolism and Its Regulation" (S. Numa, ed.), pp. 181–202. Elsevier, Amsterdam.

Kindl, H. (1987). *Eur. J. Cell Biol.* (in press).

Kindl, H., and Kruse, C. (1983). *In* "Methods in Enzymology" (S. Fleischer and B. Fleischer, eds.), Vol. 96, Part J, pp. 700–715, Academic Press, New York.

Kindl, H., Koller, W., and Frevert, J. (1980). *Hoppe-Seyler's Z. Physiol. Chem.* **361,** 465–467.

Kionka, C., Löffler, H.-G., and Kunau, W.-H. (1985). *J. Bacteriol.* **161,** 153–157.

Kirsch, T., and Kindl, H. (1986). *J. Biol. Chem.* **261,** 8570–8575.

Köller, W., and Kindl, H. (1977). *Arch. Biochem. Biophys.* **181,** 236–248.

Köller, W., and Kindl, H. (1978). *Z. Naturforsch.* **C33c,** 962–968.

Köller, W., and Kindl, H. (1980). *Hoppe-Seyler's Z. Physiol. Chem.* **361,** 1437–1444.

Kruse, C., and Kindl, H. (1982). *Ann. N.Y. Acad. Sci.* **386,** 499–502.

Kruse, C., and Kindl, H. (1983a). *Arch. Biochem. Biophys.* **223,** 618–628.

Kruse, C., and Kindl, H. (1983b). *Arch. Biochem. Biophys.* **223,** 629–638.

Kruse, C., Frevert, J., and Kindl, H. (1981). *FEBS Lett.* **129,** 36–38.

Lazarow, P. B., and DeDuve, C. (1973a). *J. Cell Biol.* **59,** 491–506.

Lazarow, P. B., and DeDuve, C. (1973b) *J. Cell Biol.* **59,** 507–524.

Ledeboer, A. M., Edens, L., Maat, J., Visser, C., Bos, J. W., Verrips, C. T., Janowicz, Z., Eckart, M., Roggenkamp, R., and Hollenberg, C. P. (1985). *Nucleic Acids Res.* **13,** 3063–3082.

Liu, K. D. F., and Huang, A. H. C. (1977). *Plant Physiol.* **59,** 777–782.

Longo, G. P., Pedretti, M., Rossi, G., and Longo, C. P. (1979). *Planta* **145,** 209–217.

Macey, M. J. K. (1983). *Plant Sci. Lett.* **30,** 53–60.

Macey, M. J. K., and Stumpf, P. K. (1983). *Plant Sci. Lett.* **28,** 207–212.

Martin, C., Beeching, J. R., and Northcote, D. H. (1984). *Planta* **162,** 68–76.

Mettler, J. J., and Beevers, H. (1980). *Plant Physiol.* **66,** 555–560.

Miernyk, J. A., and Trelease, R. N. (1981a). *Plant Physiol.* **67,** 341–346

Miernyk, J. A., and Trelease, R. N. (1981b). *Plant Physiol.* **67,** 875–881.

Miernyk, J. A., and Trelease, R. N. (1981c). *FEBS Lett.* **129,** 139–141.

Moreau, R. A., and Huang, A. H. C. (1979). *Arch. Biochem. Biophys.* **194,** 422–430.

Moreno de la Garza, M., Schultz-Borchard, J., Crabb, J. W., and Kunau, W. H. (1985). *Eur. J. Biochem.* **148,** 285–291.

Newcomb, E. H., Tandon, S. R., and Kowal, R. R. (1985). *Protoplasma* **125,** 1–12.

Nguyen, T., Zelechowski, M., Foster, V., Bergmann, H., and Verma, D. P. S. (1985). *Proc. Natl. Acad. Sci. U.S.A.* **82,** 5040–5044.

Ohnishi, J., Yamozaki, M., and Kanai, R. (1985). *Plant Cell Physiol.* **26,** 797–803.

Osumi, T., Ishii, N., Hijikata, M., Kamijo, K., Ozawa, H., Furuta, S., Miyazawa, S., Kondo, K., Inoue, K., Kagamiyama, H., and Hashimoto, T. (1985). *J. Biol. Chem.* **260,** 8905–8910.

Patel, T. R., and McFadden, B. A. (1977). *Arch. Biochem. Biophys.* **183,** 24–30.

Rachubinski, R. A., Fujiki, Y., Mortensen, R. M., and Lazarow, P. B. (1984). *J. Cell Biol.* **99,** 2241–2246.

Ramsey, R. R., and Singer, T. P. (1984). *Biochem. J.* **221,** 489–497.

Roa, M., and Blobel, G. (1983). *Proc. Natl. Acad. Sci. U.S.A.* **80,** 6872–6876.

Roberts, L. M., and Lord, J. M. (1982). *Eur. J. Biochem.* **119,** 43–49.

Schiefer, S., Teifel, W., and Kindl, H. (1976). *Hoppe-Seyler's Z. Physiol. Chem.* **357,** 63–175.

Schopfer, P., Bajracharya, D., Bergeld, R., and Falk, H. (1976). *Planta* **133,** 73–80.

Tanaka, A., Osumi, M., and Fukui, S. (1982). *Ann. N.Y. Acad. Sci.* **386,** 183–199.

Thomas, D. R., and Wood, C. (1986). *Planta* **168,** 261–266.

Titus, D. E., and Becker, W. M. (1985). *J. Cell Biol.*

Tolbert, N. E. (1971). *Annu. Rev. Plant Physiol.* **22,** 45–74.

Tolbert, N. E. (1981). *Annu. Rev. Biochem.* **50,** 133–157.

Tolbert, N. E. (1982). *Ann. N.Y. Acad. Sci.* **386,** 254–268.

Trelease, R. N., Becker, W. M., Gruber, P. J., and Newcomb, E. H. (1971). *Plant Physiol.* **48,** 461–475.

Trelease, R. N. (1984). *Annu. Rev. Plant Physiol.* **35,** 321–347.

Visser, N., Opperdoes, F. R., and Borst, P. (1981). *Eur. J. Biochem.* **118,** 521–527.

Walk, R.-A., and Hock, B. (1976). *Eur. J. Biochem.* **71,** 25–32.

Wanner, G., and Theimer, R. R. (1982). *Ann. N.Y. Acad. Sci.* **386,** 269–284.

Wood, C., Noh Hj Jalil, M., McLaren, I., Yong, B. C. S., Ariffin, A., McNeil, P. H., Burgess, N., and Thomas, D.-R. (1984). *Planta* **161,** 255–260.

Wood, C., Burgess, N., and Thomas, D.-R. (1986). *Planta* **167,** 54–57.

Yamaguchi, J., Nishimura, M., and Akazawa, T. (1984). *Proc. Natl. Acad. Sci. U.S.A.* **81,** 4809–4813.

Yang, S. Y., and Schulz, H. (1983). *J. Biol. Chem.* **258,** 9780–9785.

Zehler, H., Thomson, K.-S., and Schnarrenberger, C. (1984). *Physiol. Plant.* **60,** 1–8.

Zwart, K. B., Veenhuis, M., and Harder, W. (1983). *Antonie van Leeuwenhoek* **49,** 369–375.

Oxidative Systems for Modification of Fatty Acids: The Lipoxygenase Pathway

3

BRADY A. VICK
DON C. ZIMMERMAN

I. INTRODUCTION

There are four major enzyme systems operative in plants by which fatty acids are oxidatively modified: α-oxidation, β-oxidation, ω-oxidation, and the lipoxygenase (LOX) pathway. Chapter 2 covers the topic of β-oxidation of fatty acids in plants, and this chapter will present an overview of recent advances in plant LOX research. The α- and ω-oxidation pathways will not be reviewed here.

The existence of plant enzymes that catalyze the oxidation of lipids by molecular oxygen has been recognized for more than 50 years (André and Hou, 1932). Historically, the enzyme responsible for the oxygenation reaction has been called carotene oxidase, fat oxidase, lipoxidase, and most recently, lipoxygenase (LOX). During the 1940s the oxidized lipid products were identified as fatty acid hydroperoxides (Bergström, 1945). Final characterization of the products was completed by Privett *et al.* (1955).

Despite the wealth of knowledge about the mechanism of the LOX reaction and the characterization of its products, much less progress has been made in understanding the physiological role of this enzyme in plants. Nevertheless, research interest in LOX remains high. Much attention has focused on commercial aspects of LOX action, with emphasis on the characterization of the reactions that produce flavors and odors resulting from the breakdown of lipid hydroperoxides in food products. Some of these flavors and odors are desirable, others are not.

Interest in LOX from a physiological view is increasing. Recent advances in mammalian research have demonstrated the powerful regulatory activity of oxygenated fatty acid metabolites that result from LOX and cyclooxygenase reactions in animals. The importance of fatty acid hydroperoxides, leukotrienes, prostaglandins, prostacyclin, and thromboxane in mammals has undoubtedly contributed to the continuing interest in LOX in plants.

Several reviews on plant LOX have appeared in recent years. The comprehensive review by Galliard and Chan (1980) in Vol. 4, Chapter 5 of this series encompasses most of the information through 1979 regarding the properties of LOX and the mechanism of the oxygenation reaction it catalyzes. The reader is referred to this chapter for background information on LOX. Other recent reviews are by Nicolas and Drapron (1981), Gardner (1980, 1985a, 1987), and Vliegenthart and Veldink (1982). Comparisons between LOXs from plant and mammalian sources have been made in reviews by Schewe *et al.* (1986) and Kühn *et al.* (1986). Extensive reviews by Axelrod (1974), Veldink *et al.* (1977), Eskin *et al.* (1977), Galliard (1974, 1975, 1978), and Grosch (1972), covering various aspects of plant LOX, appeared during the 1970s.

This chapter will examine the advances in plant LOX research since 1979. A special effort has been made to focus on physiological roles in which LOX may participate. A clear theme emerging from this effort is that a unique physiological role for LOX probably does not exist in plants. It likely has a more general function of responding to plant requirements by synthesizing fatty acid hydroperoxide, the central intermediate for a variety of diverging metabolic pathways. In this respect the LOX pathway in plants may be similar to the mammalian LOX or cyclo-oxygenase pathways. In animals, the hydroperoxide or endoperoxide products of these two enzymes are transformed into a variety of leukotrienes and prostaglandins, depending on the nature of the cellular stimulus.

II. THE LIPOXYGENASE REACTION

Lipoxygenase (E.C. 1.13.11.12) catalyzes the incorporation of molecular oxygen into polyunsaturated fatty acids according to the following reaction:

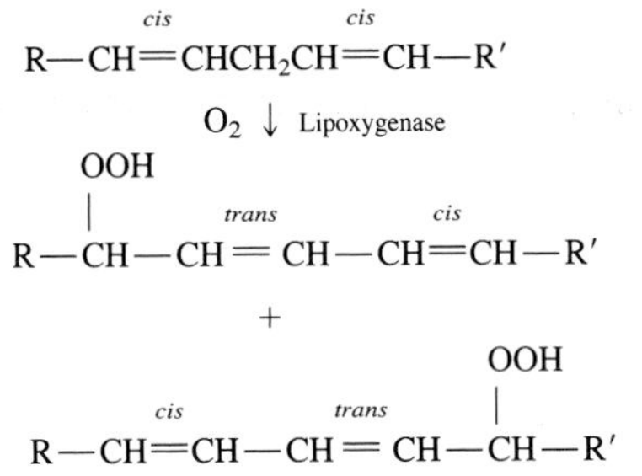

$$\overset{cis}{R-CH}=CHCH_2\overset{cis}{CH}=CH-R'$$

$$O_2 \downarrow \text{Lipoxygenase}$$

$$\underset{|}{\overset{OOH}{R-CH}}-CH\overset{trans}{=}CH-\overset{cis}{CH}=CH-R'$$

$$+$$

$$\overset{cis}{R-CH}=CH-CH\overset{trans}{=}CH-\underset{|}{\overset{OOH}{CH}}-R'$$

Substrates for LOX are fatty acids that possess a cis,cis-1,4-pentadiene structure. The most common plant fatty acids which have this structure are linoleic and (9,12,15)-linolenic acids, but other fatty acids having this structure are also effective substrates to varying degrees. Usually, but not always, oxygen is incorporated into the fatty acid at the n-6 or n-10 position. The product of the reaction is a conjugated hydroperoxydiene in which the cis double bond attacked by oxygen moves into conjugation with a neighboring cis double bond. In the process,the migrating double bond assumes the trans configuration. The carbon bearing the hydroperoxide group, whether it is the n-6 or n-10 carbon, has the S stereoconfiguration.

III. DISTRIBUTION OF LIPOXYGENASES

A. Distribution within the Plant Kingdom

Pinsky et al. (1971) surveyed 40 plant materials for LOX activity based on the uptake of oxygen by plant extracts in the presence of linoleic acid. Another list of plants that have LOX activity has been tabulated by Axelrod (1974). Together, more than 60 plant species have been reported to contain the enzyme. Other plants more recently surveyed for LOX activity are given in Table I. The assertion by Axelrod (1974) that undetected LOX in a plant probably reflects an insensitive or unreliable detection method may be correct in view of the wide distribution of the enzyme

TABLE I

Sources and Properties of Some Recently Investigated Lipoxygenases

Source	Botanical name	MW	Optimum pH	Ratio of 9:13 hydroper-oxides[a]	References
Apple	*Malus sylvestris* Mill.	206,000	6.9	18:82	Feys *et al.* (1982); Grosch *et al.* (1977a)
Baker's yeast	*Saccharomyces cerevisiae*		6.5	50:50	Shechter and Grossman (1983)
Cotton	*Gossypium hirsutum* L.	100,000	6.0–6.5	6:94	Vick and Zimmerman (1981a)
Cucumber	*Cucumis sativus* L.		5.5	15:85	Wardale and Lambert (1980); Sekiya *et al.* (1979)
Gooseberry	*Ribes uva-crispa* L.		6.5	50:50	Kim and Grosch (1978)
Grapes	*Vitis vinifera* L.		8	20:80	Cayrel *et al.* (1983)
Pear	*Pyrus communis* L.		6.0	30:70	Kim and Grosch (1978)
Pinto bean	*Phaseolus vulgaris* L.		6.9		McCurdy *et al.* (1983)
Rice embryo isoenzymes:	*Oryza sativa* L.	95,000	4–7	97:3	Yamamoto *et al.* (1980a,b)
L-1			4.5	52:48	Ida *et al.* (1983); Ohta *et al.* (1986)
L-2			5.5	51:49	
L-3			7.0	83:17	
Strawberry	*Fragaria vesca* L.		6.5	20:80	Kim and Grosch (1978)
Sunflower	*Helianthus annuus* L.	>250,000	6.2		Leoni *et al.* (1985)
Tea leaves	*Thea sinensis* L.			16:84	Kajiwara *et al.* (1980)
Winged bean isoenzymes:	*Psophocarpus tetragonolobus* L.				
FI		80,000	6.0		Truong *et al.* (1982b)
FII		80,000	5.8		

[a]Linoleic acid substrate.

already substantiated. Undoubtedly, endogenous LOX inhibitors such as chlorophyll (Cohen *et al.*, 1984) or α-tocopherol (Grossman and Waksman, 1984) have complicated the detection of LOX in plant extracts. Other LOXs, such as that from *Chlorella pyrenoidosa* (Zimmerman and Vick, 1973), may be oxygen-sensitive and have escaped detection. LOXs are also present in the fungi *Fusarium oxysporum* (Matsuda *et al.*, 1979) and *Saprolegnia parasitica* (Hamburg, 1986).

B. Distribution within Plant Organs

According to early reports, LOX is present in such a wide variety of plant organs that no meaningful conclusions about its physiological function could be made (Galliard and Chan, 1980). In recent years, however, sophisticated techniques have been employed to establish the precise location of LOX within plant organs and within plant cells.

Vernooy-Gerritsen *et al.* (1983) developed an immunofluorescence staining technique to determine which cells of various plant organs contain LOX. They applied rabbit antibodies directed against two soybean LOX isoenzymes (LOX-1 and LOX-2) to sections of soybean cotyledons, hypocotyls, and leaves at various stages of germination. With the addition of a fluorescently labeled anti-rabbit IgG, the cells that contained LOX could be distinguished by their fluorescence. Their results showed that at 20 h into germination, the predominant location of LOX in the cotyledons was in the cytoplasm of storage parenchyma cells (Fig. 1A). By 40 h, the fluorescence (i.e., LOX enzymes) in the storage parenchyma cells directly beneath the epidermis had nearly disappeared (Fig. 1B), but the abaxial epidermis (lower surface) showed high concentrations of LOX. After 67 h, the LOX was confined primarily to the epidermis, hypodermis, and cells surrounding the vascular bundle (Fig. 1C).

In hypocotyls the fluorescent zones, signifying the presence of LOX, appeared first in the parenchyma cells below the epidermis, then in the pith and cortex parenchyma, and finally in the vascular cylinder. Initially, the young leaves did not fluoresce after immunofluorescent staining, but by 20 h all the leaf cells fluoresced. One isoenzyme, LOX-2, produced especially intense fluorescence labeling in leaves. Mature leaves of 2-month-old plants showed a very weak fluorescence around chloroplasts of spongy and palisade parenchyma cells.

C. Subcellular Distribution

Information about the location of LOX within a cell would be valuable in determining a physiological role for the enzyme. Many attempts have been made to establish the subcellular location of LOX, but usually with ambiguous or conflicting results. Nevertheless, there is increasing evidence that photosynthetic leaf tissue has high LOX activity associated with the chloroplasts. Cytoplasm and vacuoles have also been reported as sites of LOX activity in other tissues.

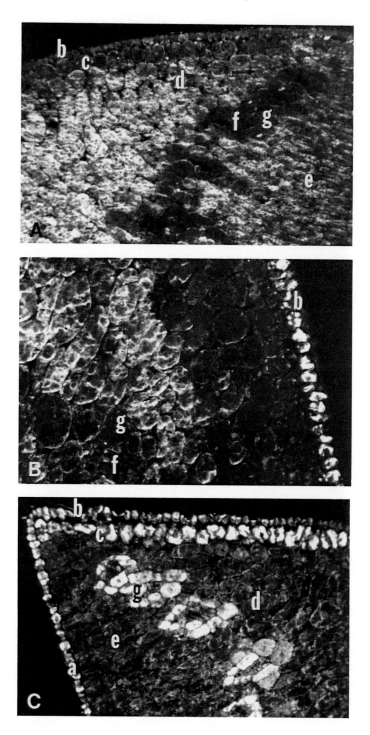

Douillard and Bergeron (1978, 1979) found that when thylakoids from wheat shoots were purified on a sucrose gradient, the LOX activity equilibrated at a density slightly higher than the chlorophyll marker of the thylakoids. They attributed the higher density, LOX-rich fractions to the presence of "juvenile" plastid membranes. Thus, in wheat shoots LOX appeared to be more active in young plastids than in mature chloroplasts. However, this interpretation did not consider the inhibitory effect of chlorophyll on LOX activity later reported by Cohen et al. (1984). It is possible that the mature chloroplasts in their native state were also high in LOX activity, but that the high concentrations of chlorophyll reduced the LOX activity in vitro. Approximately 60% of the LOX in an extract of young wheat shoots was bound to thylakoid membranes and the rest was soluble, some of which may have been in the stroma (Douillard, 1979).

Douillard and Bergeron (1981) also examined the properties of LOX in the aerial parts of 7- to 10-day-old pea shoots. Lipoxygenase that was bound tightly to the thylakoids accounted for 5–10% of the total homogenate activity. At least 60% of the activity in intact chloroplasts was in the stroma. Five isoenzymes of LOX were identified in pea shoots (Douillard et al., 1982), four of which were present in the chloroplast stroma. In pea roots, however, LOX paralleled acid phosphatase activity on sucrose or Ficoll gradients and was determined to be associated with "lysosomal" organelles, presumably vacuoles (Wardale and Galliard, 1977). In contrast, the same authors found that LOX from cauliflower florets was present in an organelle with a density similar to, but different from, plastids. Potato LOX was not associated with a particulate fraction to any significant degree. Like the enzyme from pea roots, LOX from cucumber flesh or peel coincided with acid phosphatase on a sucrose gradient and was also concluded to be located in the vacuoles (Wardale and Lambert, 1980).

Vernooy-Gerritsen et al. (1984) probed the intracellular location of LOX in tissues of germinating soybean seeds by the use of an indirect, immunolabeling procedure for electron microscopy. Ultrathin cryosections of soybean tissues were first incubated with rabbit anti-LOX-1 or anti-LOX-2. Gold-labeled staphylococcal protein A, which binds to immunoglobulin molecules, was then added to the tissue sections. The protein A–gold complex produced electron-opaque areas in the electron micrograph, presumably at the sites of LOX protein. In storage parenchyma cells of cotyledons, the gold label was randomly distributed throughout the cytoplasm during the first 7 days of germination (Fig. 2A and B). The label was not

Fig. 1. Immunofluorescence-stained cotyledons of germinating soybean seedlings exposed to anti-LOX antibodies. (A) Cotyledon after 20 h of germination showing fluorescing storage parenchyma tissue (anti-soybean LOX-1). (B) Cotyledon after 40 h of germination with fluorescent abaxial epidermis and decreased fluorescence in storage parenchyma cells lying under the epidermis (anti-soybean LOX-1). (C) Cotyledon after 67 h of germination showing high fluorescence in epidermis, hypodermis, and vascular bundle sheath (anti-soybean LOX-2). a, Adaxial epidermis; b, abaxial epidermis; c, abaxial hypodermis; d, outer spongy storage parenchyma; e, inner-palisade storage parenchyma; f, vascular tissue; g, vascular bundle sheath. [Reproduced from Vernooy-Gerritsen et al. (1983); by permission.]

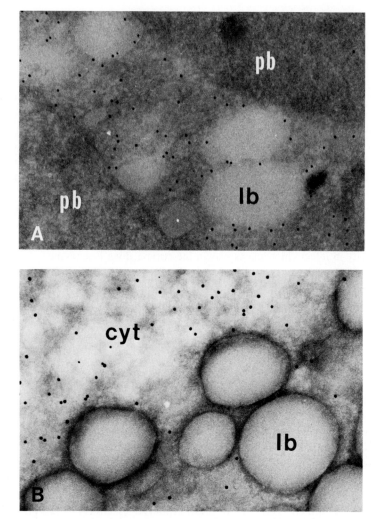

Fig. 2. Electron micrographs of storage parenchyma cells of germinating soybean cotyledons exposed to anti-soybean LOX-2 and protein A–gold complex. (A, B) Cotyledon after 26 h germination showing nonlabeled protein bodies (pb) and lipid bodies (lb) surrounded by highly labeled cytoplasm (cyt). [Reproduced from Vernooy-Gerritsen *et al.* (1984); by permission.]

present in protein bodies, lipid bodies, mitochondria, or other organelles. In cotyledons, LOX (especially type 2) in the cells surrounding the vascular bundle was predominantly associated with an aberrant, nonfusing type of protein body. In cells of cotyledon epidermis and in young leaf parenchyma cells, most of the gold label was found near regions in the cytoplasm where vacuoles were about to be formed. Cytoplasm and protein bodies were sites of labeling in cotyledon hypodermal cells.

IV. PROPERTIES OF LIPOXYGENASE ENZYMES

A. Soybean Lipoxygenase

1. Physical and Chemical Properties

Soybean LOX has been studied more extensively than any other plant LOX, and the chapter by Galliard and Chan (1980) in Vol. 4 of this series has summarized its salient properties. It should suffice to review here that there are two main types of LOX which can be distinguished primarily by differences in their pH activity profiles. Type 1 enzymes have optimum activity around pH 9 while type 2 enzymes are most active between pH 6.5 and 7. The enzyme from soybean consists of at least three isoenzymes. LOX-1 is a type 1 enzyme and is maximally active at pH 9. LOX-2 and LOX-3 are type 2 enzymes. LOX-2 has its pH optimum at pH 6.5 and LOX-3 has a broad pH activity profile centered around pH 7. The isoelectric points for LOX-1, LOX-2, and LOX-3 are 5.68, 6.25, and 6.15, respectively. All have a molecular weight of approximately 100,000 and contain one atom of iron per molecule. LOX-1 produces mainly the 13-hydroperoxy isomer during the reaction with linoleic acid, while LOX-2 yields approximately equal proportions of the 9- and 13-isomers. Ramadoss and Axelrod (1982) have shown that arachidonic acid is the more active substrate for LOX-2, whereas linoleic acid is more active with LOX-3. LOX-1 shows no difference between the two substrates. Stimulation by Ca^{2+} occurs with LOX-2, whereas Ca^{2+} inhibits LOX-3. A peculiarity of the aerobic LOX-3 reaction is the formation of oxodienoic acids in addition to hydroperoxide products (Ramadoss and Axelrod, 1982). The type 2 isoenzymes also have different pigment cooxidation properties than LOX-1 (see Section V,B).

Grosch et al. (1977b) have suggested that LOX-3 is actually a mixture of two LOX-2 isoenzymes. This now seems unlikely in view of the unique chemical properties cited above. Dreesen et al. (1982) have reported still another isoenzyme of LOX from soybeans that was stimulated by Ca^{2+} ions. The isoenzyme was apparently different from both LOX-1 and LOX-2 because its pH optimum at 7.5 and isoelectric point of 5.90 differed from published values for the other forms. However, the authors noted that such variances could be due to varietal differences.

Although several reports on the amino acid composition of LOX-1 have been published, there is no firm agreement on the number of cysteine and cystine residues present. Stevens et al. (1970) reported four cysteine and two disulfide residues, whereas Stan and Diel (1976) found 3 cysteine and 3 disulfide residues. The results of Mitsuda et al. (1967) indicated the presence of 2 disulfide groups while Axelrod (1974) has suggested that up to 12 half-cystines may be present. The differing results from the various groups reflect the difficult nature of these determinations. There is general agreement that the free sulfhydryl groups of the native enzyme are inaccessible to the common reagents that interact with sulfhydryl groups. Spaapen et al. (1980) used small organic mercurial reagents to determine the number of sulfhydryl groups in soybean LOX-1. Their results showed that the enzyme contained five free sulfhydryl groups with no disulfide bridges. Interestingly, they found that modification of LOX-1 with methylmercuric halides caused the enzyme to acquire some LOX-2-type properties. Similar results were reported by Feiters *et al.* (1986), who showed that methylmercuric iodide modified three sulfhydryl groups of LOX-1, whereas only two were modified in type 2 LOXs. Only the activity of LOX-1 was affected by methylmercuric iodide, suggesting that type 1 LOX possesses one sulfhydryl residue near the active site. Thus, the differences between the two isoenzymes may be due in part to variations in cysteine content. Tryptophan also plays an essential catalytic role in LOX action. According to Klein *et al.* (1985), there are two tryptophan residues at the active site of soybean LOX-1 that are necessary for both linoleate oxidation and pigment cooxidation activity.

2. Immunological Considerations

Immunological properties of soybean LOX isoenzymes have been compared by several groups. Rabbit polyclonal antibodies prepared with LOX-2 cross-reacted with LOX-1 in an Ouchterlony double-diffusion test (Peterman and Siedow, 1985a). The line of identity between LOX-1 and LOX-2 indicated that the two isoenzymes had at least two common antigenic determinants, and thus were structurally related. However, the two forms could be immunologically distinguished because the anti-LOX-2 inhibited the catalytic activity of LOX-2 but not of LOX-1. Similar conclusions were drawn by Ramadoss and Axelrod (1982) from observations of the cross-reactivity of LOX-1 and LOX-3 with goat anti-LOX-2.

Lipoxygenases from newly differentiated tissues such as roots, hypocotyls, and leaves were immunologically distinct from seed-derived LOX-2; that is, they did not cross-react in the double-diffusion test with anti-LOX-2 (Peterman and Siedow, 1985a). However, the seedling LOXs did form soluble, noninhibitory complexes with anti-LOX-2, demonstrating that they were structurally related to, yet different than, seed-derived LOX-2.

B.   Lipoxygenase from Other Sources

1. Physical and Chemical Properties of Isoenzymes

Isoenzymes of LOX are found in many plant species. The properties of the LOX isoenzymes from alfalfa, barley, fava bean, lupine, lentil, pea, potato tuber, and wheat flour have been reviewed by Eskin *et al.* (1977) and Galliard and Chan (1980). Among the more recently studied plants (see Table I), rice germ was reported to contain three isoenzymes of LOX (Ida *et al.*, 1983). The major rice seed LOX isoenzyme, L-3, had a pH optimum at 7.0, while the minor isoenzymes were most active at pH 4.5 and 5.5, respectively. Like the soybean isoenzymes, the rice isoenzymes have different product specificities. The L-3 isoenzyme produced mainly the 9-hydroperoxy isomer of linoleic acid, while the two minor isoenzymes catalyzed the formation of approximately equal amounts of the 9- and 13-hydroperoxy isomers. Following germination of the rice seed, the lipoxygenase activity in the seedling increased 20-fold by the third day. This was due mainly to increased synthesis of the L-2 isoenzyme (Ohta *et al.*, 1986). Borisova *et al.* (1981) reported the presence of three LOX isoenzymes in cucumber roots. Winged bean, a tropical legume, has three LOX isoenzymes that can be isolated from the seed (Truong *et al.*, 1982a).

2. Immunological Considerations

Antibodies raised against LOXs from soybean or potato cross-reacted with the other's LOX in double-diffusion tests and with eggplant LOX (Trop *et al.*, 1974). This result indicated a similarity in the structures of the LOXs from the three sources. The fact that the antibodies also cross-inhibited the LOX reactions further suggested that a close structural similarity exists among the active sites of the three enzymes. Similarly, another report concluded that a cross-reaction occurred between soybean LOX-2 (but not LOX-1) antisera and LOXs from other members of the Leguminosae family (Vernooy-Gerritsen *et al.*, 1982), indicating a structural similarity between soybean LOX-2 and the LOXs from some other legumes. However, their results showed no cross-reactivity between LOX-2 antisera and the LOXs from corn, wheat germ, flax, potato, or eggplant. The reason for the differing results obtained by the two laboratories for cross-reactivity of the potato and eggplant LOXs is unclear, but may be related to the purity of the LOXs used in antibody preparation.

V. PROPERTIES OF THE LIPOXYGENASE REACTION

A. Mechanism: The Role of Iron

1. The Aerobic Lipoxygenase Reaction

Iron is essential to the catalytic role of LOX. Each molecule of LOX contains one atom of iron that alternates between the Fe(II) and Fe(III) states during catalysis. One of the characteristics of the LOX reaction is the initial lag period which can only be overcome by the addition of product hydroperoxide. To explain this mechanistically, Vliegenthart *et al.* (1982), studying soybean LOX-1, have put forward the concept of two enzyme populations, one containing Fe(II) and the other containing Fe(III) (Fig. 3). EPR studies (Slappendel *et al.*, 1981) have shown that purified LOX-1 (the "native" enzyme) is predominantly in the Fe(II) form and contains only a very small fraction ($\sim$ 1%) of Fe(III). Recent studies by Feiters *et al.* (1985) indicate that native Fe(II)-LOX does not coordinate dioxygen, which makes it unlikely that this species catalyzes hydroperoxide formation. Consequently, it is not clear why the reaction of the native enzyme with substrate fatty acid should occur at all. It could be, however, that the small amount (1%) of Fe(III)-LOX present in the purified enzyme catalyzes the initial formation of hydroperoxide via the aerobic ferric cycle (Fig. 3). The resulting hydroperoxide product could then react with the Fe(II) enzyme and oxidize it to the "active"

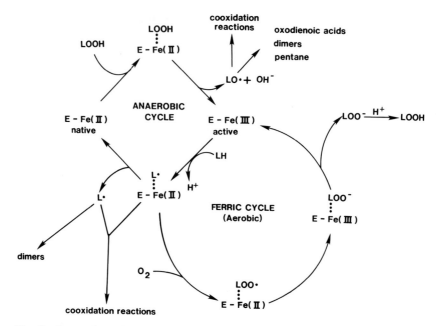

Fig. 3. Proposed reaction scheme for soybean LOX under aerobic and anaerobic conditions. [Adapted from Veldink *et al.* (1977); by permission.]

Fe(III) enzyme. As more Fe(III)-LOX became available, the reaction would accelerate. The scheme is probably more complicated than this, however, because this model would predict the elimination of the lag phase if a molar equivalent of hydroperoxide product were present to oxidize the native enzyme to the active enzyme. In fact, a large excess of hydroperoxide product is required to eliminate the lag phase completely, indicating that other steps are also operative in the activation of the enzyme (de Groot *et al.*, 1975).

There are two forms of the active enzyme that can be detected *in vitro*. The yellow form, observed when a molar equivalent of hydroperoxide is added, also shows an initial lag period in the oxygenation reaction. When an excess of 13-hydroperoxide is added (the 9-hydroperoxy isomer is ineffective), the enzyme solution turns purple. This is due to the formation of a complex between the yellow enzyme and the hydroperoxide. The purple enzyme apparently has a largely rhombic environment around the iron atom, whereas the symmetry about the iron in the yellow enzyme is predominantly axial (Slappendel *et al.*, 1981). De Groot *et al.* (1975) have speculated that the purple enzyme is the active form during the aerobic reaction. From ^{1}H NMR and magnetic susceptibility measurements of soybean LOX-1, the iron in the native enzyme was confirmed to be high-spin Fe(II) with largely axial symmetry (Cheesbrough and Axelrod, 1983; Slappendel *et al.*, 1982; Petersson *et al.*, 1985). It has been established through measurements on the yellow form of the activated enzyme that it is in the high-spin Fe(III) state (Slappendel *et al.*, 1981; Cheesbrough and Axelrod, 1983). For a discussion of the stereochemistry of hydrogen abstraction and oxygen insertion, see the review by Veldink *et al.* (1977).

The mechanism of the type 2 LOX reaction is not as well characterized as the type 1 mechanism. It is thought to be a free-radical process similar to type 1, but with some deviations. Under optimum pH conditions for the type 2 enzyme (pH 6.6), radicals dissociate more easily from the enzyme than from its type 1 counterpart. This may explain why some type 2 enzymes produce relatively high proportions of carbonyl compounds absorbing at 285 nm (oxodienes) under aerobic conditions. Under such a scheme oxygen would also be free to attack the uncomplexed fatty acid radical at either the 9- or 13-position carbon of linoleic acid without stereochemical restrictions. This proposal is consistent with the observation that type 2 LOX produces approximately equal proportions of 9- and 13-hydroperoxides, each having nearly equal amounts of the *R* or *S* configurations (Van Os *et al.*, 1979). The participation of free peroxy radicals in the type 2 reaction has been confirmed by oxygen isotope scrambling experiments (Schieberle *et al.*, 1981). Hausknecht and Funk (1984) also concluded from inhibitor studies that free-radical intermediates are integral to the type 2 mechanism, and that enzyme activation may be dependent on free-radical intermediates rather than on the 13-hydroperoxide product.

Singlet oxygen has been detected in soybean LOX reactions catalyzed by certain isoenzymes (Kanofsky and Axelrod, 1986); however, its significance is not yet understood. The reaction catalyzed by LOX-3 produces the most singlet oxygen, whereas singlet oxygen production in the LOX-1 reaction occurs only at low oxygen concentrations. Singlet oxygen is not detectable in the LOX-2 reaction.

The positional specificity of LOX appears to depend on how the enzyme recognizes the methylene group from which hydrogen is initially removed. For some enzymes, the site is determined by the distance from the methyl end of the molecule (Hamberg and Samuelsson, 1967), and for others by the distance from the carboxyl end (Kühn *et al.*, 1985).

2. *The Anaerobic Lipoxygenase Reaction*

Soybean LOX-1 also undergoes an anaerobic reaction which has been described in detail in previous reviews (Galliard and Chan, 1980; Veldink *et al.*, 1977). When the LOX reaction becomes anaerobic, fatty acid hydroperoxide produced during the aerobic phase binds to native soybean LOX-1 and activates it. As a result of enzyme activation under anaerobic conditions, the hydroperoxide is converted to hydroxyl ions and alkoxy radicals that rearrange or recombine to form oxodienoic acids, dimers, and pentane (Fig. 3). The activated enzyme combines with the substrate fatty acid in the usual manner and abstracts a hydrogen. But because oxygen is unavailable, the fatty acid dissociates as a radical that can recombine to form dimers, and the enzyme returns to the native state.

B. Cooxidation Reactions

The ability of LOX to oxidize carotene and other pigments was known prior to the recognition of its lipid-oxidizing properties (Bohn and Haas, 1928), hence its original name of carotene oxidase. For pigment cooxidation to occur, both substrate fatty acid and LOX must be present in the reaction. The LOX enzyme is an integral part of the cooxidation process because the pigment is not oxidized by the hydroperoxide product alone:

$$LH + O_2 \xrightarrow{\text{LOX}} LOOH$$
$$\underset{X_{red} \;\; X_{ox}}{}$$

Lipoxygenase isoenzymes have varying degrees of effectiveness in catalyzing the cooxidation reaction, although some confusion exists as to whether individual isoenzymes or a combination of them is most effective. Grosch and Laskawy (1979) reported that type 2 soybean LOXs, LOX-2 and LOX-3, are equally effective in cooxidation activity. In contrast, Ramadoss et al. (1978) indicated that LOX-2 and LOX-3 were poor catalysts individually, but that a combination of LOX-1 and LOX-3 or LOX-2 and LOX-3 were effective in promoting the cooxidation of pigments.

Soybean LOX-1 is not an efficient catalyst of the reaction under aerobic conditions, but under anaerobic conditions it is a strong cooxidizing agent in the presence of substrate fatty acid and product hydroperoxide (Axelrod, 1974; Klein et al., 1984). Klein et al. (1984) showed that the anaerobic LOX-1 cooxidation reaction was affected by the presence of antioxidants. Their results suggested that the enzyme–fatty acid radical complex E-Fe(II)···L· (Fig. 3, anaerobic cycle), and not an alkyl, alkoxy, or peroxy radical, was the major pigment oxidizer. In any event, free radicals are generally thought to be the agents responsible for bleaching of pigments during the LOX reaction. Since free radicals are produced in greatest abundance during the anaerobic reaction, the cooxidation ability of a particular isoenzyme may depend in part on its capacity to participate in the anaerobic cycle. It could accomplish this by releasing the radical before it could react with oxygen. This view is supported by experiments which showed that LOX-1 could be chemically altered by modification of its sulfhydryl groups with methylmercuric iodide (Grossman et al., 1984). Under aerobic conditions the modified enzyme bleached pigments in a manner similar to the anaerobic reaction of unmodified LOX-1. This suggests that chemical modification of LOX-1 sulfhydryl groups enhances the dissociation of the radicals from the enzyme complex before the complex can react with oxygen.

Despite the abundant literature on pigment bleaching by LOXs from various sources, an important aspect of this research area has not been reported. The complete characterization of a LOX-oxidized pigment would undoubtedly provide insight into the mechanism of the cooxidation reaction. It is highly probable that several cooxidation mechanisms exist, depending on the pigment substrate. Indeed, experiments by Hildebrand and Hymowitz (1982) suggest that the mechanisms of carotene and chlorophyll bleaching are different.

C. Other Catalytic Properties of Lipoxygenase

1. Oxygen Exchange

When hydroperoxide product labeled with ^{18}O was incubated with soybean LOX-1 in the presence of $^{16}O_2$, an enzyme-catalyzed exchange of oxygen occurred between the two molecules (Matthew and Chan, 1983). Stereoconfiguration of the hydroperoxide was retained during the exchange reaction which was optimal at pH 5.5. While O_2 exchange occurs naturally as a chemical phenomenon, the retention of stereoconfiguration indicated enzyme participation in the exchange process. The experiments suggested that LOXs may be capable of catalyzing reactions of the type:

$$LOOH \underset{-H}{\rightleftharpoons} LOO\cdot \rightleftharpoons L\cdot + O_2$$

2. Double Oxygenation

When fatty acids with more than two double bonds, such as arachidonic acid, are used as a substrate at high concentrations of soybean LOX-1, oxygenation can occur twice within the same molecule (Bild et al., 1977). Van Os et al. (1981) have identified two double-oxygenated isomers that resulted from the reaction with arachidonic acid. One isomer had hydroperoxy groups at carbons 8 and 15, whereas the other was oxygenated at carbons 5 and 15. The hydroperoxy groups had predominantly the S stereoconfiguration at all positions. Under favorable conditions, double-oxygenated fatty acids can undergo rearrangement to prostaglandins (Bild et al., 1978).

D. Inhibitor Studies

A review of LOX inhibitors has been presented by Eskin et al. (1977). Inhibitors such as nordihydroguaiaretic acid, n-propylgallate, and butylated hydroxytoluene are antioxidants that function by reacting with free-radical intermediates of the reaction. Hydrogen peroxide is an analog of fatty acid hydroperoxide that attacks the iron atom of the enzyme and perturbs its coordination sphere (Egmond et al., 1975). Polyunsaturated fatty acid analogs that have acetylenic bonds can also inhibit the reaction. In recent years much effort has been exerted in the study of LOX from mammalian systems. Experiments on the inhibition of mammalian LOX have often used soybean LOX as a model system. Some LOX inhibitors which have recently been investigated are listed in Table II.

VI. METABOLISM OF THE HYDROPEROXIDE PRODUCTS OF LIPOXYGENASE

A. Free-Radical Reactions

The hydroperoxide products resulting from the LOX reaction readily form fatty acid hydroperoxy radicals, which are highly reactive compounds. Lipid hydro-

TABLE II
Some Recently Investigated Inhibitors of Lipoxygenase

Inhibitor	Proposed mode of inhibition	Reference
From plant sources		
α-Tocopherol	Enzyme–inhibitor complex	Grossman and Waksman (1984)
Chlorophyll		Cohen et al. (1984)
Quercetin	Free-radical reduction	Takahama (1985)
Epicatechin		Prusky et al. (1985)
Monoenoic fatty acids	Competitive	St. Angelo and Ory (1984)
Synthetic chemicals		
Fatty acids with S substituted for CH_2	Reversible competitive	Funk and Alteneder (1983); Corey et al. (1985)
Nonsteroidal antiinflammatory drugs[a]		Sircar et al. (1983)
Nupercaine	Substrate modification	Bishop and Oertle (1983)
Ketone hydrazones		Baumann and Wurm (1982)
Phenylhydrazine derivatives		Gibian and Singh (1984)
Phenylhydrazones	Noncompetitive	Wallach and Brown (1981)
N-Alkylhydroxylamines	Reduction of Fe^{3+} to Fe^{2+}	Clapp et al. (1985)
Disulfiram	Inhibits conversion of native to active enzyme	Hausknecht and Funk (1984)

[a]Naproxen, BW 755C, indomethacin, and isoxicam were the most effective inhibitors tested.

peroxy radicals are known to damage membrane structure and organization (Tappel, 1973; Mead, 1976), as well as proteins and amino acids (Gardner, 1979a). Furthermore, DNA can be destroyed by the reaction of hydroperoxides with guanine nucleotides (Inouye, 1984). While there may be certain physiological reactions which require the presence of fatty acid hydroperoxides as coreactants, a more likely fate for these compounds is their prompt conversion to other metabolites by one of several known enzymatic pathways described below.

B. Hydroperoxide Lyase

Tressl and Drawert (1973) demonstrated that linoleic or linolenic acid hydroperoxides were cleaved by extracts of ripening bananas to form volatile aldehydes and oxoacids that contributed to the aroma of bananas. Although they were not able to isolate the enzyme system responsible for the production of aldehydes, an enzyme system catalyzing such a cleavage was later demonstrated in watermelon seedlings by Vick and Zimmerman (1975, 1976) and in cucumber fruit by Galliard and Phillips (1976). Like the banana extracts, the enzyme, hydroperoxide lyase, catalyzed the cleavage of hydroperoxy fatty acids to aldehydes and oxoacids (Fig. 4).

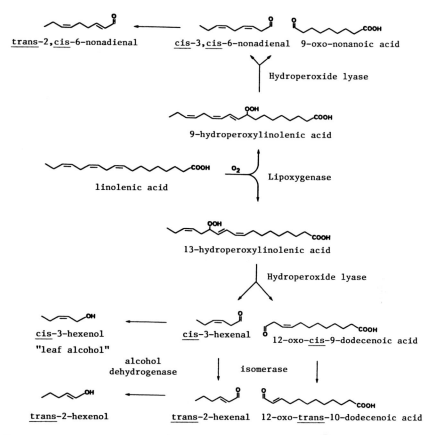

Fig. 4. Pathways for the production of aldehydes, alcohols, and ω-oxoacids from linolenic acid. Similar reactions with linoleic acid lead to the production of hexanal, hexanol, *cis*-3-nonenal, and *trans*-2-nonenal, as well as the ω-oxoacids shown.

In the majority of plants studied thus far, the 13-hydroperoxide of linoleic or linolenic acid is the most active substrate for hydroperoxide lyase. The products from 13-hydroperoxylinoleic acid are hexanal and 12-oxo-*cis*-9-dodecenoic acid. When 13-hydroperoxylinolenic acid is the substrate, the six-carbon product is *cis*-3-hexenal. In some plants, such as cucumber fruit (Galliard *et al.*, 1976), the enzyme can utilize either the 13- or the 9-hydroperoxy isomer as substrate. In the case of pear fruit (Kim and Grosch, 1981), only the 9-hydroperoxy isomer of linoleic or linolenic acid is the substrate for hydroperoxide lyase. The products of lyase action with the 9-hydroperoxy isomer of linoleic or linolenic acid are 9-oxo-nonanoic acid and *cis*-3-nonenal or *cis*-3,*cis*-6-nonadienal, respectively. In most plants, the products which have a *cis*-3-enal structure are quickly isomerized by an isomerase enzyme to the *trans*-2-enal form (Phillips *et al.*, 1979). Under certain physiological conditions, alcohol dehydrogenase can convert the aldehyde products to their corresponding alcohols. Hexanal, cis-3-nonenal, and trans-2-nonenal are components of the characteristic flavor and odor of cucumber. The alcohol product, cis-3-hexenol, is also known as leaf alcohol and contributes to the aroma of green leaves.

Other plants in which hydroperoxide lyase has been detected include tomato fruits (Galliard and Matthew, 1977), tea leaves (Hatanaka et al., 1979), alfalfa seedlings (Sekiya et al., 1979), bean leaves (Matthew and Galliard, 1978), and apple fruits (Schreier and Lorenz, 1982). Sekiya et al. (1983) surveyed the green leaves of various plant species and found hydroperoxide lyase activity in 28 species tested. The enzyme reportedly has a high molecular weight, more than 250,000 in watermelon seedlings (Vick and Zimmerman, 1976) and about 200,000 in cucumber fruit (Phillips et al., 1979). Mushrooms have an unusual hydroperoxide lyase that utilizes only the 10S-hydroperoxide isomer of linoleic acid as substrate, producing 8- and 10-carbon fragments, that is, 1-octen-3-ol and 10-oxo-trans-8-decenoic acid (Wurzenberger and Grosch, 1984b,c). The alcohol product, 1-octen-3-ol, is responsible for the characteristic flavor of mushrooms. Presumably, mushrooms also possess an unusual LOX that catalyzes the incorporation of dioxygen at carbon 10 of linoleic acid.

The mechanism of the hydroperoxide lyase reaction has not been studied in depth until recently. In tea leaves, hydroperoxide lyase has a stereospecific requirement for the S configuration of the hydroperoxide as a substrate (Kajiwara et al., 1982). The α-ketol products of the hydroperoxide isomerase reaction (see below) would seem to be likely candidates as intermediates in the lyase reaction because their oxo and hydroxyl groups are positioned such that scission between them would likely yield the observed products. However, attempts to convert α-ketols to aldehyde products with enzyme extracts have failed; they do not appear to be intermediates in the lyase reaction. Nevertheless, under favorable reaction conditions, nonenzymatic radical reactions can lead to the formation of hexanal from the 12,13-ketol of linoleic acid (Yabuuchi, 1978).

Gardner and Plattner (1984) have studied model chemical reactions that mimic hydroperoxide lyase activity. They were able to show that linoleate hydroperoxides could be cleaved to a six-carbon aldehyde and 12-oxo-dodecenoic acid by use of a strong Lewis acid in an aprotic solvent (Fig. 5). Hydroperoxide lyase probably catalyzes the reaction in a similar manner, with a particular amino acid acting as a Lewis acid at the active site. Oxygen isotope labeling experiments with the hydroperoxide lyase reaction (Hatanaka et al., 1985) supported the proposed mechanism because the oxo group of 12-oxo-dodecenoic acid was shown to arise from a hydroperoxide oxygen atom, whereas the oxo group of the aldehyde originated from water. Similar labeling experiments with the anomalous hydroperoxide lyase from mushrooms have shown that one hydroperoxy oxygen atom from 10-hydroperoxylinoleic acid becomes the hydroxyl group of 1-octen-3-ol, and the other probably remains as the oxo group of 10-oxo-trans-8-decenoic acid (Wurzenberger and Grosch, 1984a).

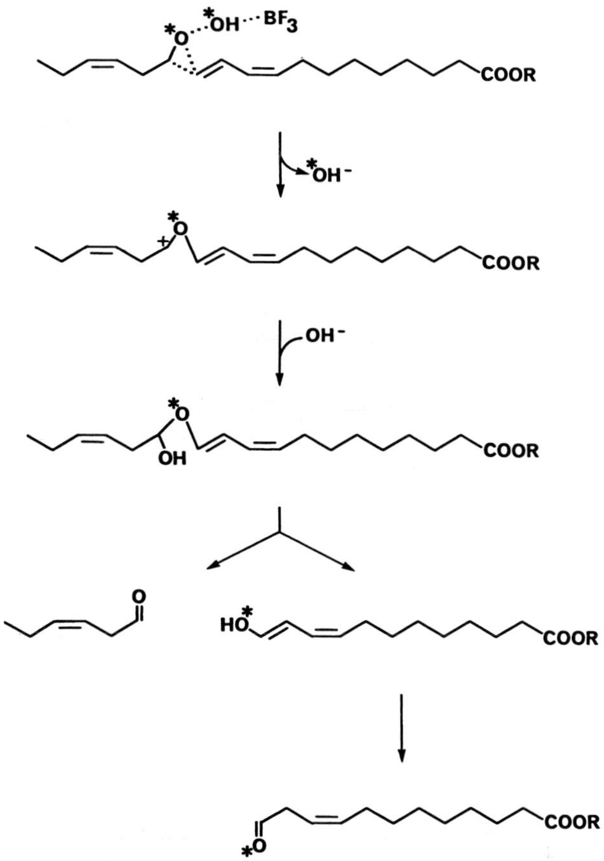

Fig. 5. Proposed mechanism for the cleavage of 13-hydroperoxylinolenic acid, catalyzed by a strong Lewis acid in an aprotic solvent. Hydroperoxide lyase may catalyze the enzyme reaction by a similar mechanism with an amino acid residue acting as a Lewis acid. Asterisks indicate the labeling patterns of the enzyme-catalyzed reaction, which were determined through ^{18}O isotope experiments. [Adapted from Gardner and Plattner (1984); by permission.]

Hydroperoxide lyase reactions are of considerable importance to food technologists who are interested in the biosynthesis of flavors and aromas in food products. Two reviews have appeared recently that describe the role and properties of hydroperoxide lyase in the biogenesis of flavors and aromas (Gardner, 1985b; Hatanaka *et al.*, 1986).

C. Hydroperoxide Isomerase

Hydroperoxide isomerase from flaxseed was the first enzyme discovered that could metabolize the hydroperoxide product of LOX (Zimmerman, 1966; Zimmerman and Vick, 1970). The products of the isomerase reaction are α- and γ-ketols:

The enzyme is widely distributed among plant species (Table III) and usually has an optimum pH between 6 and 7. Substrate specificity varies among species. The reaction with the flaxseed enzyme is about 36 times faster with the 13-hydroperoxy isomer of linoleic acid than with the 9-hydroperoxy isomer (Feng and Zimmerman, 1979). In contrast, hydroperoxide isomerase from corn germ shows no preference between the two substrates (Gardner, 1970). The hydroperoxides of linolenic acid are just as effective as the substrates derived from linoleic acid.

The mechanism of the isomerase reaction has been studied extensively. Oxygen isotope labeling experiments have established that the oxo function of the α- or γ-ketol originates from one of the hydroperoxy oxygen atoms (Gerritsen et al., 1976). The other oxygen is derived from substitution by OH^- or other nucleophiles at the hydroperoxy carbon, and involves inversion of stereoconfiguration from S to R. On the basis of this information, Gardner (1979b) has proposed a mechanism for hydroperoxide isomerase action that involves the loss of OH^- from the hydroperoxide and the formation of an epoxy-cation intermediate (Fig. 6). A bimolecular nucleophilic substitution reaction at the carbon originally bearing the hydroperoxide leads to an α-ketol product. If the cation were to distribute its charge over the three

TABLE III
Sources of Hydroperoxide Isomerase

Source	Botanical name	Optimum pH[a]	References
Barley	Hordeum vulgare L.	6.8	Yabuuchi and Amaha (1976); Lulai et al. (1981)
Corn	Zea mays L.	6.6	Gardner (1970)
Cotton	Gossypium hirsutum L.	6	Vick and Zimmerman (1981a)
Eggplant	Solanum melongena L.	6.5	Grossman et al. (1983)
Flaxseed	Linum usitatissimum L.	7.0	Zimmerman and Vick (1970)
Lettuce	Lactuca sativa L.		Vick and Zimmerman (1979b)
Oats	Avena sativa L.		Vick and Zimmerman (1979b)
Spinach	Spinacea oleracea L.		Vick and Zimmerman (1979b)
Sunflower	Helianthus annuus L.		Vick and Zimmerman (1979b)
Wheat	Triticum aestivum L.		Vick and Zimmerman (1979b)

[a] Linoleic acid substrate.

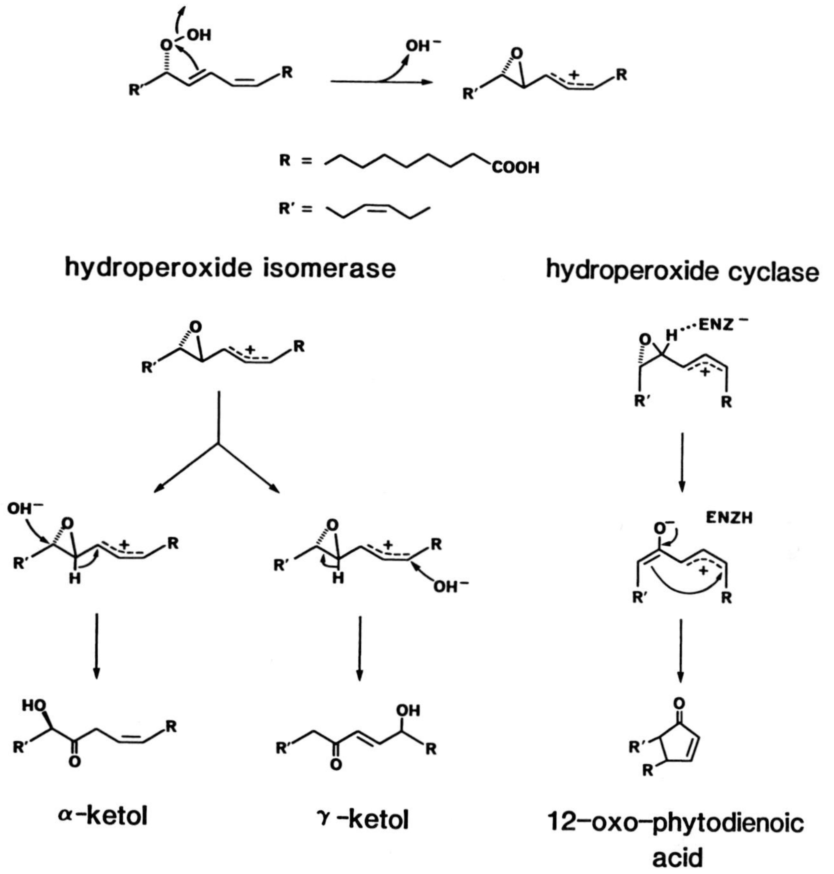

Fig. 6. Proposed mechanisms of action for hydroperoxide isomerase and hydroperoxide cyclase. [Adapted from Gardner (1979b) and Vick *et al.* (1980); by permission.]

carbons adjacent to the epoxy group to form an allylic hybrid, substitution could occur at the γ-carbon, leading to a γ-ketol. No further metabolism of α-ketol or γ-ketol fatty acids by plant enzymes has yet been reported.

D. Hydroperoxide Cyclase

Much of the early research on the LOX pathway utilized linoleic acid as the substrate for investigation. In the few experiments in which linolenic acid metabolism was studied, investigators often concluded that linolenic acid behaved similarly to linoleic acid and produced analogous products with one additional double bond. That concept changed when Zimmerman and Feng (1978) reported the formation of a cyclic, prostaglandinlike fatty acid from the reaction of linolenic acid with a flaxseed extract:

12–oxo–phytodienoic acid

An intermediate in the biosynthesis of the cyclopentenone fatty acid was 13-hydroperoxylinolenic acid, derived from the action of LOX on linolenic acid (Vick and Zimmerman, 1979a). An analogous cyclic product from linoleic acid was not formed. The enzyme responsible for catalyzing the reaction was called hydroperoxide cyclase. The product, 8-[2-(cis-2′-pentenyl)-3-oxo-cis-4-cyclopentenyl] octanoic acid, was given the common name 12-oxo-phytodienoic acid (12-oxo-PDA).

Vick and Zimmerman (1979b) surveyed 24 plant species for hydroperoxide cyclase and demonstrated its presence in 15 of the species. Additional species that possess hydroperoxide cyclase activity have been added to the list since the initial survey. The enzyme appears to be distributed throughout the plant kingdom and is present in seedlings, leaves, and fruit.

Other polyunsaturated fatty acid hydroperoxides are also substrates for the hydroperoxide cyclase reaction. Fatty acid hydroperoxides ranging in length from 18 to 22 carbons were good substrates when derived from fatty acids with n-3,6,9 unsaturation with the hydroperoxide at the n-6 carbon (Vick and Zimmerman, 1979a). Cyclization did not occur when the n-3 double bond was absent. However, a structurally similar substrate, 9-hydroperoxy-γ-linolenic acid, was cyclized despite the absence of n-3 unsaturation:

It was concluded that n-3 unsaturation was not the critical factor, but rather that a cis double bond β,γ to the hydroperoxide was necessary for the enzyme to catalyze the cyclization reaction (Vick *et al.*, 1980).

The mechanism of the hydroperoxide cyclase reaction is probably similar to the hydroperoxide isomerase mechanism. Oxygen isotope labeling studies showed that the 12-oxo oxygen atom arose from a 13-hydroperoxy oxygen atom (Vick *et al.*, 1980). The proposed mechanism involves the formation of an epoxy-cation intermediate, followed by abstraction of a proton at carbon 12 to give an enolate anion (Fig. 6). Rearrangement of this intermediate would give cyclization between carbons 9 and 13. Because the mechanism of hydroperoxide isomerase and hydroperoxide cyclase appears to be similar, and because enzymes for the two reactions have never been separated by gel filtration or ion-exchange chromatography, it has been suggested that the same protein may be responsible for the two reactions (Vick and Zimmerman, 1981a,b). However, final conclusions about dual isomerase/cyclase activity of a single protein must await more rigorous separation efforts, that is, the demonstration of both activities with a single protein following isoelectric focusing or gel electrophoresis techniques.

The absolute stereoconfiguration of the side chains of 12-oxo-PDA has not been determined unequivocally, but it is probably 9S,13S based on the known stereochemistry of its metabolite, jasmonic acid. It is known, however, that the naturally occurring form has a cis orientation of the side chains with respect to the plane of the ring (Vick *et al.*, 1979). 12-Oxo-PDA is easily isomerized by heat, acid, or base treatment to its trans configuration. The cis form of 12-oxo-PDA is an intermediate in the biosynthesis of jasmonic acid by the reactions described in the following section.

E. The Jasmonic Acid Pathway

In plants, 12-oxo-PDA is converted to jasmonic acid (JA) by a series of reactions (Fig. 7) (Vick and Zimmerman, 1983). The first step in the sequence is the reduction of the double bond in the cyclopentenone ring to form 3-oxo-2-(2'-pentenyl)cyclopentaneoctanoic acid. The reaction is catalyzed by 12-oxo-PDA reductase for which NADPH is the most active reductant (Vick and Zimmerman, 1986). The enzyme from corn kernels has a molecular weight of about 54,000 and has a broad pH range from pH 6.8 to 9.0. The K_m for 12-oxo-PDA is 190 μM and for NADPH the K_m is 13 μM. NADH with a K_m of 4.2 mM can also serve as a reductant in the reaction, but it is much less effective than NADPH.

Other intermediates that have been identified in the pathway have hexanoic and butanoic acid side chains, leading to the proposal that chain shortening to the 12-carbon product occurs by three cycles of β-oxidation. The final product, (1R,2S)-3-oxo-2-(2'-pentenyl)cyclopentaneacetic acid (jasmonic acid), retains the same cis stereoconfiguration of the side chains as the parent 12-oxo-PDA. As with 12-oxo-PDA, the cis form is easily isomerized to the trans form. In the early literature the name "jasmonic acid" was given to the compound having the trans

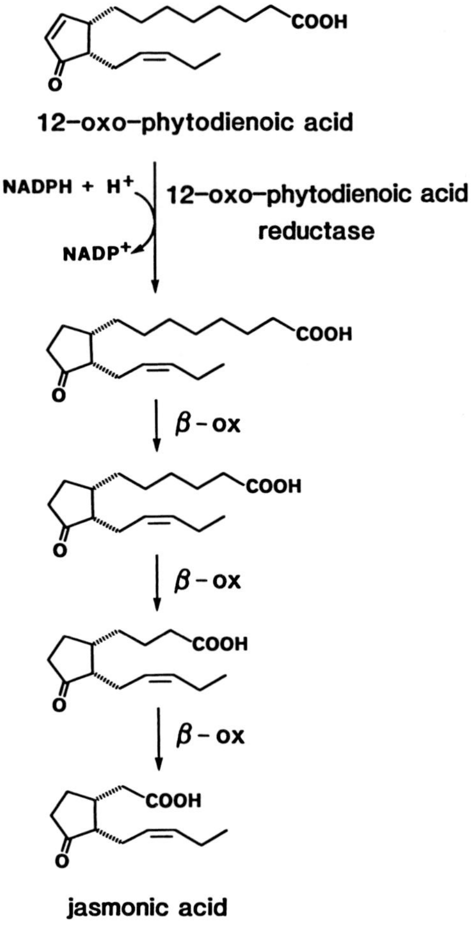

12–oxo–phytodienoic acid

Fig. 7. The jasmonic acid pathway. [Adapted from Vick and Zimmerman (1984); by permission.]

configuration of its side chains (Demole *et al.*, 1962). The trans form undoubtedly arose from isomerization of the cis form during its purification. To distinguish between the two isomers, some researchers have used the term "epijasmonic acid" or "7-iso-jasmonic acid" to refer to the naturally occurring cis form of the compound.

The JA pathway occurs in a variety of plant species. The conversion of 12-oxo-PDA to JA has been demonstrated in *Vicia faba* L. bean pods (Vick and Zimmerman, 1983) and in corn coleoptiles, flax and sunflower cotyledons, wheat and oat leaves, and eggplant fruit (Vick and Zimmerman, 1984). Consequently, JA synthesis appears to be a general metabolic pathway in plants.

F. Other Products of Hydroperoxide Metabolism

1. Divinyl Ethers

Potato tubers possess an enzyme system that catalyzes the formation of divinyl ether products from fatty acid hydroperoxides (Galliard and Chan, 1980). Potatoes appear to be unique in catalyzing this reaction because divinyl ethers have not been detected in reactions catalyzed by extracts of any other plant tissues.

2. Leukotrienes

Shimizu et al. (1984) demonstrated recently the synthesis of 6-trans-leukotriene B_4 (LTB$_4$) from arachidonic acid incubated with a potato tuber homogenate. Intermediates in the reaction were (5S)-hydroperoxy-trans-6,cis-8,11,14-icosatetraenoic acid (5-HPETE) and an epoxide, leukotriene A_4. A single enzyme, LOX, was apparently responsible for catalyzing both the formation of the hydroperoxide and its conversion to LTB$_4$:

LTB$_4$

VII. PROPOSED PHYSIOLOGICAL ROLES FOR METABOLITES OF THE LIPOXYGENASE PATHWAY

A. Fatty Acid Hydroperoxides

1. Ethylene Formation

Ethylene is a plant hormone with such diverse physiological effects that a fundamental knowledge of its mechanism of biosynthesis is crucial to understanding plant growth regulation. Ethylene formation can be classified into two categories: (1) the normal formation of ethylene at particular stages in the life of a plant, a process normally associated with high auxin levels in nonsenescing vegetative tissue, and (2) the production of stress ethylene that results from physical, chemical, or biological stress to the plant.

Lieberman et al. (1966) were the first to report that methionine is a biological precursor to ethylene in plants. An important advance in the clarification of the biosynthetic route from methionine to ethylene resulted when Adams and Yang (1979) reported that the penultimate intermediate in ethylene biosynthesis was 1-aminocyclopropane-1-carboxylic acid (ACC). However, characterization of the enzyme system responsible for the final step in the pathway, the conversion of ACC to ethylene, has presented a difficult challenge to researchers. Through the combined efforts of many laboratories, two general features of the conversion of ACC to ethylene are known: a requirement for oxygen and a possible participation by free radicals.

Because the LOX reaction has an oxygen dependence and also involves free-radical intermediates, several groups have investigated its possible participation in ethylene formation. Legge and Thompson (1983) demonstrated that in the presence of etiolated pea microsomal membranes and certain small organic hydroperoxides such as t-butyl hydroperoxide and cumene hydroperoxide, ACC could be converted to ethylene. The addition of LOX and linolenic acid to the microsomal membranes also promoted ethylene synthesis, presumably due to the formation of hydroperoxylinolenic acid. They also provided evidence for the involvement of two free radicals in the reaction, the superoxide anion O_2^- and an ACC-derived radical. Lynch and Thompson (1984) suggested that production of the superoxide anion is dependent on LOX activity. The manner in which superoxide anion might be produced by LOX-mediated reduction of O_2 is unknown, but Hamilton (1974) has proposed a mechanism for such a reaction. The hypothetical scheme involves the enzyme-mediated transfer of an electron from a fatty acid hydroperoxide anion to O_2 to form superoxide anion and a fatty acid peroxy radical; however, there is no direct evidence that such a reaction can occur.

Further evidence for the participation of LOX in ethylene biosynthesis is the parallel change in LOX activity and ethylene formation during senescence (Lynch and Thompson, 1984) or membrane injury (Kacperska and Kubacka-Zebalska, 1985). Direct evidence for the participation of LOX was provided by Bousquet and Thimann (1984), who showed that an *in vitro* soybean LOX system free of membrane components could oxidize ACC to ethylene. The system, which required manganese and pyridoxal phosphate, had properties that resembled the ethylene-synthesizing system in pea microsomal membranes and in intact oat leaves. Consequently, they suggested that the LOX system may be identical to the *in vivo* system that produces ethylene from ACC in senescent leaves. Kacperska and Kubacka-Zebalska (1985) also demonstrated the active conversion of ACC to ethylene in a model system containing purified soybean LOX-1, linoleic or linolenic acid, and ACC. The observation that an ethylene-producing fungus, *Fusarium oxysporum*, also possesses LOX activity (Matsuda *et al.*, 1979) may indicate a relationship between LOX and ethylene synthesis in fungi.

Other evidence, however, argues against the participation of LOX in ethylene biosynthesis via the "normal" pathway in intact plants. Lynch *et al.* (1985) used senescing carnation flowers to compare ethylene formation from the intact flower with ethylene synthesis from microsomal membranes isolated at various stages of senescence. The periods of maximum ethylene synthesis in the two systems did not coincide, indicating that the ethylene-producing activity in the microsomes (a "stressed" tissue) was not the same as the native, *in vivo* ethylene-forming enzyme. The synthesis of ethylene by intact tissues is highly stereoselective (Hoffman *et al.*, 1982), and such stereoselectivity is not characteristic of cell-free ethylene-forming systems (Venis, 1984), nor has it been demonstrated for ethylene formation catalyzed by purified LOX (Pirrung, 1986). However, it could be argued that spatial constraints of the membrane-bound enzyme permit high stereoselectivity which would be lost after dissociation from the cell membrane. Furthermore, LOX-mediated ethylene production is totally inhibited by aminoxyacetic acid, whereas the compound has no influence on *in vivo* ethylene biosynthesis (Pirrung, 1986). Taken as a whole, the evidence accumulated thus far suggests that LOX and its hydroperoxide products may participate in the production of stress ethylene. However, the bulk of the ethylene produced by intact plants during other stages of growth most likely involves another ACC-oxidizing system which awaits discovery.

2. Alternative Respiration

Many plants possess a supplementary mitochondrial respiratory pathway that differs from the usual route of electron transport via the cytochrome system. Unlike normal respiration, this alternative pathway is insensitive to cyanide and antimycin A, but sensitive to *n*-propyl gallate and hydroxamic acids. The pathway is thought to involve the diversion of electron flow from ubiquinone to a mitochondrial oxidase rather than to the cytochromes of the normal electron transport pathway (Laties, 1982).

The nature of the alternative, cyanide-resistant oxidase is not completely understood. Alternative respiration is stimulated by the addition of linoleic acid and is similar to LOX in its sensitivity to various inhibitors. This has prompted some speculation that O_2 consumption during mitochondrial alternative respiration is due to the presence of LOX (Goldstein *et al.*, 1981; Dupont, 1981). Peterman and Siedow (1983), however, maintain that the structural features of propyl gallate necessary for inhibition of LOX are different than those for inhibition of the alternative pathway, and consequently the two activities are distinct. In support of this view, it has been shown that disulfiram inhibits alternative respiration but not LOX (Miller and Obendorf, 1981).

Another explanation of cyanide-resistant respiration was offered by Rustin *et al.* (1983). They proposed that all respiratory substrates can donate electrons to ubiquinone to form ubiquinol. In turn, electrons can be transferred from ubiquinol to either the normal or the cyanide-resistant pathway. When electron transfer to the cyanide-resistant pathway occurs, ubiquinol donates electrons to unsaturated fatty acid hydroperoxy radicals to form water and fatty acid radicals. Oxygen combines with the fatty acid radicals (the observed respiration) to re-form hydroperoxy radicals and complete the R·/ROO· cycle. Under such a scheme the hydroperoxy radicals are required only in catalytic amounts and could arise from reactions other than those catalyzed by LOX. The proposed mechanism for alternative respiration would account for the observed effects of linoleic acid and various free-radical inhibitors. However, a troublesome feature of this scheme is the proposed formation of fatty acid radicals by transfer of electrons to fatty acid hydroperoxy radicals. The expected product from the reduction of fatty acid hydroperoxy radicals would be fatty acid hydroperoxide, not fatty acid radicals. Thus, the absence of fatty acid radicals from the cycle would preclude the uptake of oxygen. Collectively, the information available on this topic at present suggests that alternative respiration is a distinct metabolic phenomenon, but that LOX is not the terminal oxidase of that process.

3. Senescence

Some experimental evidence suggests that LOX and its hydroperoxide products might participate in reactions that lead to plant senescence. For example, LOX activity in pea leaves increases during leaf senescence, but the senescence can be retarded by the addition of LOX inhibitors such as cytokinin or α-tocopherol (Leshem *et al.*, 1979). Apparently, cytokinin acts as a free-radical scavenger to inhibit LOX and decrease the rate of fatty acid hydroperoxide production.

Fatty acid hydroperoxides reportedly can cause senescence by several different mechanisms, including inactivation of protein synthesis (Dhindsa, 1982) and the inhibition of photochemical activity in chloroplasts (Köckritz *et al.*, 1985). The latter mechanism appears to be due to a chlorophyll-catalyzed reaction of hydroperoxylinolenic acid with certain components of the electron transport chain. Fatty acid hydroperoxides also promote the deterioration of cellular membranes (Pauls and Thompson, 1984). In senescing cotyledons of *Phaseolus vulgaris,* LOX activity and lipid peroxidation products in microsomal membranes show a concomitant increase as the cotyledons age. Concurrently, the oxidized membrane lipids acquire rigid gel-phase properties, become permeable, and the membranes eventually disintegrate. Leshem (1984) and Sridhara and Leshem (1986) have suggested that this process involves Ca^{2+}-stimulated calmodulin that activates phospholipase A_2 and LOX. The resulting hydroperoxidized fatty acids are proposed to act as specific Ca^{2+} ionophores to enable even more Ca^{2+} to penetrate the cell and activate more calmodulin. This cycle is repeated until the membranes become progressively senescent.

A physiological role for LOX as a mediator of cell membrane disruption has some attractive features. Although cell membrane disruption is usually associated with plant senescence, there are times when such events are required for growth. For example, during the early stages of seedling growth, reserve lipids or carbohydrates must be rapidly broken down and the metabolites transported to the growing axis. During this process, the permeability of the storage cell membranes increases to allow movement of the metabolites to the developing embryo. In many plants LOX activity is at its maximum during this period and is probably localized in storage cells as in soybean cotyledons (Fig. 1A). Perhaps the high activity of LOX during germination results in the formation of lipid hydroperoxides that accelerate the disruption of cellular membranes and facilitate the transport of metabolites to the developing embryo (Pauls and Thompson, 1984).

However, the senescence hypothesis is not consistent with some observations. For example, the high LOX activity observed during early germination in many plants is usually accompanied by high activities of other enzymes, such as hydroperoxide isomerase, lyase, or cyclase, which metabolize fatty acid hydroperoxides. These enzymes are presumably available to convert hydroperoxides to other products before serious cellular damage occurs. Moreover, the high LOX activities are usually not confined to the senescing storage tissues. High LOX activity is also observed in newly differentiated and rapidly growing tissues such as hypocotyls, roots, coleoptiles, and leaves. Finally, LOX activity in senescing wheat and rye leaves (Kar and Feierabend, 1984) declined, not increased, during senescence. Lipid peroxidation was detectable, but only in light-grown leaves, and was a result of photochemical reactions, not LOX. The light-grown leaves (with higher peroxide levels) actually senesced slower than the dark-grown leaves. Similarly, LOX activity in soybean cotyledons decreased during substrate mobilization and subsequent senescence, but increased upon rejuvenation of the cotyledons (Peterman and Siedow, 1985b). In both examples LOX activity and lipid peroxidation appeared to be unrelated to senescence. Peterman and Siedow (1985b) have suggested that LOX-1 may function as a seed storage protein in soybean cotyledons rather than serve a catalytic function.

4. Regulation of Enzyme Activity

Fatty acid hydroperoxides are highly reactive with sulfhydryl groups. Douillard (1981) proposed that this property may enable hydroperoxides to regulate the activity of several Calvin cycle enzymes in the chloroplast. According to the hypothesis, thioredoxin maintains the sulfhydryl groups of the chloroplast enzymes in their reduced, active states. If the sulfhydryl groups of thioredoxin are oxidized by hydroperoxides or by oxidized glutathione (which can result from hydroperoxide oxidation of glutathione), thioredoxin will no longer be able to supply reducing equivalents to the thiol groups of the susceptible carbon fixation cycle enzymes, and enzyme inactivation results.

B.   Aldehydes and ω-Oxoacids

The physiological roles of the aldehydes and ω-oxoacids that result from the action of hydroperoxide lyase are not well understood. It is not even clear whether one or both products have functional roles in plant metabolism. The 12-carbon product, 12-oxo-*trans*-10-dodecenoic acid, has been identified as the active component of the wound hormone, traumatin (Zimmerman and Coudron, 1979). The compound mimics the effect of wounding in certain bioassays by stimulating the formation of callous tissue. It is easily oxidized nonenzymatically to the dicarboxylic acid *trans*-2-dodecenedioic acid, commonly called traumatic acid. Consequently, bioassays with the oxoacid are difficult to conduct; it cannot be easily determined whether the compound penetrates the cell as the oxoacid or as a dicarboxylic acid. The transformation of oxoacid to dicarboxylic acid is not known to occur enzymatically in plants, but such a possibility cannot be ruled out. Traumatic acid was identified originally as the active component of traumatin (English *et al.*, 1939), but it now appears that traumatic acid was derived from the oxidation of the oxoacid during its purification. In addition to a wound response, traumatic acid also stimulates adventitious bud initiation (Tanimoto and Harada,

1984). No metabolic role has yet been demonstrated for 9-oxo-nonanoic acid, which results from the 9-hydroperoxide isomer of linoleic or linolenic acid.

The physiological role of the aldehyde products is also unknown. The synthesis of hexanal and hexenal in tea leaves varies with the season, being higher in summer than winter (Sekiya *et al.*, 1984). The variation is due to increased LOX activity in summer leaves rather than to hydroperoxide lyase activity, which remains relatively constant all year. While the seasonal change in aldehyde biosynthesis has a pronounced effect on the flavor of green tea, its physiological effect is less clear, but may be related to temperature acclimation of the leaves. Another possibility is the role of aldehydes in the defense mechanism of the plant. For example, *trans*-2-hexenal is an inhibitor of fungal growth (Major *et al.*, 1960) and has bactericidal activity (Schildknecht and Rauch, 1961). When leaves are injured or intentionally macerated, LOX and hydroperoxide lyase are activated, leading to the production of hexanal and hexenal. The same response could result from invasion by insects, bacteria, or other plant pathogens. The toxicity of α,β-unsaturated aldehydes is due to the high reactivity of the enal function with sulfhydryl, amino, and hydroxyl groups through addition reactions or formation of Schiff bases (Schauenstein *et al.*, 1977). At subtoxic levels, α,β-unsaturated aldehydes may regulate certain pathways by inhibiting critical sulfhydryl-requiring enzymes.

C. Ketols

No specific metabolic function has yet been ascribed to the α-ketol or γ-ketol products of hydroperoxide isomerase. In our experience, neither compound seems to be enzymatically metabolized to any extent by cell-free homogenates or tissue slices. In the metabolism of 13-hydroperoxylinolenic acid, the ketols are always present as products along with 12-oxo-PDA in approximately the same ratios from all plant enzyme sources, about $5:2:1$ for α-ketol, 12-oxo-PDA, and γ-ketol, respectively.

The α-ketol concentration in young, developing seedlings does not parallel the rise and decline of LOX and hydroperoxide isomerase activities commonly observed in germinating seedlings (Vick and Zimmerman, 1982). Ketol levels (as free acids) in these tissues are very low despite the high activity of the enzymes responsible for their synthesis. This probably reflects a cellular compartmentation of the enzymes apart from the substrates, linoleic or linolenic acid, a rapid turnover to some unidentified compound, or an irreversible binding to an insoluble material. Interaction between substrate and enzyme could be triggered by some cellular stimulus such as wounding. Maceration of corn leaves results in a 3- to 5-fold increase in α-ketol concentration within 10 min (Vick and Zimmerman, 1982). One function of hydroperoxide isomerase in plants could be to convert excess fatty acid hydroperoxides to innocuous ketol products as a protective measure.

A significant feature of the γ-ketols is their structural similarity to 4-hydroxy-alkenals. These aldehyde compounds are extremely reactive, particularly with sulfhydryl groups, and lead to the inactivation of enzymes (Schauenstein *et al.*,

1977). It is possible that γ-ketols function in this manner to regulate enzyme activity *in vivo*.

D. 12-Oxo-Phytodienoic Acid and Jasmonic Acid

The similarity in structure between 12-oxo-PDA in plants and prostaglandin A_1 from animals suggests that 12-oxo-PDA may have a regulatory function in plants, just as the prostaglandins are potent metabolic regulators in animals. Compounds with structures related to 12-oxo-PDA have been isolated from various plants, including *Chromolaena morii* K. et R. (Bohlmann *et al.*, 1981) and *Eleocharis microcarpa* Torr. (Van Aller *et al.*, 1985). The oxygenated fatty acids from the latter organism have allelochemical effects against blue-green algae.

While 12-oxo-PDA may have physiological activity of its own, its most likely function in plants is its role as a precursor to JA. In some plants, JA is also present as the methyl ester (JA-Me). JA-Me was first isolated and characterized from the essential oil of *Jasminum grandiflorum* L. (Demole *et al.*, 1962) and it has become an important ingredient in the perfume industry. Both JA and JA-Me have been of considerable research interest during the past 5 years because of their growth-regulating properties.

JA was first identified as a plant growth inhibitor after its isolation from cultures of the fungus *Lasiodiplodia theobromae* (Aldridge *et al.*, 1971). JA or JA-Me have now been detected in several plant species (Ueda and Kato, 1980, 1982; Yamane *et al.*, 1981, 1982; Dathe *et al.*, 1981; Meyer *et al.*, 1984; Anderson, 1985a). Generally, JA and JA-Me affect plants by retarding growth or promoting senescence. In tomato fruit, JA-Me inhibited the synthesis of lycopene, the pigment responsible for the characteristic red color, and stimulated the production of ethylene in red-ripe fruits (Saniewski and Czapski, 1983, 1985). In many respects, the two compounds act similarly to abscisic acid (ABA). JA-Me is known to induce stomatal closure (Satler and Thimann, 1981) and JA inhibits the IAA- or light-induced opening of the pulvinules of *Mimosa pudica* L. plants (Tsurumi and Asahi, 1985). But there are also differences between JA and ABA. JA has been suggested as an endogenous regulator of *Camellia* pollen germination, a process in which ABA had little activity (Yamane *et al.*, 1981, 1982). JA was a stronger inhibitor of kinetin-induced soybean callus growth than ABA (Ueda and Kato, 1982). Conversely, JA was less effective than ABA acid as an inhibitor of lettuce germination (Yamane *et al.*, 1981), wheat seedling growth (Dathe *et al.*, 1981), and in the inhibition of growth or chlorophyll production in cucumber cotyledons (Fletcher *et al.*, 1983).

A limiting factor in most JA growth bioassays has been the lack of availability of the pure, naturally occurring cis isomer, (1*R*,2*S*)-JA. Most studies have used a synthetic mixture of JA containing at least four stereoisomers in which the trans form predominates. The synthetic mixture typically contains only about 5% in the cis form. Only limited information is available on the relative efficacy of the four isomers in plant bioassays. Studies by Miersch *et al.* (1986) showed that in some bioassays the cis isomer of JA had significantly higher biological activity than the trans isomer. In odor activity tests, only (1R,2S)-JA-Me exhibited fragrant properties (Acree et al., 1985), and in insect attractant bioassays only the same isomer had biological activity (Baker et al., 1981). Consequently, if the 1R,2S-isomer of JA also has higher regulatory activity than the other isomers in plants, then many previous studies may have underestimated the potency of the natural compound.

The biosynthesis of JA is perhaps the best example available of a metabolic function for plant LOX. Evidence of a physiological effect for JA is much better than for any of the other LOX metabolites. There is no question that JA has potent growth-regulating properties. Researchers are now suggesting that JA may be a previously unrecognized plant hormone. Experiments have demonstrated the presence of JA in phloem exudates (Anderson, 1985b), indicating that it is probably translocated in the phloem. Although JA has been implicated in promoting senescence in various bioassays, it is usually detected in young, actively growing plant tissues (Meyer et al., 1984), as are the enzymes in its biosynthetic pathway (Vick and Zimmerman, 1984). Consequently, JA may be involved in other phases of plant development which are not yet recognized. In support of this view, both jasmonic acid and 12-oxo-phytodienoic acid at 0.1 mM concentrations were shown to stimulate significantly adventitious root formation in mung bean seedlings pretreated with 5 μM indolebutyric acid (Zimmerman and Vick, 1983).

E. Oxidized Pigments: Abscisic Acid Synthesis

Although there is good evidence that abscisic acid (ABA) is synthesized from mevalonic acid in the isoprenoid biosynthetic pathway, an alternate pathway may also exist in which ABA is derived from a carotenoid, specifically violaxanthin. In a LOX-catalyzed cooxidation reaction with linoleic acid, violaxanthin was cleaved to xanthoxin (Firn and Friend, 1972). The conversion of xanthoxin to ABA occurs in the shoots of tomato and dwarf bean (Taylor and Burden, 1973). Whether this pathway is of any significance in other plants is not yet clear, but it offers yet another possible physiological role for LOX. In this regard, LOX activity in developing maize embryos shows two peaks of maximum activity, one at 14 days after pollination and the second at 21 days after pollination (Belefant and Fong, 1985). Changes in the ability of developing maize embryos to take up ABA closely paralleled the changes in LOX activity (Fong et al., 1985). Other experiments (Henson, 1984) have shown that the application of inhibitors of carotenoid synthesis, such as norflurazon, also results in decreased levels of ABA.

VIII. PERSPECTIVE: THE OCTADECANOIDS

Lipoxygenase research has acquired considerable attention during the past decade as new oxygenated metabolites of linoleic or linolenic acid are added to the metabolic charts and new theories concerning the physiological function of LOX are put forward. The interest in the plant LOX pathway has been inspired in part because of the rapid advances made in prostaglandin and leukotriene research in mammalian biochemistry. These potent metabolic regulators in mammals are members of a large group of oxygenated fatty acids known collectively as the icosanoids because they are derived from arachidonic acid or other 20-carbon fatty acids. It now appears that plants have a similar version of the arachidonic acid cascade. In the plant polyunsaturated fatty acid cascade, the 18-carbon fatty acids predominate. Consequently, it might be appropriate to refer to plant oxygenated fatty acids as octadecanoids, meaning metabolites derived from linoleic or linolenic acid via the LOX reaction. The central intermediate in this scheme is a fatty acid hydroperoxide, which can be directed to one of several alternative pathways.

It is clear from a review of recent literature that LOX participates in more than one physiological process. This view is reinforced by considering the example of JA. JA is the final product of a metabolic pathway initiated by the formation of the 13-hydroperoxide of linolenic acid. No other fatty acid hydroperoxide can serve as an intermediate in the pathway. Yet many plants have lipoxygenases that are highly specific for the formation of 9-hydroperoxy fatty acids which are not involved in JA biosynthesis. These hydroperoxides apparently follow a different course of metabolism, most likely through hydroperoxide isomerase or hydroperoxide lyase reactions to products whose functions are not yet clear.

Most of the oxygenated fatty acid metabolites of LOX are not available commercially. This has impeded progress in LOX research because nearly all of the intermediates must be synthesized enzymatically for further metabolism studies. A challenge to organic chemists would be the synthesis of the naturally occurring stereoisomers of 12-oxo-phytodienoic acid and (1R,2S)-JA. Another limiting factor is the availability of reasonably priced, uniformly radiolabeled linolenic acid with high specific activity. Despite the above limitations, significant progress has been made in the field of LOX research, and more advances are undoubtedly soon forthcoming. In view of the important role of oxygenated fatty acids in mammalian metabolism, we should not be surprised to discover physiological functions of comparable significance for these metabolites in plants.

REFERENCES

Acree, T. E., Nishida, R., and Fukami, H. (1985). *J. Agric. Food Chem.* **33**, 425–427.
Adams, D. O., and Yang, S. F. (1979). *Proc. Natl. Acad. Sci. U.S.A.* **76**, 170–174.
Aldridge, D. C., Galt, S., Giles, D., and Turner, W. B. (1971). *J. Chem. Soc. C* pp. 1623–1627.
Anderson, J. M. (1985a). *J. Chromatogr.* **330**, 347–355.
Anderson, J. M. (1985b). *Plant Physiol.* **77S**, 75.
André, E., and Hou, K.-W. (1932). *C. R. Hebd. Seances Acad. Sci.* **194**, 645–647.
Axelrod, B. (1974). "Food Related Enzymes," Advances in Chemistry Series No. 136, pp. 324–348. Am. Chem. Soc., Washington, D.C.
Baker, T. C., Nishida, R., and Roelofs, W. L. (1981). *Science* **214**, 1359–1361.
Baumann, J., and Wurm, G. (1982). *Agents Actions* **12**, 360–364.
Belefant, H., and Fong, F. (1985). *Plant Physiol.* **77S**, 78.

Bergström, S. (1945). *Ark. Kemi, Mineral. Geol.* **21A**, 1–8.

Bild, G. S., Ramadoss, C. S., Lim, S., and Axelrod, B. (1977). *Biochem. Biophys. Res. Commun.* **74**, 949–954.

Bild, G. S., Bhat, S. G., Ramadoss, C. S., and Axelrod, B. (1978). *J. Biol. Chem.* **253**, 21–23.

Bishop, D. G., and Oertle, E. (1983). *Plant Sci. Lett.* **31**, 49–53.

Bohlmann, F., Gupta, R. K., King, R. M., and Robinson, H. (1981). *Phytochemistry* **20**, 1417–1418.

Bohn, R. M., and Haas, L. W. (1928). *Cited in* "Chemistry and Methods of Enzymes" (J. B. Sumner and G. F. Somers, eds.), 3rd Ed., p. 311. Academic Press, New York, 1958.

Borisova, I. G., Gordeeva, N. T., and Budnitskaya, E. V. (1981). *Dokl. Biochem. (Engl. Transl.)* **261**, 404–407.

Bousquet, J.-F., and Thimann, K. V. (1984). *Proc. Natl. Acad. Sci. U.S.A.* **81**, 1724–1727.

Cayrel, A., Crouzet, J., Chan, H. W.-S., and Price, K. R. (1983). *Am. J. Enol. Vitic.* **34**, 77–82.

Cheesbrough, T. M., and Axelrod, B. (1983). *Biochemistry* **22**, 3837–3840.

Clapp, C. H., Banerjee, A., and Rotenberg, S. A. (1985). *Biochemistry* **24**, 1826–1830.

Cohen, B.-S., Grossman, S., Pinsky, A., and Klein, B. P. (1984). *J. Agric. Food Chem.* **32**, 516–519.

Corey, E. J., Dalarcao, M., and Kyler, K. S. (1985). *Tetrahedron Lett.* **26**, 3919–3922.

Dathe, W., Rönsch, H., Preiss, A., Schade, W., Sembdner, G., and Schreiber, K. (1981). *Planta* **153**, 530–535.

de Groot, J. J. M. C., Garssen, G. J., Veldink, G. A., Vliegenthart, J. F. G., and Boldingh, J. (1975). *FEBS Lett.* **56**, 50–54.

Demole, E., Lederer, E., and Mercier, D. (1962). *Helv. Chim. Acta* **45**, 675–685.

Dhindsa, R. S. (1982). *Phytochemistry* **21**, 309–313.

Douillard, R. (1979). *Physiol. Veg.* **17**, 457–476.

Douillard, R. (1981). *Physiol. Veg.* **19**, 533–542.

Douillard, R., and Bergeron, E. (1978). *C. R. Hebd. Seances Acad. Sci., Ser. D* **286**, 753–755.

Douillard, R., and Bergeron, E. (1979). *In* "Advances in the Biochemistry and Physiology of Plant Lipids" (L.-Å. Appelqvist and C. Liljenberg, eds.), pp. 159–164. Elsevier/North-Holland, Amsterdam.

Douillard, R., and Bergeron, E. (1981). *Plant Sci. Lett.* **22**, 263–268.

Douillard, R., Bergeron, E., and Scalbert, A. (1982). *Physiol. Veg.* **20**, 377–384.

Dreesen, T. D., Dickens, M., and Koch, R. B. (1982). *Lipids* **17**, 964–969.

Dupont, J. (1981). *Physiol. Plant.* **52**, 225–232.

Egmond, M. R., Finazzi-Agrò, A., Fasella, P. M., Veldink, G. A., and Vliegenthart, J. F. G. (1975). *Biochim. Biophys. Acta* **397**, 43–49.

English, J., Jr., Bonner, J., and Haagen-Smit, A. J. (1939). *Science* **90**, 329.

Eskin, N. A. M., Grossman, S., and Pinsky, A. (1977). *CRC Crit. Rev. Food Sci. Nutr.* **9**, 1–40.

Feiters, M. C., Aasa, R., Malmström, B. G., Slappendel, S., Veldink, G. A., and Vliegenthart, J. F. G. (1985). *Biochim. Biophys. Acta* **831**, 302–305.

Feiters, M. C., Veldink, G. A., and Vliegenthart, J. F. G., (1986). *Biochim. Biophys. Acta* **870**, 367–371.

Feng, P., and Zimmerman, D. C. (1979). *Lipids* **14**, 710–713.

Feys, M., de Mot, R., Naesens, W., and Tobback, P. (1982). *Z. Lebensm.-Unters.-Forsch.* **174**, 360–365.

Firn, R. D., and Friend, J. (1972). *Planta* **103**, 263–266.

Fletcher, R. A., Venkatarayappa, T., and Kallidumbil, V. (1983). *Plant Cell Physiol.* **24**, 1057–1064.

Fong, F., Smith, J. D., and Resler, S. (1985). *Plant Physiol.* **77S**, 75.

Funk, M. O., and Alteneder, A. W. (1983). *Biochem. Biophys. Res. Commun.* **114**, 937–943.

Galliard, T. (1974). *Recent Adv. Phytochem.* **8**, 209–241.

Galliard, T. (1975). *In* "Recent Advances in the Chemistry and Biochemistry of Plant Lipids" (T. Galliard and E. I. Mercer, eds.), pp. 319–357. Academic Press, New York.

Galliard, T. (1978). *In* "Biochemistry of Wounded Plant Storage Tissues" (G. Kahl, ed.), pp. 155–201. de Gruyter, Berlin.

Galliard, T., and Chan, H. W.-S. (1980). *In* "The Biochemistry of Plants: A Comprehensive Treatise Vol. 4, Lipids: Structure and Function" (P. K. Stumpf and E. E. Conn, eds.), pp. 131–161. Academic Press, New York.

Galliard, T., and Matthew, J. A. (1977). *Phytochemistry* **16**, 339–343.

Galliard, T., and Phillips, D. R. (1976). *Biochim. Biophys. Acta* **431**, 278–287.

Galliard, T., Phillips, D. R., and Reynolds, J. (1976). *Biochim. Biophys. Acta* **441**, 181–192.

Gardner, H. W. (1970). *J. Lipid Res.* **11**, 311–321.

Gardner, H. W. (1979a). *J. Agric. Food Chem.* **27**, 220–229.

Gardner, H. W. (1979b). *Lipids* **14**, 208–211.

Gardner, H. W. (1980). *In* "Autoxidation in Food and Biological Systems" (M. G. Simic and M. Karel, eds.), pp. 447–504. Plenum, New York.

Gardner, H. W. (1985a). *In* "Chemical Changes in Food during Processing" (T. Richardson and J. W. Finley, eds.), pp. 177–203. AVI, Westport, Connecticut.

Gardner, H. W. (1985b). *In* "Flavor Chemistry of Fats and Oils" (D. B. Min and T. H. Smouse, eds.), pp. 189–206. American Oil Chemists' Society, Champaign, Illinois.

Gardner, H. W. (1987). *In* "Advances in Cereal Science and Technology" (Y. Pomeranz, ed.). American Association of Cereal Chemists, St. Paul, Minnesota, in press.

Gardner, H. W., and Plattner, R. D. (1984). *Lipids* **19**, 294–299.

Gerritsen, M., Veldink, G. A., Vliegenthart, J. F. G., and Boldingh, J. (1976). *FEBS Lett.* **67**, 149–152.

Gibian, M. J., and Singh, K. (1984). *Biochemistry* **23**, 3354.

Goldstein, A. H., Anderson, J. O., and McDaniel, R. G. (1981). *Plant Physiol.* **67**, 594–596.

Grosch, W. (1972). *Fette, Seifen, Anstrichm.* **74**, 375–381.

Grosch, W., and Laskawy, G. (1979). *Biochim. Biophys. Acta* **575**, 439–445.

Grosch, W., Laskawy, G., and Fischer, K.-H. (1977a). *Z. Lebensm.-Unters.-Forsch.* **163**, 203–205.

Grosch, W., Laskawy, G., and Kaiser, K.-P. (1977b). *Z. Lebensm.-Unters.-Forsch.* **165**, 77–81.

Grossman, S., and Waksman, E. G. (1984). *Int. J. Biochem.* **6**, 281–289.

Grossman, S., Bergman, M., and Sofer, Y. (1983). *Biochim. Biophys. Acta* **752**, 65–72.

Grossman, S., Klein, B. P., Cohen, B., King, D., and Pinsky, A. (1984). *Biochim. Biophys. Acta* **793**, 455–462.

Hamberg, M. (1986). *Biochim. Biophys. Acta.* **876**, 688–692.

Hamberg, M., and Samuelsson, B. (1967). *J. Biol. Chem.* **424**, 5329–5335.

Hamilton, G. A. (1974). *In* "Molecular Mechanisms of Oxygen Activation" (O. Hayaishi, ed.), pp. 405–451. Academic Press, New York.

Hatanaka, A., Kajiwara, T., Sekiya, J., and Fujimura, K. (1979). *Agric. Biol. Chem.* **43**, 175–176.

Hatanaka, A., Kajiwara, T., and Sekiya, J. (1985). *Abstr. Pap. Am. Chem. Soc. Meet.*, **190th**, Chicago, Ill. No. 57.

Hatanaka, A., Kajiwara, T., and Sekiya, J. (1986). *In* "Biogeneration of Aromas" (T. H. Parliment and R. Croteau, eds.), *ACS Symp. Ser.*, No. 317, pp. 167–175. American Chemical Society, Washington D.C.

Hausknecht, E. C., and Funk, M. O. (1984). *Phytochemistry* **23**, 1535–1539.

Henson, I. E. (1984). *Z. Pflanzenphysiol.* **114**, 35–43.

Hildebrand, D. F., and Hymowitz, T. (1982). *J. Agric. Food Chem.* **30**, 705–708.

Hoffman, N. E., Yang, S. F., Ichihara, A., and Sakamura, S. (1982). *Plant Physiol.* **70**, 195–199.

Ida, S., Masaki, Y., and Morita, Y. (1983). *Agric. Biol. Chem.* **47**, 637–641.

Inouye, S. (1984). *FEBS Lett.* **172**, 231–234.

Kacperska, A., and Kubacka-Zebalska, M. (1985). *Physiol. Plant.* **64**, 333–338.

Kajiwara, T., Nagata, N., Hatanaka, A., and Naoshima, Y. (1980). *Agric. Biol. Chem.* **44**, 437–438.

Kajiwara, T., Sekiya, J., Asano, M., and Hatanaka, A. (1982). *Agric. Biol. Chem.* **46**, 3087–3088.

Kanofsky, J. R., and Axelrod, B. (1986). *J. Biol. Chem.* **261**, 1099–1104.

Kar, M., and Feierabend, J. (1984). *Planta* **160**, 385–391.

Kim, I.-S., and Grosch, W. (1978). *Z. Lebensm.-Unters.-Forsch.* **167**, 324–326.

Kim, I.-S., and Grosch, W. (1981). *J. Agric. Food Chem.* **29**, 1220–1225.

Klein, B. P., Grossman, S., King, D., Cohen, B.-S., and Pinsky, A.(1984). *Biochim. Biophys. Acta* **793**, 72–79.

Klein, B. P., Cohen B.-S., Grossman, S., King, D., Malovany, H., and Pinsky, A. (1985). *Phytochemistry* **24**, 1903–1906.

Köckritz, A., Schewe, T., Hieke, B., and Hass, W. (1985). *Phytochemistry* **24**, 381–384.

Kühn, H., Heydeck, D., Wiesner, R., and Schewe, T. (1985). *Biochim. Biophys. Acta* **830**, 25–29.

Kühn, H., Schewe, T., and Rapoport, S. M. (1986). *In* "Advances in Enzymology " (A. Meister, ed.) pp. 273–311. John Wiley and Sons, New York.

Laties, G. G. (1982). *Annu. Rev. Plant Physiol.* **33**, 519–555.

Legge, R. L., and Thompson, J. E. (1983). *Phytochemistry* **22**, 2161–2166.

Leoni, O., Iori, R., and Palmieri, S. (1985). *J. Food Sci.* **50**, 88–92.

Leshem, Y. Y. (1984). *In* "Structure, Function and Metabolism of Plant Lipids" (P.-A. Siegenthaler and W. Eichenberger, eds.), pp. 181–188. Elsevier, Amsterdam.

Leshem, Y. Y., Grossman, S., Frimer, A., and Ziv, J. (1979). *In* "Advances in the Biochemistry and Physiology of Plant Lipids" (L.-Å. Appelqvist and C. Liljenberg, eds.), pp. 193–198. Elsevier/North-Holland, Amsterdam.

Lieberman, M., Kunishi, A., Mapson, L. W., and Wardale, D. A. (1966). *Plant Physiol.* **41**, 376–382.

Lulai, E. C., Baker, C. W., and Zimmerman, D. C. (1981). *Plant Physiol.* **68**, 950–955.

Lynch, D. V., and Thompson, J. E. (1984). *FEBS Lett.* **173**, 251–254.

Lynch, D. V., Sridhara, S., and Thompson, J. E. (1985). *Planta* **164**, 121–125.

McCurdy, A. R., Nagel, C. W., and Swanson, B. G. (1983). *Can. Inst. Food Sci. Technol. J.* **16**, 179–184.

Major, R.T., Marchini, P., and Sproston, T. (1960). *J. Biol. Chem.* **235**, 3298–3299.

Matsuda, Y., Beppu, T., and Arima, K. (1979). *Agric. Biol. Chem.* **43**, 189–190.

Matthew, J. A., and Chan, H. W.-S. (1983). *J. Food Biochem.* **7**, 1–6,

Matthew, J. A., and Galliard, T. (1978). *Phytochemistry* **17**, 1043–1044.

Mead, J. F. (1976). *In* "Free Radicals in Biology" (W. A. Pryor, ed.), Vol. 1, p. 51. Academic Press, New York.

Meyer, A., Miersch, O., Büttner, C., Dathe, W., and Sembdner, G. (1984). *J. Plant Growth Regul.* **3**, 1–8.

Miersch, O., Meyer, A., Vorkefeld, S., and Sembdner, G. (1986). *J. Plant Growth Regul.* **5**, 91–100.

Miller, M. G., and Obendorf, R. L. (1981). *Plant Physiol.* **67**, 962–964.

Mitsuda, H., Yasumoto, K., and Yamamoto, A. (1967). *Agric. Biol. Chem.* **31**, 853–860.

Nicolas, J., and Drapron, R. (1981) *Sci. Aliments* **1**, 91–168.

Ohta, H., Ida, S., Mikami, B., and Morita, Y. (1986). *Plant Cell Physiol.* **27**, 911–918.

Pauls, K. P., and Thompson, J. E. (1984). *Plant Physiol.* **75**, 1152–1157.

Peterman, T. K., and Siedow, J. N. (1983). *Plant Physiol.* **71**, 55–58.

Peterman, T. K., and Siedow, J. N. (1985a). *Arch. Biochem. Biophys.* **238**, 476–483.

Peterman, T. K., and Siedow, J. N. (1985b). *Plant Physiol.* **78**, 690–695.

Petersson, L., Slappendel, S., and Vliegenthart, J. F. G. (1985). *Biochim. Biophys. Acta* **828**, 81–85.

Phillips, D. R., Matthew, J. A., Reynolds, J., and Fenwick, G. R. (1979). *Phytochemistry* **18**, 401–404.

Pinsky, A., Grossman, S., and Trop, M. (1971). *J. Food Sci.* **36**, 571–572.

Pirrung, M. C. (1986). *Biochemistry* **25**, 114–119.

Privett, O. S., Nickell, C., Lundberg, W. O., and Boyer, P. D. (1955). *J. Am. Oil Chem. Soc.* **32**, 505–511.

Prusky, D., Kobiler, I., Jacoby, B., Sims, J. J., and Midland, S. L. (1985). *Physiol. Plant Pathol.* **27**, 269–279.

Ramadoss, C. S., and Axelrod, B. (1982). *Anal. Biochem.* **127**, 25–31.

Ramadoss, C. S., Pistorius, E. K., and Axelrod, B. (1978). *Arch. Biochem. Biophys.* **190**, 549–552.

Rustin, P., Dupont, J., and Lance, C. (1983). *Trends Biochem. Sci.* **8**, 155–157.

St. Angelo, A. J., and Ory, R. L. (1984). *Lipids* **19**, 34–37.

Saniewski, M., and Czapski, J. (1983). *Experientia* **39**, 1373–1374.

Saniewski, M., and Czapski, J. (1985). *Experientia* **41**, 256–257.

Satler, S. O., and Thimann, K. V. (1981). *C. R. Hebd. Seances Acad. Sci. Ser. D* **293**, 735–740.

Schauenstein, E., Esterbauer, H., and Zollner, H. (1977). "Aldehydes in Biological Systems." Pion, London.

Schewe, T., Rapoport, S. M., and Kühn, H. (1986). *In* "Advances in Enzymology" (A. Meister, ed.), pp. 191–272. John Wiley and Sons, New York.

Schieberle, P., Grosch, W., Kexel, H., and Schmidt, H.-L. (1981). *Biochim. Biophys. Acta* **666**, 322–326.

Schildknecht, H., and Rauch, G. (1961). *Z. Naturforsch., B* **16b**, 422–429.

Schreier, P., and Lorenz, G. (1982). *Z. Naturforsch., C* **37c**, 165–173.

Sekiya, J., Kajiwara, T., and Hatanaka, A. (1979). *Agric. Biol. Chem.* **43**, 969–980.

Sekiya, J., Kajiwara, T., Munechika, K., and Hatanaka, A. (1983). *Phytochemistry* **22**, 1867–1869.

Sekiya, J., Kajiwara, T., and Hatanaka, A. (1984). *Plant Cell Physiol.* **25**, 269–280.

Shechter, G., and Grossman, S. (1983). *Int. J. Biochem.* **15**, 1295–1304.

Shimizu, T., Radmark, O., and Samuelsson, B. (1984). *Proc. Natl. Acad. Sci. U.S.A.* **81**, 689–693.

Sircar, J. C., Schwender, C. F., and Johnson, E. A. (1983). *Prostaglandins* **25**, 393–396.

Slappendel, S., Veldink, G. A., Vliegenthart, J. F. G., Aasa, R., and Malmström, B.G. (1981). *Biochim. Biophys. Acta* **667**, 77–86.

Slappendel, S., Malmström, B. G., Petersson, L., Ehrenberg, A., Veldink, G. A., and Vliegenthart, J. F. G. (1982). *Biochem. Biophys. Res. Commun.* **108**, 673–677.

Spaapen, L. J. M., Verhagen, J., Veldink, G. A., and Vliegenthart, J. F. G. (1980). *Biochim. Biophys. Acta* **618**, 153–162.

Sridhara, S., and Leshem, Y. Y. (1986). *New Phytol.* **103**, 5–16.

Stan, H.-J., and Diel, E. (1976). *Proc. World Congr. Int. Soc. Fat Res., Marseille*, pp. 15–22.

Stevens, F. C., Brown, D. M., and Smith, E. L. (1970). *Arch. Biochem. Biophys.* **136**, 413–421.

Takahama, U. (1985). *Phytochemistry* **24**, 1443–1446.

Tanimoto, S., and Harada, H. (1984). *Biol. Plant.* **26**, 337–341.

Tappel, A. L. (1973). *Fed. Proc., Fed. Am. Soc. Exp. Biol.* **32**, 1870–1874.

Taylor, H. F., and Burden, R. S. (1973). *J. Exp. Bot.* **24**, 873–880.

Tressl, R., and Drawert, F. (1973). *J. Agric. Food Chem.* 560–565.

Trop, M., Grossman, S., and Veg, Z. (1974). *Ann. Bot. (London)* **38**, 783–794.

Truong, V. D., Raymundo, L. C., and Mendoza, E. M. T. (1982a). *Food Chem.* **8**, 187–201.

Truong, V. D., Raymundo, L. C., and Mendoza, E. M. T. (1982b). *Food Chem.* **8**, 277–312.

Tsurumi, S., and Asahi, Y. (1985). *Physiol. Plant.* **64**, 207–211.

Ueda, J., and Kato, J. (1980). *Plant Physiol.* **66**, 246–249.

Ueda, J., and Kato, J. (1982). *Agric. Biol. Chem.* **46**, 1975–1976.

Van Aller, R. T., Pessoney, G. F., Rogers, V. A., Watkins, E. J., and Leggett, H. G. (1985). *In* "The Chemistry of Allelopathy: Biochemical Interactions among Plants" (A. C. Thompson, ed.), ACS Symp. Ser., No. 268, pp. 387–400. Am. Chem. Soc., Washington, D.C.

Van Os, C. P. A., Rijke-Schilder, G. P. M., and Vliegenthart, J. F. G. (1979). *Biochim. Biophys. Acta* **575**, 479–484.

Van Os, C. P. A., Rijke-Schilder, G. P. M., van Halbeek, H., Verhagen, J., and Vliegenthart, J. F. G. (1981). *Biochim. Biophys. Acta* **663**, 177–193.

Veldink, G. A., Vliegenthart, J. F. G., and Boldingh, J. (1977). *Prog. Chem. Fats Other Lipids* **15**, 131–166.

Venis, M. A. (1984). *Planta* **162**, 85–88.

Vernooy-Gerritsen, M., Veldink, G. A., and Vliegenthart, J. F. G. (1982). *Biochim. Biophys. Acta* **708**, 330–334.

Vernooy-Gerritsen, M., Bos, A. L. M., Veldink, G. A., and Vliegenthart, J. F. G. (1983). *Plant Physiol.* **73**, 262–267.

Vernooy-Gerritsen, M., Leunissen, J. L. M., Veldink, G. A., and Vliegenthart, J. F. G. (1984). *Plant Physiol.* **76**, 1070–1079.

Vick, B. A., and Zimmerman, D. C. (1975). *Plant Physiol.* **56S**, 84.

Vick, B. A., and Zimmerman, D. C. (1976). *Plant Physiol.* **57**, 780–788.

Vick, B. A., and Zimmerman, D. C. (1979a). *Plant Physiol.* **63,** 490–494.
Vick, B. A., and Zimmerman, D. C. (1979b). *Plant Physiol.* **64,** 203–205.
Vick, B. A., and Zimmerman, D. C. (1981a). *Plant Physiol.* **67,** 92–97.
Vick, B. A., and Zimmerman, D. C. (1981b). *Plant Physiol.* **67S,** 161.
Vick, B. A., and Zimmerman, D. C. (1982). *Plant Physiol.* **69,** 1103–1108.
Vick, B. A., and Zimmerman, D. C. (1983). *Biochem. Biophys. Res. Commun.* **111,** 470–477.
Vick, B. A., and Zimmerman, D. C. (1984). *Plant Physiol.* **75,** 458–461.
Vick, B. A., and Zimmerman, D. C. (1986). *Plant Physiol.* **80,** 202–205.
Vick, B. A., Zimmerman, D. C., and Weisleder, D. (1979). *Lipids* **14,** 734–740.
Vick, B. A., Feng, P., and Zimmerman, D. C. (1980). *Lipids* **15,** 468–471.
Vliegenthart, J. F. G., and Veldink, G. A. (1982). *In* "Free Radicals in Biology" (W. A. Pryor, ed.), Vol. 5, pp. 29–64. Academic Press, New York.
Vliegenthart, J. F. G., Veldink, G. A., Verhagen, J., Slappendel, S., and Vernooy-Gerritsen, M. (1982). *In* "Biochemistry and Metabolism of Plant Lipids" (J. F. G. M. Wintermans and P. J. C. Kuiper, eds.), pp. 265–274. Elsevier, Amsterdam.
Wallach, D. P., and Brown, V. R. (1981). *Biochim. Biophys. Acta* **663,** 361–372.
Wardale, D. A., and Galliard, T. (1977). *Phytochemistry* **16,** 333–338.
Wardale, D. A., and Lambert, E. A. (1980). *Phytochemistry* **19,** 1013–1016.
Wurzenberger, M., and Grosch, W. (1984a). *Biochim. Biophys. Acta* **794,** 18–24.
Wurzenberger, M., and Grosch, W. (1984b). *Biochim. Biophys. Acta* **794,** 25–30.
Wurzenberger, M., and Grosch, W. (1984c). *Biochim. Biophys. Acta* **795,** 163–165.
Yabuuchi, S. (1978). *Nippon Nogei Kagaku Kaishi* **52,** 417–425.
Yabuuchi, S., and Amaha, M. (1976). *Phytochemistry* **15,** 387–390.
Yamamoto, A., Fujii, Y., Yasumoto, K., and Mitsuda, H. (1980a). *Agric. Biol. Chem.* **44,** 443–445.
Yamamoto, A., Fujii, Y., Yasumoto, K., and Mitsuda, H. (1980b). *Lipids* **15,** 1–5.
Yamane, H., Takagi, H., Abe, H., Yokota, T., and Takahashi, N. (1981). *Plant Cell Physiol.* **22,** 689–697.
Yamane, H., Abe, H., and Takahashi, N. (1982). *Plant Cell Physiol.* **23,** 1125–1127.
Zimmerman, D. C. (1966). *Biochem. Biophys. Res. Commun.* **23,** 398–402.
Zimmerman, D. C., and Coudron, C. A. (1979). *Plant Physiol.* **63,** 536–541.
Zimmerman, D. C., and Feng, P. (1978). *Lipids* **13,** 313–316.
Zimmerman, D. C., and Vick, B. A. (1970). *Plant Physiol.* **46,** 445–453.
Zimmerman, D. C., and Vick, B. A. (1973). *Lipids* **8,** 264–266.
Zimmerman, D. C., and Vick, B. A. (1983). *Plant Physiol.* **72S,** 108.

Lipases | 4

ANTHONY H. C. HUANG

I. INTRODUCTION

Information on lipolytic enzymes in higher plants is important in understanding their physiological roles, their action in agricultural products during storage, and their potential use in biochemistry and industry. In postgermination of oilseeds, the mobilization of oil reserves is essential in providing energy and carbon skeletons for embryonic growth. Lipolytic enzymes catalyze the initial steps of lipid mobilization, and thus may be rate controlling in germination and postgerminative growth. The turnover of membrane lipids in various tissues is dependent on lipolytic enzymes; knowledge concerning their action is important for understanding cellular development and differentiation, as well as changes in agricultural products during storage. In agriculture, the crushing or storage of seeds or other agricultural products may lead to an increase in lipolytic activity. This increase results in an accumulation of free fatty acids in oilseeds, and the removal of fatty acids from food oil products adds extra costs. Also, lipolytic activities during seed storage may cause a loss of seed vigor or ability of the seeds to germinate rapidly. Furthermore, lipolytic activities cause rancidity, which renders agricultural products unsuitable for human consumption. Some of the enzymes are highly active and are suitable for biochemical study of their catalytic mechanisms. For their specific catalysis, lipolytic enzymes have been used as biochemical tools to probe membrane surface structure (e.g., phospholipase D). Currently, lipase represents about 3% of all enzymes used in industry. The use of lipases to obtain fatty acid and glycerol from triacylglycerols eliminates the cost of energy in the traditional method of hydrolysis under high temperature and pressure, and the mild enzymatic reaction generates fatty acids and glycerol which can be recovered more economically. Lipases have also been used in other industrial processes (Posorske, 1984). Although lipases used in industry have been obtained from microbes, the unique substrate specificity of plant lipases, not found in microbes or mammalian systems, may be of special value in industrial utilization.

It is apparent that the study of plant lipolytic activities is important. However, our knowledge in this area is still very limited, especially when compared with information on mammalian lipases. The nomenclature for plant lipolytic enzymes, especially with reference to substrate specificity, is a common source of confusion. This confusion, together with technical difficulties in the assay of lipase activities, has discouraged researchers. However, many of the problems have been recognized and overcome recently.

This chapter is designed to describe advances in plant lipase research since Galliard's review in 1980 in Vol. 4 of "The Biochemistry of Plants"(Chapter 3). Information already described in this previous article will not be mentioned in the current article, except for clarification, alternate interpretation, or substantial differences in organization. Two topics will be described in the current article. The first topic deals with true lipases (E.C. 3.1.1.3) in oilseeds. Substantial advances have been made in recent years such that a major overhaul of our knowledge can be made. The other topic deals with a group of apparently similar enzymes occurring in diverse tissues that possess the combined catalytic capacity of phospholipases A_1, (E.C. 3.1.1.32), A_2 (E.C. 3.1.1.4), B (E.C. 3.1.1.5), glycolipase, sulfolipase, and monoacylglycerol lipase. Because of the clarification of the literature on true lipases, these nonspecific acyl hydrolases can now be assessed further. Recent reviews on plant lipases include those on cereal lipases (Galliard, 1983), plant lipases (Huang, 1984), and lipid bodies including details on lipase assays (Huang, 1985). A multiauthored book on lipolytic enzymes has also been published (Borgström and Brockman, 1984).

II.   LIPASES (E.C. 3.1.1.3)

A.   Gluconeogenesis in Oilseeds

True lipases, which attack the fatty acyl linkage of water-insoluble triacyl-glycerols, are known to occur in oilseeds and cereals. Oilseeds generally contain 20–50% of their dry weight as storage triacylglycerols. In postgerminative growth, the oil reserve is rapidly mobilized to provide energy and carbon skeletons for the growth of the embryo (Beevers, 1980). Highly active lipases are found to catalyze the hydrolysis of reserve triacylglycerols. The triacylglycerols are localized in subcellular organelles called lipid bodies (oleosomes, oil bodies, spherosomes) which are bounded by a half-unit membrane (Fig. 1; Yatsu and Jacks, 1972). Depending on the plant species, the lipase may be localized on the membrane of the

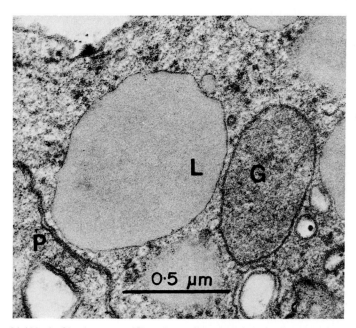

Fig. 1.   Lipid body (L), glyoxysome (G), and a small portion of plastid (P), representing organelles surrounded by "half-unit" membrane, single membrane (one "unit" membrane, tripartite), and double membrane, respectively, in the subapical zone of a 1-day-old shoot apex of maize seedling (courtesy of R. N. Trelease, 1969).

lipid bodies, or in other subcellular compartments. With only one well-documented exception (castor bean), lipase activity is absent in ungerminated seeds and increases rapidly in postgermination (Huang, 1984). The fatty acids released by lipase activity are further metabolized in the glyoxysomes (Cooper and Beevers, 1969; Hutton and Stumpf, 1969). The lipid bodies and glyoxysomes *in vivo* are in close proximity or direct physical contact with one another (Frederick *et al.*, 1975); presumably, this proximity could facilitate transport of fatty acids from the lipid bodies to the glyoxysomes.

Naturally, the studies of lipases in plants center on those in oilseeds, especially in postgermination when lipolysis is active. In fact, high activities of true lipases in tissues or organs other than seeds have not been well documented.

B.   Technical Difficulties in the Assay of Lipase Activities

In the study of lipases, technical difficulties in the assay of the enzyme activities need to be overcome. Some of the difficulties are not unusual, and are common among enzyme assays dealing with water-insoluble substrates. In the assay of lipase activity, the substrate micelles are generally prepared by sonicating triacylglycerol in the presence of an emulsifying agent, such as Acacia (gum arabic), or a detergent such as Triton X-100. The fatty acid released is quantitated by a pH stat continuously, or at time intervals by colorimetry or radioactive analysis. The assay is time-consuming. The substrate emulsion is unstable, and the sizes of the micelles are nonuniform and are not totally reproducible from experiment to experiment. Therefore, some researchers use artificial triacylglycerols containing shorter fatty acid moieties so that the emulsion is much more stable and uniform. Some other researchers measure the esterase activity of the lipase by using an artificial substrate which upon enzymatic hydrolysis generates a fluorescent product. As will be explained, these shortcuts actually generate more problems than those they are intended to solve. An account of the technical problems in measuring lipase activities is itemized below, and some of them are especially critical when dealing with plant tissues:

1. As will be described in Section II,C, a lipase from a certain seed species is likely to be relatively specific on the native storage triacylglycerols, or artificial triacylglycerols containing the major fatty acid components of the storage triacyl-glycerols in that same species. Thus, when assaying seed lipase activity, a suitable triacylglycerol should be used. For example, based on information shown in Table I (Section II,C), it is inappropriate to use tristearin or tripalmitin to study maize lipase, or to use triolein to study elm lipase.

2. Highly active nonspecific acyl hydrolases are present in many plant tissues (Section III). These hydrolases are active on various acyl lipids other than triacylglycerols. Under ideal assay conditions, their activities are several orders of magnitude higher than the lipase activity in the same tissues. Thus, using an artificial ester substrate (generally monoester of a fatty acid and a fluorescent

moiety) to study seed lipases generates uncertain results. The problem may be relatively less important in the study of lipases in other organisms, but is very critical in plant tissues due to the presence of the ubiquitous and highly active acyl hydrolases.

3. The problem described in item 2 is compounded by the general unawareness of the purity of commercial triacylglycerol preparations used as substrates. Impurity of a small amount of monoacylglycerols or other acyl lipids would lead to the assay of the dominating acyl hydrolases mentioned above. It has been reported that commercially available tributyrin preparation contains monobutyrin as a contaminant (Brockerhoff and Jensen, 1974), and trilinolein preparation, labeled as 99% pure, contains a small amount of monoacylglycerols (from one company) or fatty acids (from another company) (Lin and Huang, 1983). A simple preparative thin-layer chromatography has been described to remove fatty acid, di- and monoacylglycerols from triacylglycerol preparations (Huang, 1986). Alternatively, a one-step silicic acid column chromatography using hexane-diethyl ether as the eluting solvent also works well (Christie, 1973).

4. The optimal pH for lipase activity on the native triacylglycerols may be quite different from that on an artificial substrate. For example, rapeseed lipase has an optimal activity at pH 8.5 on N-methylindoxylmyristate, but at pH 6.5 on internal lipid (autolipolysis of the lipid bodies), trilinolein, or trierucin (Lin and Huang, 1983). It is inappropriate to assay the optimal pH for activity using an artificial substrate (or impure substrate), and then use this pH to study substrate specificity and other enzyme properties.

5. Lipase inhibitors are present in some seeds before and after germination (Widmer, 1977; Satouchi and Matsushita, 1976; Gargouri et al., 1984; Wang and Huang, 1984). These inhibitors are not the classical enzyme inhibitors which bind to or act on the enzyme molecule. Instead, the lipase inhibitors are proteins that bind to the surface of the substrate micelles in an in vitro enzyme assay. The binding prevents the normal functioning of the lipase which acts on the interfacial area between the aqueous medium and the micelle surface. If the seed extract does not contain an overwhelming amount of the protein inhibitors, as is the case with peanut cotyledons, lipase activity can be detected and measured by simply adding more substrate micelles to the assay system (Wang and Huang, 1984). However, in other seed tissues, such as soybean cotyledons, the amount of protein inhibitors is in great excess and a simple increase in the substrate micelles is not sufficient to overcome the inhibition. Methods should be designed to remove these proteins prior to assay of lipase activities. The methods of removal could take advantage of the fact that these protein inhibitors are of amphipathic nature.

In an overall assessment, items 1–4 are minor technical difficulties that can be overcome easily. The substrate triacylglycerol should be the native triacylglycerol extracted from the ungerminated seed, or artificial triacylglycerols containing the major storage fatty acid moieties. The release of fatty acid is monitored by a pH stat continuously or at time intervals by colorimetry or radioactive assay (if the triacylglycerol is available in radioactive form). The triacylglycerol obtained commercially, irrespective of the purity claimed by the chemical company, should be checked for purity, and if necessary, purified by thin-layer or column chromatography. Artificial substrates such as those used in fluorescence measurement could be employed in routine enzyme assays such as in enzyme purification, but should be used only after the lipase has been fairly well studied and separated from the nonspecific acyl hydrolases. Overcoming the lipase inhibitors described in item 5 is more difficult, and procedures to remove them will have to be designed and are likely to be species-specific. So far, no such procedure has been published.

C. Substrate Specificity

A distinct feature of plant lipases is their substrate specificity. Seed lipase from a certain plant species is relatively specific for the native triacylglycerols or triacylglycerols containing the major fatty acids of the storage triacylglycerols of that same species (Table I). For example, maize lipase is most active on triacylglycerols of linoleic acid and oleic acid, which are the major fatty acid constituents of maize oil (Lin *et al.*, 1983). Rapeseed and mustard seed (storage triacylglycerols containing substantial amount of erucic acid) lipases are more active on triacylglycerols of erucic acids than of stearic acid or behenic acid (Lin and

TABLE I
Hydrolysis of Various Triacylglycerols by Lipases from Various Sources.[a]

Substrate	Porcine pancreas[b]	Castor bean[b]	Maize[b]	Rapeseed[b]	Erucic acid-free rapeseed[c]	Elm[b]	Mustard[c]	Palm[d]
					Relative activity (%)			
Tricaprin	207	43	127	89	—	100[e]	—	100[e]
Trilaurin	92	60	0	31	—	4	—	60
Trimyristin	50	26	0	92	—	3	—	15
Tripalmitin	5	46	0	27	51	0	39	35
Tristearin	0	62	0	36	89	0	40	—
Triolein	97	55	38	44	138	4	96	—
Trilinolein	100[e]	57	100[e]	89	116	6	89	—
Triricinolein	53	100[e]	0	83	—	0	—	—
Tribehenin	—	—	0	16	—	0	—	—
Trierucin	92	36	45	100[e]	100[e]	0	100[e]	—

[a] Glyoxysomal lipases from soybean (Lin *et al.*, 1982) and castor bean (Maeshima and Beevers, 1985) are also very active on trilinolein and triricinolein, respectively.

[b] From Lin *et al.* (1986). Similar results on castor bean lipase were reported earlier (Ory *et al.*, 1962).

[c] Lin and Huang (1983).

[d] Oo and Stumpf (1983).

[e] The standard (100%) used to calculate the relative activity of the lipase on various substrates.

Huang, 1983). The elm seed lipase (Lin *et al.*, 1986) and palm kernel lipase (Oo and Stumpf, 1983) (both containing high percentages of capric acid in the storage triacylglycerols) are more active on tricaprin than on trilaurin, tripalmitin, or trilinolein. The above substrate specificity has been observed in direct comparisons of different seed lipases using individual triacylglycerol as well as mixed tri-acylglycerol preparations. The pattern of fatty acyl specificity can also be observed on diacylglycerols, monoacylglycerols, and fatty acyl 4-methylumbelliferone, although the pattern becomes less distinct (Lin *et al.*, 1986). This gradual loss in the fatty acyl specificity of the seed lipases from tri- to di- to monoacylglycerols may be of physiological significance. Each storage triacylglycerol molecule generally is not composed of only the major fatty acid, but rather of the major fatty acid and other fatty acids in the same molecule (Hilditch and Williams, 1964). Thus, after the lipase has hydrolyzed the major fatty acid from a triacylglycerol molecule, it should still have high capacity to hydrolyze the remaining diacylglycerol and monoacylglycerol containing the other fatty acids.

Very few lipases from mammalian and microbial sources have high fatty acyl specificity (Jensen *et al.*, 1983). A notable exception is the lipase from *Geotrichum candidum*, which is partially specific for fatty acids with *cis*-9 configuration (Jensen, 1974). In this and in other reported cases, the physiological relevance of fatty acyl specificity is unknown. On the contrary, the physiological significance of fatty acyl specificity in seed lipases is obvious. The plant species can afford to produce more specific lipases for a higher efficient catalysis because the fatty acid compositions of the storage triacylglycerol also are well defined and largely inherited. This fatty acyl specificity of plant lipases could be exploited in lipid biotechnology.

D. Differences in Properties of Lipases among Seed Species

A common feature among lipases from diverse seeds, with the sole well-documented exception of castor bean, is that the enzyme activities are absent in ungerminated seeds and increase in postgermination (Huang, 1984). Other than this similarity, oilseed lipases from diverse species exhibit differences in their proper-ties. These differences include substrate specificity, pH for optimal activity, reactivity toward sulfhydryl reagents, hydrophobicity of the molecule, and sub-cellular location. In the last aspect, depending on the species, three subcellular locations of lipases have been reported; namely, the lipid bodies, the glyoxysomes, and the cytosol. The lipases that are associated with lipid bodies have been studied most extensively.

E. Lipid Body Lipases

Lipases in several seed species, including castor bean, maize, rapeseed, mustard seed, and jojoba, are known to be present in the lipid bodies, and specifically, on the surrounding membrane of the organelles. In subcellular fractionations, about 50–80% of the lipase activity could be recovered in the lipid body fraction. Isolated lipid bodies can be subjected to triacylglycerol removal by ether extraction, and the remaining fraction contains the membranes of the organelles (Fig. 2). Because the membrane is composed of a monolayer of phospholipids (Yatsu and Jacks, 1972), the lipid body ghosts tend to adhere to one another. The membrane contains the lipase, presumably as a minor protein, and other proteins; the overall protein pattern as resolved in sodium dodecyl sulfate–polyacrylamide gel electrophoresis (SDS–P-AGE) is very different among species (Qu *et al.*, 1986). Thus, the properties of the lipase, as well as other lipid body proteins especially the abundant proteins, are unique to each plant species.

1. Castor Bean Acid Lipase

The classic work of Ory and his colleagues (Ory, 1969) on castor bean acid lipase has been reviewed in Galliard's 1980 article, and will not be discussed again here. Among lipases from different seed species, the castor bean lipase is unique in several aspects, including the acidic pH for optimal activity, its presence in an active form in the ungerminated seeds, its lack of substrate specificity on diverse triacylglycerols, and its being equally active on tri-, di-, and monoacylglycerols

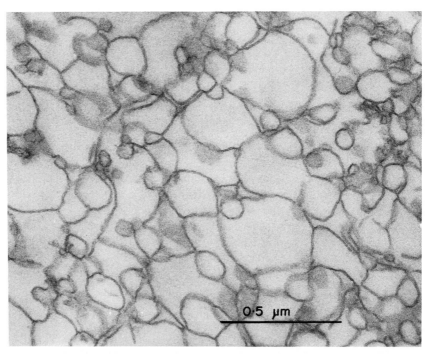

0·5 μm

Fig. 2. Lipid body membrane fraction obtained after the triacylglycerols in isolated maize lipid bodies had been extracted by diethyl ether (courtesy of R. N. Trelease, 1969).

(Table I; Ory *et al.*, 1962; Lin *et al.*, 1986). In postgermination of castor bean, the activity of the lipid body lipase decreases concomitant with the disappearance of triacylglycerol (Moreau *et al.*, 1980).

2. Maize Lipase

Maize lipase from the scutella of seedlings has been extensively studied (Lin *et al.*, 1983; Lin and Huang, 1984; Lin *et al.*, 1986). Similar to the castor bean lipase, the maize enzyme is tightly associated with the lipid body membrane, and resists solubilization by repeated washing with dilute buffer or salt solution. Lipase activity is present in germinated seed and not in ungerminated seed. Isolated lipid bodies undergo autolipolysis, releasing free fatty acids. The enzyme has an optimal activity at pH 7.5 in the autolipolysis of lipid bodies, or on trilinolein or *N*-methyl-indoxylmyristate. The activity is not greatly affected by salt or pretreatment of the enzyme with *p*-chloromercuribenzoate or mersalyl.

Among the various triacylglycerols examined, maize lipase is active only on those containing linoleic acid and oleic acid (and the short-carbon chain capric acid) which are the major fatty acid constituents of maize oil (Table I). The enzyme is inactive on lecithins. It is more active on tri- than di- or monolinolein. It releases linoleic acid from both primary and secondary positions, although it exerts some degree of preference in releasing fatty acid from the primary rather than the secondary position of a triacylglycerol (Lin *et al.*, 1986). At the primary positions, it is more active on oleyl ester than stearyl ester.

Maize lipase apparently catalyzes acyl transfer (Lin *et al.*, 1986). During its catalysis on the hydrolysis of 1,2-dilinolein, trilinolein is produced in addition to linoleic acid and monolinolein. No trilinolein is produced if 1,3-dilinolein is the substrate. Presumably, the esterification of linoleic acid is much more active on the primary position of 1,2-dilinolein than on the secondary position of 1,3-dilinolein. Similar catalysis of acyl transfer has been observed in pancreatic lipase and other lipases (Brockerhoff and Jensen, 1974).

Maize lipase has been purified 272-fold to apparent homogeneity as evidenced by SDS–PAGE and double immunodiffusion (Lin and Huang, 1984). In SDS–PAGE, the enzyme has an approximate molecular mass of 65 kDa. This molecular mass is not reduced by pretreatment of the enzyme with 2-mercaptoethanol, and the molecule contains no cysteine. The amino acid composition as well as a biphasic partition using Triton X-114 reveal the enzyme to be a hydrophobic protein. This hydrophobicity is in accord with the enzyme being tightly associated with the membrane.

3. Rape and Mustard Seed Lipase

Rape and mustard belong to the family Cruciferae, whose species generally contain a high amount of erucic acid in the seed storage triacylglycerols. In rapeseed, lipase activity is absent in the ungerminated seed and increases after germination (Rosnitschek and Theimer, 1980; Lin and Huang, 1983). About 50% of the activity can be recovered in the lipid bodies after subcellular fractionation.

However, unlike the castor bean and maize lipases, rapeseed lipase can be washed away easily from the lipid bodies by dilute buffer. The enzyme detached from the lipid body fraction (which contains all the triacylglycerol and floats during centrifugation) is still attached to membrane fragments. These fragments were proposed to be appendices of the lipid bodies and the location for the synthesis of nascent lipase (Wanner and Theimer, 1978). However, the fragments seem more likely to represent the ghosts, or part of the ghosts, of the lipid bodies after the triacylglycerol has been consumed (Bergfeld *et al.*, 1978).

The rapeseed lipase has an optimal activity at pH 6.5 on the native triacylglycerol, trilinolein, or trierucin, but at pH 8.5 on the artificial substrate *N*-methy-lindoxylmyristate. Among the acylglycerols examined, the lipase is most active on trierucin and trilinolein (Table I), and is less active on the corresponding di- and monoacylglycerols (Lin and Huang, 1983; Lin *et al.*, 1986). This substrate specificity is meaningful, since rapeseed contains roughly 40% erucic acid and 20% linoleic acid in the storage triacylglycerol. In the hydrolysis of trierucin, erucic acid is the dominant product. However, at the very early stage of hydrolysis when only a very small amount of the trierucin has been hydrolyzed, a fatty acid tentatively identified by gas chromatography as oleic acid is dominant (Lin *et al.*, 1986). This oleic acid may be the hydrolytic product of the remaining internal triacylglycerols in the enzyme preparation, or a secondary fatty acid generated from the initial product erucic acid by an unknown enzymatic reaction. Depending on how the substrate is prepared, the enzymatic activity may be inhibited by erucic acid, or promoted by NaCl and mild detergents (Rosnitschek and Theimer, 1980; Lin and Huang, 1983).

Several varieties of rapeseed containing no erucic acid in the seed oil have been produced through breeding. The lipase obtained from one of these varieties (Tower) possesses properties very similar to those of the normal rapeseed lipase (Lin and Huang, 1983), including having relatively higher activity on trierucin and trilinolein (Table I). This high activity on trierucin is not unexpected, since erucic acid-free varieties were obtained by breeding through the alterations of only a few genes (Downey and Craig, 1964). Presumably the lipase gene was not affected.

Lipase from the seed of mustard possesses properties very similar to the rapeseed lipase (Lin and Huang, 1983). These similarities include the pH for optimal activity on the native substrates and *N*-methylindoxylmyristate, the substrate specificity (highest on trierucin and trilinolein) (Table I), subcellular location, and the appearance of activity in the seed only after germination. The similarity is quite expected in view of the close relationship between the two species. A similarity also exists in the protein pattern of the isolated lipid bodies as resolved by SDS–PAGE (Qu *et al.*, 1986).

4. Oil Palm Lipases

Lipase in the fruit mesocarp of oil palm (*Elaeis guineensis*) is present in the lipid bodies (Abigor *et al.*, 1985). The partially purified enzyme has an optimal activity at pH 4.5, and is active on its native substrate (palm oil, which contains mostly oleic acid and palmitic acid), triolein, and tripalmitin. Sodium cyanide, resorcinal, cholesterol, lecithin, and glycylglycine strongly inhibit the enzyme activities, whereas phenol, L-cysteine, and EDTA enhance its activities. In another study, lipase activity was not detected in the extracts of palm kernel and haustorium in postgermination, but was detected in the shoot extract (Oo and Stumpf, 1983). The shoot lipase is most active on tridecanoin and trilaurin (Table I).

5. Jojoba Lipase

A lipid body wax ester hydrolase in jojoba seed has been studied (Moreau and Huang, 1981). Jojoba seed is unique among oilseeds as it contains intracellular wax esters instead of triacylglycerols as food storage. The wax ester hydrolase activity is absent in ungerminated seed and increases in postgermination. The enzyme is associated with the membrane of the lipid bodies and has an optimal activity at pH 8.5 (on N-methylindoxylmyristate). It has highest activities on monoacylglycerols, wax esters, jojoba wax ester, and low activities on dipalmitin, tripalmitin, and triolein. Since the enzyme hydrolyzes triacylglycerols, it can be called a true lipase (E.C. 3.1.1.3), although its native substrates are monoesters. The enzyme is inactivated by p-chloromercuribenzoate, and the inactivation is reversed by subsequent addition of dithiothreitol.

6. Soybean Lipase (Acyl Hydrolase)

Soybean lipid bodies isolated from ungerminated or germinated seed do not undergo autolipolysis (Lin et al., 1982). However, the membranes of lipid bodies isolated from ungerminated seed contain a monoacylglycerol hydrolase. This hydrolase has an optimal activity at pH 6.5 (on N-methylindoxylmyristate) and hydrolyzes monolinolein but not trilinolein. In postgermination, the enzyme activity declines before the total triacylglycerol. The above observations suggest that the enzyme is not involved in triacylglycerol hydrolysis in postgermination. Of all the oilseeds examined, only soybean possesses lipid bodies having the activity of this monoacylglycerol hydrolase. Its physiological role is unknown. Whether it is a nonspecific acyl hydrolase as described in Section III requires further studies.

F. Glyoxysomal Lipases

A lipase/acyl hydrolase is present in the glyoxysomes of diverse seed species in postgermination (Muto and Beevers, 1974; Huang, 1975b; Huang and Moreau, 1978). The enzyme is tightly bound to the membrane of the organelles, and resists solubilization from the membrane by buffer or high-salt solutions. Treatment of intact glyoxysomes with trypsin strongly diminishes the enzyme activity but does not affect the glyoxysomal matrix enzyme activities (Maeshima and Beevers, 1985). The enzyme can be solubilized by deoxycholate and KCl, and has been purified to apparent homogeneity. The molecular mass of the purified enzyme, as determined by SDS–PAGE is 62 kDa.

Originally, the glyoxysomal lipase activity was assayed as a fatty acyl hydrolase using N-methylindoxylmyristate in a fluorescence assay (Muto and Beevers, 1974).

The enzyme has an optimal activity at alkaline pH values on this artificial substrate. It is also active on monoacylglycerols, but is relatively inactive on triacylglycerols. Nevertheless, it does hydrolyze triacylglycerols, especially those containing the dominant fatty acid of the storage triacylglycerol in the same seed species (Lin *et al.*, 1982; Maeshima and Beevers, 1985). Even so, the enzyme from soybean hydrolyzes trilinolein at a rate of only about 10% of that on monolinolein. Since the enzyme can hydrolyze triacylglycerols, it can be termed a true lipase. Its role in triacylglycerol hydrolysis in postgermination remains unclear in view of its activity being much higher on monoacylglycerol than on triacylglycerol, its physical separation from the triacylglycerols of the lipid bodies, and its low activity in comparison with the true lipase in the same cell.

G. Soluble Lipases

In some oilseeds, lipase activity is detected mostly in the soluble fraction in subcellular fractionations. The enzyme is truly soluble in the sense that it is not associated with small membrane fragments. The soluble nature of these lipases is probably not due to a harsh fractionation procedure for the following two reasons. First, under the same fractionation procedure carried out in the same laboratory, other seed species yield lipase activity that is associated with lipid bodies (Lin *et al.*, 1986). Second, in the latter seed species, the lipases cannot be truly solubilized easily. These lipases are either tightly associated with the lipid bodies and not easily removed (e.g., castor bean, maize, and jojoba), or still bound to membrane fragments after its relatively easy removal from the lipid bodies (e.g., rapeseed and mustard). If the soluble lipases are indeed those involved in lipolysis *in vivo*, they should be amphipathic proteins that can associate with the membrane of the lipid bodies in order to carry out lipolysis. None of these soluble lipases, to be described in the following paragraphs, has been purified or even partially purified.

In elm seed, lipase activity is present in germinated but not ungerminated seed (Lin *et al.*, 1986). The lipase has an optimal activity at pH 9.0 on tricaprin or trilinolein. It is highly specific on tricaprin, and hydrolyzes trilaurin, trilinolein, and triolein at rate less than 10% of that on tricaprin (Table I). In subcellular fractionation, more than 90% of the lipase activity can be recovered in the 100,000-*g*, 90-min supernatant, whereas the lipid body fraction has no detectable activity.

Similarly, cotton cotyledon lipase is present only in germinated seed, and most of the activity appears in the soluble fraction in subcellular fractionation (Huang, 1984). About 10–15% of the lipase activity is associated with the lipid body fraction, and it can be removed from the fraction easily by washing the fraction with dilute buffer.

In Douglas fir, two lipase activities, measured at pH 5.1 and 7.1, are present in the storage gametophytes of seed (Ching, 1968). The activities increase in post-germination. With native neutral lipids as the substrate for enzyme assays, about 80% of the lipase activity is present in the soluble fraction, and the rest are distributed among lipid body, mitochondrial, and protein body fractions. The lack of lipase activity in fir lipid bodies is similar to that in pine (Huang, 1984), in which the lipid bodies of ungerminated and germinated seed are unable to undergo autolipolysis.

H.   Other Oilseed Lipases

Lipase activities, besides those described above, have been reported in other oilseeds, but their subcellular locations are unknown. In many cases, acetone powders of the seeds were used as enzyme sources.

In peanut, a lipase was partially purified from an acetone powder of maturing seeds, using tributyrin as the substrate (Sanders and Pattee, 1975). The enzyme activity increases during seed maturation. It has a higher activity on tributyrin than on maize oil. Similar findings were observed from germinated peanut, although the data were not presented. It is unknown whether this partially purified lipase is the same as the peanut glyoxysomal lipase (Huang and Moreau, 1978); they both share the same alkaline pH for optimal activity and increase in activity in post-germination. Even though this alkaline glyoxysomal "lipase" activity on N-methyl-indoxylmyristate is dominant over activities at pH 5 or 7 throughout postger-mination, its involvement in storage triacylglycerol hydrolysis in postgermination is still uncertain. An earlier report shows that the maximal lipolytic (autolytic) activity in crude extracts of germinated peanut occurs at pH 4–5 (St. Angelo and Altschul, 1964).

A lipase was partially purified from ungerminated seed of *Veronica anthel-mintica* (Olney *et al.*, 1968), an oilseed in which the major storage lipid is trivernolin (vernolic acid is *cis*-12,13-epoxy-*cis*-9-octadecenoic acid). The enzyme has a molecular mass of >200 kDa and an optimal activity at pH 7.5–8.0. It hydrolyzes both primary and secondary ester bonds of triacylglycerols and is equally active on saturated (palmitate, stearate) and unsaturated (oleate) triacylglycerols. Its activity on trivernolin was not examined.

An acid lipase (pH 5) is present in the acetone powder of germinated seed of *Cucumeriopsis edulis* (Opute, 1975). The enzyme activity increases in postgermina-tion. Since the enzyme indiscriminately releases fatty acids from *rac*-glyceryl-1-palmitate-2-oleate-3-stearate or *rac*-glyceryl-1-stearate-2-palmitate-3-oleate, its ac-tivity is quite nonspecific.

In acetone powder of apple seed, two lipase activities are present, with optimal activities at pH 5 and 7.5 (Somolenska and Lewak, 1974). The activities are low in ungerminated or dormant seed and increase in postgermination.

I.   Lipases in Cereal Grains

Cereal grains are not considered oilseeds because of their low content of lipids which are not utilized commercially (except maize oil and rice bran oil). They generally contain 2–10% lipids, depending on the species and varieties. Most of the lipids are triacylglycerols, of which 80–90% of the fatty acids are linoleic acid and oleic acid (Hilditch and Williams, 1964). The lipids are usually located in the embryo (germ) and the aleurone layer (the bran, which also includes the pericarp, testa, and some endosperm) (Morrison, 1983; Hammond, 1983). In the aleurone layer of wheat (Jelsema *et al.*, 1977) and barley (Firn and Kende, 1974), the triacylglycerols are present in lipid bodies. In maize, the lipids are present in lipid bodies of scutella and are mobilized in postgermination (see Section II,E).

Lipolytic activities are present in different parts of cereal grains. Some research focus has been placed on these enzymes because of their potential in causing rancidity during grain or bran storage. The physiological role of these enzymes is unclear, and in some cases, their exact tissue location is still unknown. Furthermore, the lipolytic activities might have been contributed by contaminating microbes (Galliard, 1983). These microbes proliferate upon grain storage, especially under humid conditions, and utilize the grain lipids and other food reserves for growth.

Lipases in cereal grains have been covered in Galliard's 1980 article, and there has been little progress since. They will not be discussed in the current article. Other review articles covering cereal lipases include those by Galliard (1983) and Huang (1984).

J.   Lipases in Nonseed Tissues

True lipase in nonseed tissues or organs has been reported occasionally. In maize roots, lipase activity with an optimal activity at pH 5 on olive oil occurs mostly in the zone of elongation (Heimann-Matile and Pilet, 1977). A lipase with a molecular mass of 77 kDa has been separated from other acyl hydrolases in potato tuber extract (Hasson and Laties, 1976a). The potato lipase releases fatty acids preferentially from triolein instead of diolein or monoolein and has no activity on phospholipids or galactolipids. It has an optimal activity at pH 7.5–8.0 on *N*-methylindoxyl esters and preferentially hydrolyzes *N*-methylindoxyl esters of longer chain length.

K.   Intracellular Transport of Lipolytic Products

In seeds, triacylglycerols are localized in the lipid bodies, and the hydrolyzed fatty acids are activated and β-oxidized in the glyoxysomes. The other product, glycerol, is metabolized initially in the cytosol (Stumpf, 1955; Huang, 1975a). Fatty acid is water-insoluble, and its removal from the lipid bodies is desirable and probably essential in the continuation of lipolysis. Little fatty acid accumulates in seeds *in vivo* in postgermination (St. Angelo and Altschul, 1964). The mode of transport of fatty acid from the lipid bodies to the glyoxysomes remains unknown.

There are several possible mechanisms for the above-mentioned fatty acid transport. Phospholipid transfer proteins are present in mammalian and plant systems (Boussange *et al.*, 1980), but their ability to transfer free fatty acid in oilseeds has not been established. In those seeds where soluble instead of lipid body-bound lipases are present, the enzymes may carry out lipolysis as well as transport of fatty acid; however, there is no information to support this notion. As observed *in situ* under the electron microscope, lipid bodies and glyoxysomes occasionally are in physical contact with each other (Frederick *et al.*, 1975). Even in those electron micrographs where the two organelles are observed to be in close proximity but not in direct contact, a consideration of the three-dimensional structure of the organelles may suggest that there is a point of contact at other locations on the organelles. If so, it is possible that the fatty acid is transported from the lipid body to the glyoxysome by a membrane flow mechanism. Fatty acid released by the lipase on the lipid body membrane would flow along the membrane by a simple diffusion, due to a concentration gradient, through the contact junction of the two organelles, to the membrane of the glyoxysome. Such a flow of fatty acid along the membranes of different organelles (and cells) has been proposed to occur in mammalian systems (Scow *et al.*, 1979). Finally, it should not be forgotten that the glyoxysomal membrane contains an acyl hydrolase/true lipase of unknown function (Section II,F). The physiological role of the glyoxysomal lipase in lipolysis has not been substantiated, and there is always a possibility that somehow it is involved in the transport of fatty acid.

L.   Control of Lipase Activity

Since storage triacylglycerols are used for gluconeogenesis during germination and in postgerminative growth of seeds, lipase activity which is involved in the first reaction of the gluconeogenic pathway should be a useful target for physiological control. In most seeds, lipase activity increases during germination and in postgermination. There are a few studies where hormones applied externally to the seeds or to excised seed parts enhance lipase activity and lipid consumption. However, when such a stimulation occurs, it is still unclear whether the hormone effect is specific on the increase in lipase activity preceding other developmental processes or whether it is merely a general influence on cell development or differentiation.

In castor bean, excised endosperm provided with water appears to undergo the normal postgerminative development including gluconeogenesis from lipids independent of the embryonic axis or cotyledons (Huang and Beevers, 1974). Gibberellin added externally enhances lipolysis slightly (Marriott and Northcote, 1975). Thus, if any hormonal influence on lipolysis is involved, the hormone must have been present in the endosperm.

In the embryo of dormant apple seeds, two lipase activities at pH 5 and pH 7.5 are present (Somolenska and Lewak, 1974). In excised embryo, light acting through the phytochrome system enhances the internal gibberellin content as well as the alkaline lipase activity (Zarska-Maciejewska, 1980). Externally applied gibberellin also exerts similar effects. These two treatments allow the embryo, originally from dormant seeds, to undergo germination, though slowly. Cold treatment has a similar effect on enhancing internal gibberellin content and lipase activity and it also initiates normal germination. AMO-1618, a gibberellin biosynthetic inhibitor, blocks the effect of light or gibberellin application. In contrast, light, gibberellin, or

AMO-1618 application generally causes a decrease in the acid lipase activity. It has been proposed that the alkaline lipase is involved in lipolysis during germination and that light and gibberellin are part of the control mechanism by which low temperature breaks dormancy.

In wheat grain, most of the lipase activities are present in the endosperm and bran, and they increase rapidly in postgermination (Clarke *et al.*, 1983). The activities in excised endosperm or bran do not increase unless embryo diffusate or growth regulators are supplied. Application of hydroxylamine or glutamine at 1 m*M* induces the activity increase in the endosperm but not in the bran. Lipase activity in the excised bran increases only after the addition of both hydroxylamine and indoleacetic acid. In excised aleurone layers of barley, application of gibberellin causes no change in the amount of neutral lipids in the first 12 h, even though the development of other processes is induced (Firn and Kende, 1974). Neutral lipid consumption is only evident after 24 h of incubation. Apparently, lipolysis is not one of the very first events in the hormone-induced development of the aleurone layer.

As mentioned in Section II,B, lipase inhibitors of protein nature are present in some oilseeds. They are present in the soluble fraction after subcellular fractionation. They bind to the substrate micelles in *in vitro* enzyme assays, thereby preventing the access of substrate to the enzyme. Such an inhibitory action is different from that of the classical enzyme inhibitors which act on the enzyme molecules directly. It is doubtful that the lipase inhibitors in oilseeds exert a regulatory role. *In vivo*, the triacylglycerol is packed inside the lipid bodies, which are surrounded by a half-unit membrane. Thus, the substrate triacylglycerol is not accessible to the inhibitors. It is inconceivable that a lipase regulatory mechanism is built on a massive encasement of the triacylglycerol by proteins. It is more likely that some proteins, such as special storage proteins in massive amount, are amphipathic and fortuitously bind to the triacylglycerol micelles in an *in vitro* lipase assay and thus prevent lipolysis.

M. Biosynthesis of Seed Lipases

In all oilseeds studied except castor bean, lipase activity is absent in the ungerminated seed and appears in postgermination. The mode of lipase biosynthesis is currently unknown. Undoubtedly, the lipase in castor bean is synthesized in seed maturation rather than in postgermination (Moreau *et al.*, 1980). Most likely, the castor bean lipase, which is bound to the lipid bodies, is synthesized as an active enzyme together with the lipid bodies in seed maturation. Although the lipase extracted from maturing castor bean is highly active, it is apparently inactive *in vivo*. Whether it is synthesized in the rough endoplasmic reticulum, as is the lipid body (Schwarzenbach, 1971), is unknown. In other oilseeds, the lipase may be synthesized as an inactive precursor during seed maturation. Alternatively, the enzyme may be synthesized in postgermination when activity is detected. In the latter possibility, the newly synthesized enzyme would have to be transported from the polysomes (free or bound) to the lipid bodies. Since the lipase molecule is hydrophobic or amphipathic, its transport from the polysomes to the lipid bodies may involve membrane vesicles. The enzyme may be synthesized initially as a precursor and then processed on the rough endoplasmic reticulum. These are important questions that need to be answered.

The maize lines, Illinois High Oil and Illinois Low Oil, containing about 18% and 0.5%, respectively, of kernel lipids, have been obtained through continuous breeding for oil quantity (Dudley *et al.*, 1974). The lipase activities in the kernels of the above two lines, as well as their F_1 generation, in postgermination are proportional to their lipid content (Wang *et al.*, 1984). On the contrary, the glyoxysomal enzymes remain the same in the three maize lines irrespective of the lipid content. Thus, there is a difference in the genetic control of lipase and the other gluconeogenic enzymes, and a coselection for high lipid content and high lipase activity through breeding. The mechanism of coordinate selection for both high lipid and high lipase activity is unknown. There are about 50 genes for the expression of high lipid content in Illinois High Oil (Dudley *et al.*, 1974), and apparently they are expressed only in seed maturation and not in seedling growth. Yet lipase activity is absent in the maturing and ungerminated seeds, and appears only after germination. It is unlikely that the genes for high-lipid content are tightly linked to the lipase gene(s). It is possible that the high-lipid genes and the lipase gene(s) are both expressed in seed maturation, but the latter expression results in the production of prelipase protein or mRNA, which is processed to active lipase in postgermination. Alternatively, the lipase may be synthesized or degraded in postgermination in proportion to the availability of specific binding surface on the lipid bodies or substrate and thus metabolic need.

III. LIPID ACYL HYDROLASES

A. Introduction

An enzyme that hydrolyzes acyl groups from several classes of lipids, including glycolipids, phospholipids, sulfolipids, and mono- and diacylglycerols, but is inactive on triacylglycerols, is present in many plant tissues. The acyl hydrolase releases both fatty acids from diacyl glycerolipids, and in many cases, there is no preference for either the 1- or 2-position of the acyl ester linkage. Thus, the enzyme possesses a combined catalytic capacity of phospholipase A_1, A_2 and B, as well as glycolipase, sulfolipases, and monoacylglycerol lipase. Similarities of the enzymes from various tissues include the following: (1) they exert a similar pattern of substrate specificity as described above; (2) they occur as isozymes in each tissue and they have fairly similar patterns of substrate specificity; (3) they have a similar pH for optimal activity on a particular substrate and generally the pH shifts to a more alkaline value in the presence of detergent; and (4) they catalyze acyl transferase reactions.

The hydrolytic activities on various classes of acyl lipids are apparently carried out by a single protein. The following evidence with occasional exceptions has been obtained using purified or partially purified enzymes on, usually but not exclusively, galactolipids and phosphatidylcholine.

1. The activity ratio of the enzyme preparation on galactolipid and phospholipid remains fairly constant throughout an enzyme purification procedure.

2. The activity ratio is also similar after treatment of the enzyme with high or low temperature and chemical-modifying reagents.

3. The enzyme carries out acyltransferase reactions with each of the substrates.

4. Each substrate at a similar high concentration inhibits the activity.

5. The optimal activity is at an acidic pH and shifts to a high pH in the presence of detergent.

6. Competition of the two substrates for enzyme activity exists, suggesting that the activities reside not only in a single protein, but also within the same active site.

7. Potato tuber of a special variety in which the acyl hydrolase activity is very low contains a proportional reduction of activity on each of the substrates (Galliard, 1980).

Historically, a galactolipase was first observed in bean leaves in 1964 (Sastry and Kates, 1964), and it was subsequently purified to homogeneity (Helmsing, 1969). In 1971, a partially purified enzyme from potato tuber was shown to possess glycolipase, phospholipase, and other acyl hydrolytic activities (Galliard, 1971), and this enzyme was purified to homogeneity in 1975 (Hirayama et al., 1975). Reexamination of the leaf galactolipase revealed that the enzyme also catalyzes fatty acid removal from phospholipids and other acyl lipids (Burns et al., 1977). The enzyme from rice bran was also purified (Matsuda and Hirayama, 1979a). In the following description, the properties of the potato tuber enzyme will be given in some detail. Then, the enzymes from leaves and rice bran will be mentioned, with emphasis on the major similarities and differences between them and the potato tuber enzyme. The enzymes in other tissues known to hydrolyze at least one class of the acyl lipids will be reviewed briefly. Finally, the physiological role of the enzyme will be discussed.

B. Potato Tuber Acyl Hydrolase

The potato tuber acyl hydrolase was the first acyl hydrolase identified in plants, and is perhaps the best studied. There were some interesting uncertainties in the earlier studies, and they are apparently clarified by a recent study. Galliard (1971) partially purified (5-fold) an acyl hydrolase from potato tuber. He was unable to attain a higher fold of purification by various procedures; the enzyme apparently was copurified with the bulk of the soluble proteins. Hirayama et al. (1975) achieved a 350-fold purification of an acyl hydrolase from potato tuber to apparent homogeneity. Even though the fold of purification of the enzymes by Galliard and by Hirayama et al. were enormously different, the specific activities of the two enzyme preparations were quite similar. Since the potato tuber acyl hydrolase exists in isozyme forms (Galliard and Dennis, 1974b), it is possible that Galliard's enzyme (13% yield) represents a mixture of the isozymes, whereas Hirayama *et al.*'s enzyme (1% yield) represents one of the isozymes. Presumably, Hirayama *et al.*'s procedure selectively excluded or inactivated some of the isozymes. Nevertheless, the two enzyme preparations were remarkably similar in their substrate specificity (to be described).

A report by Racusen (1984) apparently clarifies the above discrepancies. The potato tuber acyl hydrolase is identified as the storage glycoprotein termed patatin. Patatin can be separated into 6–10 closely associated protein bands in isoelectric focusing, and all of these protein bands exhibit esterase activities. Since patatin represents about 20% of the soluble protein in potato tuber, the findings can explain the unsuccessful attempt of Galliard (1971) to purify the enzyme to more than 5-fold. Because the enzyme is apparently identical to patatin, the molecular properties of the fairly well-characterized patatin should also be those of the acyl hydrolase. The 6–10 patatin isoproteins are immunologically indistinguishable, and are almost identical in at least the 22 amino-terminal amino acids (Park *et al.*, 1983). The molecular mass is 88 kDa—compared to 107 kDa of Galliard's (1971) and 70 kDa of Hirayama *et al.*'s (1975) enzymes—and the protein can be separated into two identical subunits of 44 kDa by SDS. The acyl hydrolase of Hirayama *et al.* (1975) has a molecular mass of 70 kDa, and can be separated into two subunits of molecular mass of 4 and 17 kDa. The subunits reassociate in the presence of Ca^{2+} to produce a molecule of the original molecular mass (i.e., 70 kDa) possessing part of the enzymatic activity. Chemical modification and photooxidation of the protein show that histidine, serine, and tyrosine residues are important for enzyme activity.

The potato tuber acyl hydrolase hydrolyzes fatty acyl groups from various glycolipids, phospholipids, sulfolipids, monoolein, and diolein (Galliard, 1971; Hirayama *et al.*, 1975). It has lower activities on short-chain esters and does not act on triolein, tristearin, and wax esters. Table II shows the relative activities of the enzyme obtained from two different laboratories on various lipids. It should be emphasized that there are many factors affecting the activity on each individual substrate, and a comparison of activities toward different substrates under one particular assay condition is only a rough approximation. These factors include the pH of the assay, the substrate concentration and method of preparation, the presence of an activator such as Triton X-100, and the purity of the enzyme preparation. Nevertheless, the data show a similar pattern of substrate specificity reported by two different laboratories, with the exception of acylated steryl glycoside. In addition to the substrates listed in Table II, the enzyme also hydrolyzes methyl esters and *p*-nitrophenyl esters of long-chain fatty acids (Galliard, 1971). In the hydrolysis of monogalactodiacylglycerol, fatty acid is released from the 1-position first, producing a monogalactodiacylglycerol as an intermediate and further hydrolysis generates another fatty acid and a galactoglycerol (Galliard, 1971; Hirayama *et al.*, 1975). In the hydrolysis of phosphatidylcholine, digalactodiacylglycerol, or diolein, no

TABLE II

Comparison of Substrate Specificities of Lipid Acyl Hydrolases Purified from Potato Tubers, Leaves of Two *Phaseolus* Species and Potato, and Rice Bran[a,b]

	Relative activity (%)					
	Potato tubers		Leaves			Rice bran
Substrate	(1)	(2)	*P. multiflora* (3)	*P. vulgaris* (4)	Potato (5)	(6)
Monogalactodiacylglycerol	100	100	100	100	100	100
Monogalactomonoacylglycerol	—	68	—	132	394	110
Digalactodiacylglycerol	56	62	60	39	44	169
Digalactomonoacylglycerol	—	26	—	49	146	—
Phosphatidylcholine	42	100	23	31	14	27
Lysophosphatidylcholine	233	304	142	89	26	33
Phosphatidylglycerol	40	151		47	77	
Lysophosphatidylglycerol	—	20	—	93	89	—
Phosphatidylethanolamine	35	33	—	11	1	Trace
Lysophosphatidylethanolamine	—	71	—	32	6	—
Phosphatidylinositol	—	36	—	—	—	—
Phosphatidylic acid	40	16	—	7	8	—
Monoolein	325	—	161	—	—	—
Diolein	68	—	—	—	—	—
Triacylglycerol	<1	—	<1	<0.1	<0.1	Trace
Tributyrin	<2	—	—	—	—	
Sulfoquinovosldiacylglycerol	—	52	13	14	43	—
Acylated steryl glucoside	<2	500	—	—	—	—

[a] The activities are relative to that on monogalactodiacylglycerol. There are many factors (e.g., pH, detergent inclusion) affecting the activity on each individual substrate, and a comparison of activities toward different substrates under one particular assay condition is only a rough approximation.

[b] References: 1. Galliard (1971); 2. Hirayama *et al.* (1975); 3. Burns *et al.* (1977); 4. Matsuda *et al.* (1980); 5. Matsuda and Hirayama (1979b); 6. Matsuda and Hirayama (1979a).

monoacyl intermediate was found, presumably because the first acyl group removal is much slower than the second or because the intermediate does not leave the active site. The enzyme also hydrolyzes potato tuber mitochondrial membranes (Hasson and Laties, 1976b).

The pH for optimal activity of the enzyme varies with individual substrates. For example, the optimal activity is at pH 5 in citrate buffer for monogalacto-diacylglycerol and pH 8.5 in Tris buffer for phosphatidylcholine (Hirayama *et al.*, 1975). Generally speaking, the optimal pH is around 5 in the absence of detergent, and shifts to a more alkaline value in the presence of Triton X-100 (Galliard, 1971).

The activity on phosphatidylcholine is stimulated by Ca^{2+} at pH >7.5 and inhibited by Ca^{2+} at lower pH values (Hasson and Laties, 1976b). This calcium stimulation does not appear to be mediated by calmodulin (Moreau *et al.*, 1985), unlike the potato leaf enzyme (next section). Triton X-100 enhances the activity on phosphatidylcholine but has relatively little effect on glycolipids (Galliard, 1971). The activity toward membrane phospholipids is enhanced by detergents, especially unsaturated fatty acids. In fact, the totally purified enzyme is inactive in the absence of a detergent. Thus, an "autocatalysis" was suggested to occur in tissue extracts (Galliard, 1980) because the fatty acids released by the enzyme alter the membrane structure to allow access of the enzyme to more substrates. An unusual effect of detergents on the enzyme activity has been reported recently (Moreau *et al.*, 1985). Deoxycholate at low concentrations (10–50 μM) inhibits the rate of autolysis of phosphatidycholine in tuber homogenates, whereas at higher concentrations (100–1000 μM) it stimulates the autolysis. Three common calmodulin inhibitors (dibucaine, chlorpromazine, and trifluoperazine) exhibit similar effects. The effects of the inhibitors are probably due to their detergent-like properties rather than their interaction with calmodulin.

Several isozymes of the enzyme are present in potato tuber (Galliard and Dennis, 1974b; Hasson and Laties, 1976a; Shepard and Pitt, 1976b). There are slight differences in the substrate specificities of the various isozymes. The isozyme pattern also varies, depending on the variety of the species. In potato tuber, besides this enzyme, there is a true lipase (Section II,J) and a low molecular mass (23 kDa) esterase that hydrolyzes monoolein, phosphatidylcholine, and *N*-methylindoxyl-butyrate, but does not hydrolyze triolein, diolein, glycolipids, and *N*-methylindoxylmyristate (Hasson and Laties, 1976a).

The enzyme also catalyzes an acyltransferase reaction (Galliard and Dennis, 1974a):

$$RCOOX + YOH \leftrightarrow RCOOY + XOH$$

When Y = H the reaction is a hydrolysis, and when Y = CH_3 a fatty acid methyl ester is produced. The affinity of the enzyme for methanol is about 10 times that for water.

The acyl hydrolase is always present in the soluble fraction in potato tuber extract and does not appear to be associated with any cytoplasmic organelles. However, in view of its acid pH for optimal activity and its apparent lack of activity *in vivo*, it may be localized in the cell vacuoles, which are generally broken in cell fractionation. Also, it may be present in the plastids (amyloplasts), which are also extremely fragile in cell fractionation. This latter possibility is raised in view of the suggested chloroplastic location of the leaf enzyme (Section III,C). Such a possibility has been discounted based on the lack of enzyme activity in isolated amyloplasts, although information on the integrity of the amyloplasts was not presented (Wardale, 1980).

C. Leaf Acyl Hydrolase

The earliest reference to this enzyme appeared in 1964 when Sastry and Kates reported on "galactolipases" in the leaves of *Phaseolus multiflorus*. The enzyme is present in the leaves of diverse species (Galliard, 1980). It has been purified to homogeneity from the leaves of two *Phaseolus* species (Helmsing, 1969; Matsuda *et al.*, 1979) and potato (Matsuda and Hirayama, 1979b). An important purification step is the use of palmitoylated gauze chromatography, which is supposed to stabilize the enzyme activity (Matsuda and Hirayama, 1979b). The enzymes from *P. multiflorus* (Helmsing, 1969) and potato (Matsuda and Hirayama, 1979b) have a molecular mass of 110 kDa while the *P. vulgaris* enzyme has a molecular mass of 90 kDa (Matsuda *et al.*, 1979). Whereas about 50% of the *P. multiflorus* enzyme (galactolipase) activity is inhibited by 5–10 mM 2-mercaptoethanol and completely by 1 mM cysteine (Helmsing, 1969), the *P. vulgaris* enzyme (galactolipase) activity is enhanced some 50% by the two sulfhydryl reagents at 1–10 mM (Matsuda *et al.*, 1980). Chemical-modification studies of the *P. vulgaris* enzyme suggest that histidine and tryptophan residues are important for activity.

The leaf enzymes from three different species (Burns *et al.*, 1977; Matsuda and Hirayama, 1979b; Matsuda *et al.*, 1980) exhibit a pattern of substrate specificity fairly similar to the potato tuber acyl hydrolase (Table II). Furthermore, the potato tuber enzyme and the leaf enzymes generally have an optimal activity at acidic pH and, in the presence of detergent, the optimal pH shifts to a higher value. Acyltransferase activity has been observed with the *P. vulgaris* enzyme (Matsuda *et al.*, 1980). Unlike the potato tuber enzyme, the *P. vulgaris* (Matsuda *et al.*, 1980) and potato (Matsuda *et al.*, 1980) leaf enzymes do not produce detectable amounts of monoacyl lipid intermediates from either monogalactodiacylglycerol or phosphatidylcholine.

Like the potato tuber enzyme, acyl hydrolase isozymes of fairly similar substrate specificity are present in leaves (Matsuda and Hirayama, 1979b). Two hydrolases from *P. multiflorus* have been partially separated (Burns *et al.*, 1979, 1980). One hydrolase is active on monoolein and phosphatidylcholine and the other hydrolase is active on monoolein and galactolipid. Whereas competition between monoolein and monogalactodiacylglycerol for activity occurs with the "glycolipase", little competition is present between phosphatidylcholine and digalactodiacylglycerol for the enzyme activity. The relationship of the two enzymes to the acyl hydrolase isozyme systems in the leaves of other species and in potato tuber and rice bran is unknown.

Whereas the potato tuber acyl hydrolase is a "soluble" enzyme, the leaf enzyme has been reported to be present in the chloroplasts of *P. multiflorus* (Sastry and Kates, 1964), *P. vulgaris* (Anderson *et al.*, 1974), and potato (Matsuda and Hirayama, 1979b). Washing and sonication are ineffective in releasing the enzyme from the chloroplasts; the release can be achieved by acetone extraction. A recent report shows that the enzyme activity in potato leaf extracts is much higher than those previously reported and that the enzyme activity is present largely in the soluble fraction (Moreau, 1985). Apparently, the enzyme activity is regulated by calcium and calmodulin (Moreau and Isett, 1985), and by reversible protein phosphorylation (Moreau, 1986).

D. Rice Bran Acyl Hydrolase

The enzyme from rice bran has been purified to apparent homogeneity using a purification procedure that includes palmitoylated-gauze chromatography (Matsuda and Hirayama, 1979a). The enzyme has a molecular mass of 40 kDa, which is significantly lower than that of the potato tuber or leaf enzyme. Serine and cystine residues are important for enzyme activity. It exhibits a pattern of substrate specificity fairly similar to the potato tuber and leaf enzymes (Table II). Like the leaf enzyme, no detectable monoacyl lipid intermediate is produced from either monogalactodiacylglycerol or phosphatidylcholine.

E. Other Acyl Hydrolases

Washing of isolated potato tuber mitochondria with $CaCl_2$ induces an enzymatic hydrolysis of internal membrane phospholipids or externally supplied phospholipids with the release of free fatty acids (Bligny and Douce, 1978). As a result, the mitochondrial oxidative and phosphorylative properties are damaged. The enzyme has been identified as a membrane-bound acyl hydrolase which is unmasked by $CaCl_2$. The potato tuber nonspecific acyl hydrolase described in Section III,B is a soluble enzyme and is activated by Ca^{2+}. Whether this latter enzyme is related to the mitochondrial enzyme is unknown.

Several forms of a lysophospholipase have been purified from barley endosperm in postgermination (Fujikura and Baisted, 1985). The enzyme has a polypeptide with a molecular mass of 36 kDa and an extra mass of 10–12% carbohydrates. The enzyme can release fatty acids from different lysophospholipids, with the highest activity on lysophosphatidylcholine containing palmitic acid. It has no activity on p-nitrophenyl palmitate, no phospholipase A activity on phosphatidycholine, and no transacylation activity between two lysophospholipids. The metabolic role of the enzyme appears to be in the hydrolysis of lysophospholipid starch during germination.

Acyl hydrolases are present in other plant tissues. Since their substrate specificity on different classes of lipids has not been probed, it is unknown if they are related to the nonspecific acyl hydrolases of potato tuber, leaves, and rice bran. Hydrolase activity that releases fatty acids from phosphatidylcholine and lysophosphatidylcholine is present in barley grains (Von Rebmann and Acker, 1973). The enzyme activity increases and decreases in accordance with the metabolic activity of the seed during maturation and germination. An acyl hydrolase that releases fatty acid from sulfolipids has been detected in alfalfa leaves and roots, and in maize roots (Yagi and Benson, 1962).

F. Physiological Role

The activities of the nonspecific acyl hydrolases in many tissues are extremely high. For example, in potato tuber, if all the internal phospholipids and glycolipids were available to the enzyme under the most suitable condition for activity, all the lipids would be deacylated within 1 sec at 25° (Galliard, 1980). Despite this high activity, the physiological role of the enzyme is still unknown. The situation is equivalent to phospholipase D in various plant tissues, in which the enzyme is extremely active *in vitro* but its physiological role is also unclear (Galliard, 1980).

Speculation on the role of acyl hydrolases has mainly been centered on their involvement in the turnover of membrane lipids. In green leaves, galactolipids (Douce and Joyard, 1980) and sulfolipids (Harwood, 1980) are the dominant lipid components of the chloroplasts which are the intracellular sites of acyl hydrolases. Thus, the enzyme may be involved in the turnover or rearrangement of the chloroplast membrane system, especially during leaf greening and senescence. In leaf greening, the internal membrane system of a chloroplast undergoes a vigorous structural alteration from a prolamellar body to the final thylakoid membrane system. During leaf senescence, the membranes are broken down and some of the components are translocated to other parts of the plant for reutilization. In senescing rose petal, there is a decrease in phospholipids and an increase in acyl hydrolase activity (Borochov *et al.*, 1982). In pea leaves, senescence is promoted by the activation effect of calmodulin on acyl hydrolase (Leshem *et al.*, 1984). Since acyl hydrolase also catalyzes acyl-transferring reactions, it could participate in the biosynthetic pathway of acyl lipids (Harwood, 1979). In tissues other than leaves, alterations of membrane lipids occur in development and differentiation, such as during seed maturation and germination and fruit ripening. The free fatty acids released from the enzymatic reaction may undergo oxidation catalyzed by several known oxidases to produce nonvolatile and volatile metabolites that may be of hormonal nature (Galliard and Chan, 1980). In injured potato tuber cells, the hydrolytic and acyl-transferring activities of acyl hydrolase, together with lipoxygenase, may release cytotoxic, oxidized fatty acid derivatives and water-insoluble waxes that inhibit microbial invasion (Racusen, 1984). In cereals, during germination the acyl hydrolase may be involved in the mobilization of starch-bound lysophosphatidylcholine (Fujikura and Baisted, 1983). All the above-mentioned physiological processes may require the involvement of acyl hydrolase activities. However, there is no sufficient evidence for these involvements.

Some evidence is available on the involvement of acyl hydrolases in the hydrolysis of internal lipids under pathological conditions. Mechanical or freezing injury disrupts subcellular compartmentation. Acyl hydrolases, which otherwise would be separated from the substrates physically, will be in direct contact with various cytoplasmic organelles such that deacylation occurs readily. The fatty acids released could inhibit the activities of cellular organelles; this inhibition has been observed *in vitro* on chloroplasts (McCarty and Jagendorf, 1965), mitochondria (Hasson and Laties, 1976b), and microsomes (Ben Abdelkader *et al.*, 1973). Alternatively, the fatty acids released could be oxidized to volatile and nonvolatile components which may be the sources of rancidity (Galliard and Chan, 1980); the involvement of acyl hydrolase in the development of rancidity production in rice bran during storage has been demonstrated (Hirayama and Matsuda, 1975). During the "aging" of potato tuber slices, acyl hydrolase is thought to be involved in a major turnover of the internal membrane lipids (Hasson and Laties, 1976b). During the formation of leaf galls of oak, there is a lesser amount of phospholipids and an enhancement of acyl hydrolase activity (Bayer, 1983). In fungal infection, it has been suggested that the fungus simply secretes a few enzymes to disrupt the subcellular compartmentation, and all subsequent reactions of cell disintegration are accomplished by internal acyl hydrolases and other enzymes (Shepard and Pitt, 1976a). For example, during the infection of potato tubers by *Botrytis cinerea*, the fungus produces a phospholipase which does not attack isolated protoplasts but is able to enter the cells and hydrolyze the tonoplast.

IV. PERSPECTIVE

Research on oilseed lipases has not progressed rapidly due to various experimental difficulties described earlier in this article. However, these difficulties are being overcome, and the potential of future research to yield valuable information is tremendous. Most of the oilseed lipases that have been studied in some detail are associated with the membranes of the lipid bodies. Further studies on these membrane-associated lipases should be carried out with purified enzymes as well as with enzymes still attached to the membrane. The latter aspect is especially important in that the substrate triacylglycerols have to interact with the lipase-containing membrane. The effect of association of the enzyme with the membrane on the catalytic mechanism should be analyzed. The positional specificity of the enzyme with reference to the types of fatty acids attached to different carbon atoms of the glycerol and the mechanisms of progressive hydrolysis of one triacylglycerol to three fatty acids are important information. As far as the purified maize lipid body lipase (or castor bean glyoxysomal lipase) is concerned, one single enzyme can catalyze the removal of all the three fatty acids. How the product fatty acids move from the lipid bodies to the glyoxysomes, where fatty acid activation and β-oxidation occur, remains to be elucidated. The storage tissues do not accumulate free fatty acids during postgermination, even though the lipase activity as measured *in vitro* is much higher than the rate of *in vivo* lipolysis. The control of lipase activity by internal factors is apparent and requires elucidation.

In some oilseeds, the lipases do not associate with the lipid bodies but remain in the soluble fraction in subcellular fractionation. The properties of these soluble lipases are unknown. To play a role in triacylglycerol hydrolysis, the enzyme must come into contact with the lipid bodies. It is likely that the enzyme is an amphipathic protein, and is actually associated with the lipid body membrane permanently but loosely *in vivo*. Alternatively, the enzyme may be associated with the membrane only transiently during catalysis. If so, it may also be involved in the transport of fatty acids from the lipid bodies to the glyoxysomes (unless the lipid bodies and the glyoxysomes are in direct contact). There are plenty of intriguing questions waiting to be answered.

The physiological role of the glyoxysomal lipase remains a puzzle. Its physical separation from the substrate triacylglycerol and its low activity on triacylglycerols relative to monoacylglycerols argue against its role in triacylglycerol hydrolysis. Furthermore, there is always a highly active true triacylglycerol-hydrolyzing lipase present in the lipid bodies or cytosol. Since the glyoxysomal lipase is facing the cytosol, there is a possibility that its real physiological role is to aid the reception of fatty acids from the cytosol or from the contact point between the glyoxysomes and lipid bodies.

Research on the biosynthesis of lipase is just at the beginning. With the availability of modern cell biology techniques, the mode of lipase biosynthesis, as outlined earlier in this article, can be pinpointed quite easily. Also, the time is ripe to apply modern molecular biology techniques to study the detailed structure of the lipase and its gene and the regulation of gene expression.

The activity of nonspecific acyl hydrolase is extremely high in many tissues. Although its action in causing rancidity in stored agricultural products and in damaged or infected tissues has been quite well documented, its *in vivo* physiological role is still unknown. Circumstantial evidence has suggested that the enzyme participates in membrane lipid turnover or net synthesis; however, direct evidence is lacking. Investigations should be carried out with tissues or organs that have a high rate of *in vivo* turnover, net synthesis, or degradation of acyl lipids. Examples are senescing leaves, and greening leaves in which the prolamellar bodies of the plastids are actively being converted to thylakoids. The subcellular location of the enzyme and its isozymes should be studied. In view of its hydrolytic nature and its optimal activity being at an acidic pH, the possible localization of the enzyme, or a fraction thereof, in the vacuoles should be examined. At least part of the extractable enzyme activity is associated with the chloroplasts; information on its exact subchloroplast location, especially in relation to the acyl lipids, may give hints to its physiological role. The catalytic mechanism of the enzyme, which can act on so many different types of acyl lipids but not triacylglycerols, should be studied. The isozymic nature of the enzyme and the relative specificity of different isozymes toward various substrates should be reexamined carefully in view of a report suggesting that the activities on galactolipids and phospholipids in *Phaseolus* leaves belong to two different enzymes (Burns *et al.*, 1980).

REFERENCES

Abigor, D. R., Opute, F. I., Opoku, A. R., and Osagie, A. U. (1985). *J. Sci. Food. Agric.* **36,** 599–606.
Anderson, M. M., McCarty, R. E., and Zimmer, E. A. (1974). *Plant Physiol.* **53,** 699–704.
Bayer, M. H. (1983). *Plant Physiol.* **73,** 179–181.
Beevers, H. (1980). *In* "The Biochemistry of Plants" (P. K. Stumpf and E. E. Conn, eds.), Vol. 4, pp. 117–130. Academic Press, New York.
Ben Abdelkader, A., Cherif, A., Demandre, C., and Mazliak, P. (1973). *Eur. J. Biochem.* **32,** 155–165.
Bergfeld, B., Hong, Y. N., Kuhnl, T., and Schopfer, P. (1978). *Planta* **143,** 297–307.

Bligny, R., and Douce, R. (1978). *Biochim. Biophys. Acta* **529**, 419–428.
Borgström, B., and Brockman, H. L. (1984). "Lipases." Elsevier, Amsterdam.
Borochov, A., Halevy, A. H., and Shinitzky, M. (1982). *Plant Physiol.* **69**, 296–299.
Boussange, J., Dounady, D., and Kader, J. (1980). *Plant Physiol.* **65**, 355–358.
Brockerhoff, H., and Jensen, R. G. (1974). "Lipolytic Enzymes." Academic Press, New York.
Burns, D. D., Galliard, T., and Harwood, J. L. (1977). *Biochem. Soc. Trans.* **5**, 1302–1304.
Burns, D. D., Galliard, T., and Harwood, J. L. (1979). *Phytochemistry* **18**, 1793–1797.
Burns, D. D., Galliard, T., and Harwood, J. L. (1980). *Phytochemistry* **19**, 2281–2285.
Ching, T. M. (1968). *Lipids* **3**, 482–488.
Christie, W. W. (1973). "Lipid Analysis," pp. 158–161, Pergamon, Oxford.
Clarke, N. A., Wilkinson, M. C., and Laidman, D. L. (1983). *In* "Lipids in Cereal Technology" (P. J. Barnes, ed.), pp. 57–92. Academic Press, London.
Cooper, T. G., and Beevers, H. (1969). *J. Biol. Chem.* **244**, 3507–3513.
Douce, R., and Joyard, J. (1980). *In* "The Biochemistry of Plants" (P. K. Stumpf and E. E. Conn, eds.), Vol. 4, pp. 321–362. Academic Press, New York.
Downey, R. K., and Craig, B. M. (1964). *J. Am. Oil Chem. Soc.* **41**, 475–478.
Dudley, J. W., Lambert, R. J., and Alexander, D. E. (1974). *In* "Seventy Generations of Selection for Oil and Protein in Maize" (J. W. Dudley, ed.), pp. 181–212. Crop Sci. Soc. Am., Madison, Wisconsin.
Firn, R. D., and Kende, H. (1974). *Plant Physiol.* **54**, 911–915.
Frederick, S. E., Gruber, P. J., and Newcomb, E. H. (1975). *Protoplasma* **84**, 1–29.
Fujikura, Y., and Baisted, D. (1983). *Phytochemistry* **22**, 865–868.
Fujikura, Y., and Baisted, D. (1985). *Arch. Biochem. Biophys.* **234**, 570–578.
Galliard, T. (1971). *Biochem. J.* **121**, 379–390.
Galliard, T. (1980). *In* "The Biochemistry of Plants" (P. K. Stumpf and E. E. Conn, eds.), Vol. 4, pp. 85–116. Academic Press, New York.
Galliard, T. (1983). *In* "Lipids in Cereal Technology" (P. J. Barnes, ed.), pp. 111–147. Academic Press, New York.
Galliard, T., and Chan, H. W. S. (1980). *In* "The Biochemistry of Plants" (P. K. Stumpf and E. E. Conn, eds.), Vol. 4, pp. 131–161. Academic Press, New York.
Galliard, T., and Dennis, S. (1974a). *Phytochemistry* **13**, 1731–1735.
Galliard, T., and Dennis, S. (1974b). *Phytochemistry* **13**, 2463–2468.
Gargouri, Y., Julien, R., Pieroni, G., Verger, R., and Sarda, L. (1984). *J. Lipid Res.* **25**, 1214–1221.
Hammond, E. G., (1983). *In* "Lipids in Cereal Technology" (P. J. Barnes, ed.), pp. 331–352. Academic Press, New York.
Harwood, J. L. (1979). *Prog. Lipid Res.* **18**, 55–86.
Harwood, J. L. (1980). *In* "The Biochemistry of Plants" (P. K. Stumpf and E. E. Conn, eds.), Vol. 4, pp. 301–320. Academic Press, New York.
Hasson, E. P., and Laties, G. G. (1976a). *Plant Physiol.* **57**, 142–147.
Hasson, E. P., and Laties, G. G. (1976b). *Plant Physiol.* **57**, 148–152.
Heimann-Matile, J., and Pilet, P. E. (1977). *Plant Sci. Lett.* **9**, 247–252.
Helmsing, P. J. (1969). *Biochim. Biophys. Acta* **178**, 519–533.
Hilditch, T. P., and Williams, P. N. (1964). "The Chemical Constitution of Natural Fats." Wiley, New York.
Hirayama, O., and Matsuda, H. (1975). *Nippon Nogei Kagaku Kaishi* **49**, 569–576.
Hirayama, O., Matsuda, H., Takeda, H., Maenaka, K., and Takatsuka, H. (1975). *Biochim. Biophys. Acta* **384**, 127–137.
Huang, A. H. C. (1975a). *Plant Physiol.* **55**, 555–558.
Huang, A. H. C. (1975b). *Plant Physiol.* **55**, 870–874.
Huang, A. H. C. (1984). *In* "Lipases" (B. Borgström and H. L. Brockman, eds.), pp. 419–442. Elsevier, Amsterdam.
Huang, A. H. C. (1985). *In* "Modern Methods of Plant Analyses" (J. F. Jackson and H. F. Linskens, eds.), pp. 145–151. Springer-Verlag, Berlin and New York.

Huang, A. H. C., and Beevers, H. (1974). *Plant Physiol.* **54,** 277–279.
Huang, A. H. C., and Moreau, R. A. (1978). *Planta* **141,** 111–116.
Hutton, D., and Stumpf, P. K. (1969). *Plant Physiol.* **44,** 508–516.
Jelsema, C., Morre, D. J., Ruddat, M., and Turner, C. (1977). *Bot. Gaz.* **138,** 138–149.
Jensen, R. G. (1974). *Lipids* **9,** 149–157.
Jensen, R. G., De Jong, F. A., and Clark, R. M. (1983). *Lipids* **18,** 239–252.
Leshem, Y. Y., Sridhara, S., and Thompson, J. E. (1984). *Plant Physiol.* **75,** 329–335.
Lin, Y. H., and Huang, A. H. C. (1983). *Arch. Biochem. Biophys.* **225,** 360–369.
Lin, Y. H., and Huang, A. H. C. (1984). *Plant Physiol* **76,** 719–722.
Lin, Y. H., Moreau, R. A., and Huang, A. H. C. (1982). *Plant Physiol.* **70,** 108–112.
Lin, Y. H., Wimer, L. T., and Huang, A. H. C. (1983). *Plant Physiol.* **73,** 460–463.
Lin, Y. H., Yu, C., and Huang, A. H. C. (1986). *Arch. Biochem. Biophys.* **244,** 346–356.
McCarty, R. E., and Jagendorf, A. T. (1965). *Plant Physiol.* **40,** 725–735.
Maeshima, M., and Beevers, H. (1985). *Plant Physiol.* **79,** 489–493.
Marriott, K. M., and Northcote, D. H. (1975). *Biochem. J.* **152,** 65–70.
Matsuda, H., and Hirayama, O. (1979a). *Agric. Biol. Chem.* **43,** 463–469.
Matsuda, H., and Hirayama, O. (1979b). *Biochim. Biophys. Acta* **573,** 155–165.
Matsuda, H., Tanaka, G., Morita, K., and Hirayama, O. (1979). *Agric. Biol. Chem.* **43,** 563–570.
Matsuda, H., Morita, K., and Hirayama, O. (1980). *Agric. Biol. Chem.* **44,** 783–790.
Moreau, R. A. (1985). *Phytochemistry* **24,** 411–414.
Moreau, R. A. (1986). *Plant Science* **47,** 1–9.
Moreau, R. A., and Huang, A. H. C. (1981). *In* "Methods in Enzymology"(J. M. Lowenstein, ed.), Vol. 71, Part C, pp. 804–813. Academic Press, New York.
Moreau, R. A., and Isett, T. F. (1985). *Plant Science* **40,** 95–98.
Moreau, R. A., Liu, K. D. F., and Huang, A. H. C. (1980). *Plant Physiol.* **65,** 1176–1180.
Moreau, R. A., Isett, T. F., and Piazza, G. J. (1985). *Phytochemistry* **24,** 2555–2558.
Morrison, W. R. (1983). *In* "Lipids in Cereal Technology" (P. J. Barnes, ed.), pp. 11–32. Academic Press, New York.
Muto, S., and Beevers, H. (1974). *Plant Physiol.* **54,** 23–28.
Olney, C. E., Jensen, R. G., Sampugna, J., and Quinn, J. G. (1968). *Lipids* **3,** 498–502.
Oo, K. C., and Stumpf, P. K. (1983). *Plant Physiol.* **73,** 1028–1032.
Opute, F. I. (1975). *J. Exp. Bot.* **26,** 379–386.
Ory, R. L. (1969). *Lipids* **4,** 177–185.
Ory, R. L., St. Angelo, A. J., and Altschul, A. M. (1962). *J. Lipid Res.* **3,** 99–105.
Park, W. D., Blackwood, C., Mignery, G. A., Hermodson, M. A., and Lister, R. (1983). *Plant Physiol.* **71,** 156–160.
Posorske, L. H. (1984). *J. Am. Oil Chem. Soc.* **61,** 1758–1760.
Qu, R., Wang, S. M., Lin, Y. H., Vance, V. B., and Huang, A. H. C. (1986). *Biochem. J.* **235,** 57–65.
Racusen, D. (1984). *Can. J. Bot.* **62,** 1640–1644.
Rosnitschek, I., and Theimer, R. R. (1980). *Planta* **148,** 193–198.
St. Angelo, A. J., and Altschul, A. M. (1964). *Plant Physiol.* **39,** 880–883.
Sanders, T. H., and Pattee, H. E. (1975). *Lipids* **10,** 50–54.
Sastry, P. S., and Kates, M. (1964). *Biochemistry,* **3,** 1280–1287.
Satouchi, K., and Matsushita, S. (1976). *Agric. Biol. Chem.* **40,** 889–897.
Schwarzenbach, A. M. (1971). *Cytobiologie* **4,** 145–147.
Scow, R. O., Desnuelle, P., and Verger, R. (1979). *J. Biol. Chem.* **254,** 6456–6463.
Shepard, D. V., and Pitt, D. (1976a). *Phytochemistry* **15,** 1465–1470.
Shepard, D. V., and Pitt, D. (1976b). *Phytochemistry* **15,** 1471–1474.
Somolenska, G., and Lewak, S. (1974). *Planta* **116,** 361–370.
Stumpf, P. K. (1955). *Plant Physiol.* **30,** 55–58.
Von Rebmann, H., and Acker, L. (1973). *Fette, Seifen Anstrichm.* **75,** 409–411.

Wang, S. M., and Huang, A. H. C. (1984). *Plant Physiol.* **76,** 929–934.

Wang, S. M., Lin, Y. H., and Huang, A. H. C. (1984). *Plant Physiol.* **76,** 837–839.

Wanner, G., and Theimer, R. R. (1978). *Planta* **140,** 163–169.

Wardale, D. A. (1980). *Phytochemistry* **19,** 173–177.

Widmer, F. (1977). *J. Agric. Food Chem.* **25,** 1142–1145.

Yagi, T., and Benson, A. A. (1962). *Biochim. Biophys. Acta* **57,** 601–603.

Yatsu, L. Y., and Jacks, T. J. (1972). *Plant Physiol.* **49,** 937–943.

Zarska-Maciejewska, B., Sinska, I., Witkowska, E., and Lewak, S. (1980). *Physiol. Plant.* **48,** 532–535.

The Biosynthesis of Saturated Fatty Acids

5

P. K. STUMPF

I. INTRODUCTION

Since 1980, rapid progress has been made in elucidating the molecular events in the synthesis of saturated fatty acids in higher plants. Nevertheless, intriguing questions continue to arise as to the origin of acetyl-CoA, the biosynthesis of medium-chain and very long-chain fatty acids, and the mechanisms involved in regulating the composition of fatty acids. In this chapter we will examine the information which has been brought forth since Chapter 7 of Volume 4 of "The Biochemistry of Plants" was written. The reader should check that chapter for the literature prior to 1980.

The Biochemistry of Plants, Vol. 9

121

II. ORIGIN OF ACETYL-CoA

It is well recognized that free acetate is an effective substrate for fatty acid synthesis (FAS) (Stumpf, 1980). Probably this effectiveness relates to the observation that free acetate readily diffuses into organelles (Jacobson and Stumpf, 1972), is nontoxic, and is metabolically inert. Thus acetate can enter the target organelle, there to be rapidly converted to acetyl CoA by acetyl-CoA synthetase, which in most leaf tissue is exclusively localized in the chloroplast (Kuhn et al., 1981). In seeds the enzyme is distributed both in the proplastid and in the cytosol (Dennis and Miernyk, 1982).

Acetate does occur endogenously in plant cells. Recently, employing mature spinach leaves from local markets, Kuhn et al. (1981) estimated acetate concentrations as high as 1 mM whereas Liedvogel, working with young greenhouse-grown spinach leaves, observed a lower concentration of 0.065 mM (Liedvogel, 1985). Although these results reflect a steady-state level, the rate of acetate flow from its source to its utilization is not known. A good discussion on the role of acetate is presented by Given in his excellent review on this subject (Givan, 1983).

There is no question that acetate derives from the oxidative decarboxylation of pyruvic acid to acetyl-CoA by pyruvate dehydrogenase complex (PDC) (Camp and Randall, 1985), and thence its hydrolysis to acetate by a suitable hydrolase (Liedvogel and Stumpf, 1982). In fact, Liedvogel and Stumpf (1982) demonstrate the presence of an acetyl-CoA hydrolase in highly purified spinach leaf mitochondria but its absence in spinach chloroplasts. In contrast, Givan and Hodgson (1983) were not able to detect its activity in pea leaf mitochondria.

PDC has been well studied in leaf and seed mitochondria (Dennis and Miernyk, 1982). Recently PDC has been documented both in pea chloroplasts and in spinach chloroplasts though at somewhat lower levels than in corresponding leaf mitochondria (Liedvogel, 1985; Trude and Heise, 1985). Of considerable interest is the observation by Camp and Randall (1985) that whereas mitochondrial PDC is regulated by both end product inhibition and covalent modification by a phosphorylation–dephosphorylation cycle, the chloroplast PDC resembles the *Escherichia coli* PDC in that there is no evidence of any covalent modification.

As to the source of pyruvic acid, there appears to be no question that the glycolytic breakdown of glucose to pyruvic acid is the principal source (Miernyk and Dennis, 1982, 1983). In proplastids, as well as in the cytosol, all the enzymes necessary for the degradation of glucose to pyruvate have been demonstrated. However, the generation of pyruvate in the chloroplast is poorly understood. Although most of the enzyme required for glycolysis exists in the chloroplast stroma, the evidence for the presence of phosphoglyceromutase is weak (Givan, 1983; Stitt and Ap Rees, 1979; Botha and Dennis, 1986). Furthermore, in the chloroplast a considerable transport of triose phosphate into the cytosol is required to fulfill the needs for the synthesis of sucrose in the cytosol and the concomitant flow of inorganic phosphate into the chloroplast via the phosphate translocator. Moreover, pyruvate, unlike acetate, is a very reactive substrate. Its actual concentration in spinach tissue has been determined to be 0.145 mM (Liedvogel, 1985). Important reactions, which include transamination, synthesis of branched-chain amino acid, participation in C_4 plants, CO_2 fixation, and involvement in TCA cycle anaplerotic reactions, are only a partial list of pyruvate participation in general cell metabolism.

Two types of compartmentations may be involved in the generation of acetyl-CoA. The first is a one-compartment system in which pyruvate is generated and broken down to acetyl-CoA, which then is channeled into fatty acid and terpenoid biosynthesis. This system appears to function in castor bean proplastids (Dennis and Miernyk, 1982) and in pea chloroplasts (Camp and Randall, 1985). The other system involves three compartments, namely, the mitochondrion, the cytosol, and the chloroplast in some leaf tissue. Pyruvate is converted to acetate via acetyl-CoA in the mitochondrion; acetate then diffuses through the cytosol to the chloroplast where it is converted to acetyl-CoA by the acetyl-CoA synthetase. Evidence for this comes from the work by Liedvogel and Stumpf (1982), who showed that only by a combination of pure spinach mitochondria and spinach chloroplasts could pyruvate be converted to fatty acids. This system would also explain the function of acetyl-CoA synthetase, which is widespread in plant cells.

Another possibility is that citrate serves as a source of acetyl-CoA in the presence of an ATP, CoA-dependent citrate lyase. While this system is very important in animal cells, its role in plant cells appears to be limited to only a few plants (Nelson and Rinne, 1977; Fritsch and Beevers, 1979).

There is some evidence that acetyl-CoA may be transported into a plastid by a carnitine acetyltransferase mechanism. Thomas and Wood (1982) and McLaren *et al.* (1985) have detected this enzyme in pea chloroplasts and have suggested that it plays an important role in the transport of acetyl-CoA into organelles.

In conclusion, in some plant cells pyruvate serves as the direct precursor for acetyl-CoA generation; in other plants free acetate is the immediate precursor. With new techniques developed to prepare pure organelles and new procedures to detect very low levels of acetyl-CoA (Liedvogel, 1985), the solution to the problem of the origin of acetyl-CoA should be near at hand. It is quite possible that all three systems are operative depending on the developmental phase of the cell.

III. FORMATION OF MALONYL-CoA

Malonyl-CoA serves as a key substrate in the synthesis of a wide variety of important compounds in plants. Its primary function is as the C_2 unit in FAS (Stumpf, 1980). In addition, it provides C_2 units for the synthesis of cuticular waxes (Kolattukudy *et al.*, 1976), flavonoids and anthocyanins (Hahlbrock, 1981), stilbenoids (Gorham, 1980), anthroquinones (Packter, 1980), malonyl amino cyclopropane-1-carboxylic acid (Amrhein and Kionka, 1983), and free malonic acid (Stumpf and Burris, 1981).

It has been known for some time that the prime mechanism of its synthesis involves the biotinyl enzyme acetyl-CoA carboxylase, which is widely distributed in plant tissues (Stumpf, 1980). In recent years investigations have characterized the enzyme, examined its localization, considered its role in the regulation of FAS. Since 1980, a number of important publications have examined these aspects.

A. Characterization of Acetyl-CoA Carboxylase

Earlier work with wheat germ and barley carboxylase have clearly indicated the enzyme to be a soluble and somewhat unstable large complex with a M_r of approximately 6×10^5.

The enzyme has now been extensively studied in maize leaf (Nikolau *et al.*, 1981, 1984a,b), developing endosperm of castor bean seeds (Finlayson and Dennis, 1983), avocado mesocarp and spinach chloroplast (Mohan and Kekwick, 1980), cell suspension cultures of parsley (Egin-Bühler and Ebel, 1983), and rapeseed (Turnham and Northcote, 1983). A consistent picture has emerged from these studies. A range of 4.2×10^5 to 8.4×10^5 M_r is observed for the native enzyme, depending on its source. The Michaelis–Menten constants are quite similar for the various substrates. Biotin-containing subunits of M_r 240,000 for both wheat germ and parsley cell culture carboxylases, 21,000 for barley embryo, and 60,000 for maize leaf, have been reported. In the case of purified maize leaf acetyl-CoA carboxylase, polyacrylamide gel electrophoresis (PAGE) in the presence of 1% sodium dodecyl sulfate (SDS) revealed a single biotinyl subunit of 60,000 M_r (Nikolau and Hawke, 1984). With purified rapeseed carboxylase, a polypeptide isolated by avidin–agarose chromatography had a M_r of 225,000. Unfortunately, an examination of the enzymatic function of subunit structures is difficult in that treatment of the native enzyme with SDS irreversibly inactivates the proteins, thereby preventing possible reconstitution studies. Furthermore, with the exception of the total molecular weight averages of the native enzyme and the 60,000 M_r biotinyl subunit, a considerable range of subunit molecular weights have been observed from different plant sources. Hopefully, a more consistent picture will emerge in future research.

B. Regulation

Since 1961 it has been reported consistently that FAS in chloroplasts is dependent on the presence of light. Explanations for the light requirement include (1) the original suggestion of Smirnov (1960), who was the first to observe the light effect and thus suggested a photoacetylation reaction, (2) a light-activated acetyl-CoA carboxylase, and (3) a favorable shift in concentration of Mg^{2+}, ATP as well as pH changes, all of which promoted an increase in acetyl-CoA carboxylase activity in a dark-to-light transition.

A careful kinetic analysis has been carried out with highly purified acetyl-CoA carboxylase isolated from the developing endosperm of castor bean (Finlayson and Dennis, 1983). The enzyme displayed normal Michaelis–Menten kinetics. Based on the results of substrate interaction and product inhibition, a hybrid Ping-Pong mechanism for the carboxylation of acetyl-CoA was proposed. The results suggest that the active site of the enzyme is separated into two functionally distinct catalytic sites. The carboxybiotinyl arm at one site formed by the conversion of ATP to ADP and P_i, swings to the second site where acetyl-CoA is carboxylated to malonyl-CoA. The mechanism is described as a hybrid rapid-equilibrium random bi bi uni uni Ping-Pong mechanism. These results also reveal that ADP acts as a competitive inhibitor to ATP, a noncompetitive inhibitor to HCO_3^-, and an uncompetitive inhibitor toward acetyl-CoA. Similar results have been obtained with other plant acetyl-CoA carboxylases (Stumpf, 1980).

In fact, recent investigations indicate that the levels of ATP and ADP may be of importance in the regulation of activity of this enzyme (Eastwell and Stumpf, 1983). When crude extracts of wheat germ were examined for acetyl-CoA carboxylase activity, zero-order kinetics were not followed. It was then shown that an active ATPase and adenyl kinase generated considerable amounts of ADP and AMP in the assay system. Thus, not only is the availability of ATP for the carboxylation reaction limited by these competing reactions, but also the generation of ADP and AMP permits the accumulation of competitive inhibitors of the carboxylase. In the chloroplast, light would tend to lower ADP and AMP concentrations and raise ATP concentrations via photophosphorylation.

Finally, although mammalian acetyl-CoA carboxylases have two levels of regulation, that is, by a polymerization from the inactive protomer form to the active oligomeric form and by a phosphorylation–dephosphorylation cycle, there is no evidence that the plant enzyme undergoes similar changes (Eastwell and Stumpf, 1983). The avocado carboxylase has its V_{max} increased 2-fold at 3 mM level of citrate (Mohan and Kekwick, 1980), but similar effects have never been observed with other plant carboxylases.

C. Localization of Activity

While localization of acetyl-CoA carboxylase can be readily carried out with leaf organelles, namely isolated chloroplasts, similar studies with endosperm or cotyledonous tissue are difficult because of the fragility of isolated proplastids. Nevertheless, proplastids isolated from avocado mesocarp and from endosperm of developing castor bean seeds clearly contain high levels of the enzyme, although carboxylase activity in the cytosol cannot be ruled out. Indeed, indirect evidence obtained from the developing jojoba bean strongly implies both a cytosolic and a proplastidic carboxylase (Ohlrogge et al., 1978). Thus in the developing jojoba seed, the generation of oleic acid occurs in one compartment in which is housed the de novo system for the conversion of C_2 units to oleic acid; oleic acid is then transported into the cytosol, there to be elongated via a microsomal elongase system to 20:1(11) and 22:2(13) fatty acids. Acetyl-CoA carboxylase must therefore be present in both the proplastid compartment to generate malonyl-CoA for the synthesis of oleic acid and in the cytosolic compartment for the elongation reactions to take place.

In addition, other tissues, in which elongation reactions are required for conversion of C_{18} fatty acids to long-chain fatty acids, would include rapeseed (22:1), leek (22:0 to 30:0), and limnanthes (22:1) (Pollard and Stumpf, 1980).

Although the localization of acetyl-CoA carboxylase in leaf chloroplast has been known for some time, a precise determination of its distribution in the leaf cell was not available. Recently, the subcellular distribution of the enzyme was determined in mesophyll protoplasts from barley, a C_3 plant, and sorghum, a C_4 plant. In both plants all of the mesophyll cell enzyme was chloroplastic. Moreover a single biotinyl protein of 60,000 Da was identified by a modified Western-blotting procedure when both total leaf and chloroplasts were examined. This 60,000-kDa subunit has also been observed in both leek and pea extracts (Nikolau *et al.*, 1984a).

In further distribution studies, the mesophyll tissue of pea and leek leaves contained 90% of the carboxylase activity, with the epidermal cells containing the remaining 10%. A number of biotinyl proteins were observed employing a ^{125}I-streptavidin probe and Western blotting of proteins fractionated by SDS–PAGE (Nikolau *et al.*, 1984b). These included 62, 51, and 32 kDa proteins for pea and 62, 34, and 32 kDa proteins in leek. Controls were run to eliminate possible proteolytic degradation or artifactual fragmentation of these subunits. Because the only biotinyl protein to have been observed in plants is acetyl-CoA carboxylase, one possible interpretation of these results would be the presence of a number of different isozymes of acetyl-CoA carboxylase, each functioning to generate malonyl-CoA for the six or more biosynthetic pathways requiring this substrate. These small biotinyl peptides then would reflect subunits of the carboxylase(s).

Although it was reported in the early 1970s that the spinach chloroplast acetyl-CoA carboxylase was of a prokaryotic type similar to that of *E. coli* (Kannangara and Stumpf, 1972), recent experiments have not been able to demonstrate such a molecular system (Nikolau and Hawke, 1984). The data from chloroplasts from barley, sorghum, spinach, and maize show a multiunit complex of high molecular weight. It is quite probable that in the earlier studies, the biotinyl protein, namely the biotinyl carboxyl carrier protein (BCCP) subunit, had been inactivated and the addition of *E. coli* BCCP reactivated the system. Data have been presented that the castor bean carboxylase and the avocado and spinach enzyme are in part membrane-associated (Finlayson and Dennis, 1983; Mohan and Kekwick, 1980).

Obviously a number of interesting problems have emerged from these recent investigations and should be further explored in the near future. In particular, the definition of possible isozymes in different compartments synthesizing malonyl-CoA for specific requirements would be useful. In addition, because of the localization of this large multienzyme complex in organelles, studies should be directed to understand whether or not the protein(s) are derived from nuclear or organelle DNA, where biotinylation takes place, and how the subunits are organized to form an active enzyme.

IV. BIOSYNTHESIS OF SATURATED FATTY ACIDS

A. The *de Novo* System

The *de novo* system is defined as that system of enzymes which employs acetyl-ACP and malonyl-ACP as substrates for the synthesis of palmitic and stearic acids.

Although many investigators have demonstrated the capability of a wide variety of tissue extracts to convert acetyl-CoA and malonyl-CoA as ACP derivatives to C_{16} and C_{18} fatty acids (Stumpf, 1980), it was only in 1982 that laboratories from England, Denmark, Germany and the United States elucidated the molecular structure of the FAS systems from avocado (Caughey and Kekwick, 1982), barley (Høj and Mikkelsen, 1982), parsley suspension culture cells (Schuz *et al.*, 1982), safflower (Shimakata and Stumpf, 1982a), and spinach chloroplast (Shimakata and Stumpf, 1982b). The results from these laboratories complemented each other very well. By appropriate protein fractionation procedures, these workers showed that the FAS system in these plants was of the nonassociated type similar to that of *E. coli*. Table I lists the molecular weights of a number of these systems and compares them to the available MW determinations made with the *E. coli* enzymes. While there is some scattering of molecular weights, the order of magnitude appears to be quite similar.

The most detailed study was that of Shimakata and Stumpf working with spinach leaf extracts, and results of these studies are discussed briefly as well as those from other laboratories. The following enzyme activities could be separated as discrete proteins and their properties were determined.

1. Acetyl-CoA:ACP Transacetylase

$$\text{Acetyl-CoA} + \text{ACP} \rightleftharpoons \text{acetyl-ACP} + \text{CoA}$$

The pH optimum for the spinach enzyme is 8.1. The K_m values for acetyl-CoA, butyryl-CoA, and hexanoyl-CoA were, respectively, 8.00, 8.31, 8.62 μM, and the

TABLE I
Molecular Weights of Several Plant FAS Systems

Enzyme	MW × 10^3				
	Safflower	Spinach	Barley	Avocado	*E. coli*
Acetyl-CoA:ACP transacylase	—	48	87	—	—
Malonyl-CoA:ACP transacylase	22	30	41	40.5	37
β-Ketoacyl-ACP synthetase I	—	56	92	—	80
β-Ketoacyl-ACP synthetase II	—	57	—	—	85
β-Ketoacyl-ACP reductase (NADPH)	83	97	125	46	—
D-β-Hydroxyacyl-ACP dehydrase	64	85	—	—	170
Enoyl-ACP reductase (NADH)	83	115	—	62.4	90

V_{max} values (nmol/min/mg protein) were 6.18, 2.17, and 1.18, respectively. The transacylase was inhibited completely at 5 mM concentration by sodium arsenite and by p-hydroxymercuribenzoic acid (Shimakata and Stumpf, 1983a).

Acetyl-CoA:ACP transacetylase has, however, very low specific activity in extracts from developing seeds of *Cuphea lutea*, safflower, rapeseed, and from pea and spinach leaves. When the concentration of this enzyme was changed in a reconstituted system necessary for FAS, the composition of the final products of synthesis were greatly perturbed in that at low levels of this enzyme, the normal products were 70% stearate and 24–30% palmitate, but at high levels the products were now about 80% lauric acid with lesser amounts of the higher homologs. These results suggested that controlling the levels of activity of this enzyme in the cell may have major effects on the fatty acid composition of seed cells (Shimakata and Stumpf, 1983a).

In addition, because of the low specific activity of the transacetylase, it would also suggest that this enzyme may be the rate-limiting step for FAS insofar as the seven FAS enzymes were involved.

2. Malonyl-CoA:ACP Transacylase

$$\text{Malonyl-CoA} + \text{ACP} \rightleftharpoons \text{malonyl-ACP} + \text{CoA}$$

Hilt purified the avocado enzyme to homogeneity (Hilt, 1984). The pH optimum was 8.3, the M_r 42,500. The K_m values for malonyl-CoA and ACP were 8 μM and 34 μM respectively. Another study on avocado mesocarp transacylase has reported similar results with M_r of 40,500, a pH optimum of 8.0, and a K_m for malonyl-CoA of 3.6 μM at pH 7.0 (Caughey and Kekwick, 1982). The enzyme has also been purified from barley leaves (Høj and Svendsen, 1983) as a single monofunctional protein with a M_r of 34,500. When purified from barley chloroplast the M_r was 43,000. From soybean leaves (Guerra and Ohlrogge, 1986) two isoforms (I and II) were isolated based on different adsorptive properties on blue dye affinity columns and different properties to heat, inhibitors, and other factors. Both forms have a M_r of 43,000 and a pH optimum of 8.5. Interestingly, isoform I, with a K_m for malonyl-CoA of 9.4 μM is only found in seeds, but both isoforms I and II (with a K_m of 15 μM) are found in leaf extracts. Finally, a recent study with leek leaf tissue has contrasted the properties of the parenchymal and the epidermal cell transacylase (Lessire and Stumpf, 1983). The transacylase from either epidermal and parenchymal cells was quite similar with a pH optimum of 7.9. The activity was associated with proteins of M_r 38,000 and 45,000, respectively. Further studies should characterize these as possible isozymes.

Since the activity is high in a wide variety of tissues, it probably plays no role in regulation of FAS. Because of its specificity toward ACP, it has been employed as a sensitive assay for free ACP (Høj and Svendsen, 1983).

3. β-Ketoacyl-ACP Synthetase I

$$\text{Acyl ACP}_{(C_2 - C_{14})} + \text{malonyl-ACP} \rightarrow \beta\text{-ketoacyl-ACP}_{(C_4 - C_{16})} + CO_2 + \text{ACP}$$

This enzyme has been purified some 180-fold from spinach leaf extracts (Shimakata and Stumpf, 1983b). The molecular weight is 56,000 by gel filtration; the K_m for malonyl-ACP was 4 μM. The purified synthetase I was highly active with acyl-ACPs having chain lengths from C_2 to C_{14}, with hexanoyl-ACP being the most effective substrate. Palmitoyl-ACP was far less effective, and stearoyl-ACP was inactive. Cerulenin at 3 μM concentration caused 50% inhibition of the above reaction. Thus, synthetase I is an integral component of the FAS required for the formation of palmitoyl-ACP.

4. β-Ketoacyl-ACP Synthetase II

$$\text{Palmitoyl-ACP} + \text{malonyl-ACP} \rightarrow \beta\text{-ketostearoyl-ACP} + CO_2 + \text{ACP}$$

This enzyme was discovered in spinach leaf extracts when it was observed that in crude extracts palmitoyl-ACP was elongated as well as decanoyl-ACP (Shimakata and Stumpf, 1982b,c). However, when the purified synthetase I (with decanoyl-ACP as the assay substrate in the purification procedure) was tested with palmitoyl-ACP, no activity was observed. With palmitoyl-ACP as the assay substrate, a protein was isolated that specifically elongated both myristoyl- and palmitoyl-ACP, to β-ketostearoyl-ACP, but had no activity for decanoyl-ACP. Synthetase II was purified some 295-fold and has a M_r of 57,000. In sharp contrast to synthetase I, synthetase II was 50% inhibited only at 40 μM levels of cerulenin. Arsenite at 1 mM level did not inhibit synthetase I but inhibited synthetase II by 42%.

5. β-Ketoacyl-ACP Reductase

$$\beta\text{-Ketoacyl-ACP} + \text{NADPH} + H^+ \rightarrow \text{D-}\beta\text{-hydroxyacyl-ACP} + \text{NADP}^+$$

This enzyme was purified some 422-fold by employing the polyethylene glycol (PEG) fractionation procedure, blue-agarose, Sephadex G-200, and hydroxyapatite chromatographies (Shimakata and Stumpf, 1982a,b,d). The M_r as determined by Sephacryl S-300 filtration to be 97,000, whereas by SDS–PAGE, a M_r of 24,200 was observed. These results would indicate that the native enzyme exists as a tetramer. Amino acid analysis showed that 57% of the amino acids were polar and 43% nonpolar. Two cysteines were present per monomer. With acetoacetyl-ACP as substrate, NADPH was the reductant with a K_m of 25 μM. The product was the D-isomer. NADH was about 10% as effective as a reductant. While both aceto-acetyl-ACP and acetoacetyl-CoA served as substrate, the ACP derivative was the most effective. The K_m values for the ACP derivative and the CoA derivative were 3.7 and 250 μM, respectively, and the maximal velocities 16.1 and 5.4 μmol/min/mg protein, respectively. Only acetoacetyl-ACP was tested but, since the purified reductase functioned in the reconstituted system for the formation of stearoyl-ACP, it is assumed that the reductase has broad β-ketoacyl specificities. The barley reductase also showed an NADPH requirement (Høj and Mikkelsen, 1982). Two reductases have been reported to occur in avocado mesocarp extracts (Caughey and Kekwick, 1982). A NADPH-specific reductase had a sharp optimum at pH 6.5 and a K_m of 9.65 μM for acetoacetyl-ACP and a M_r of 46,000; a membrane-associated NADH reductase had, on the other hand, a broad pH optimum from pH 6.8 to 7.5, a K_m of 8.8 μM for the same substrate, and a M_r of 168,000 by sucrose density gradient centrifugation.

6. D-β-Hydroxyacyl-ACP Dehydrase:

$$\text{D-β-hydroxyacyl-ACP} \rightleftharpoons \textit{trans}\text{-2-enoyl-ACP} + H_2O$$

The spinach enzyme has been purified some 4900-fold to homogeneity (Shimakata and Stumpf, 1982d). Its M_r by gel filtration was determined to be 85,000 and on an SDS–PAGE system, 19,000. Thus, the native enzyme would appear to be a tetramer. The amino acid composition indicated 62% polar and 38% nonpolar amino acid residues. One cysteine was present per monomer. The D-isomer was active whereas the L-isomer was completely inactive. A range of 2-enoyl-ACPs was tested for activity and the single enzyme reacted with substrate ranging from C_4 to C_{16} enoyl-ACPs. No activity was observed with crotonyl-CoA.

7. Enoyl-ACP Reductase

$$\text{2-Enoyl-ACP} + \text{NADH} + H^+ \rightleftharpoons \text{acyl-ACP} + NAD^+$$

The spinach enzyme was purified some 2200-fold to homogeneity (Shimakata and Stumpf, 1982d). Its M_r as determined by gel filtration was 115,000 and 32,500 by the SDS–PAGE procedure. Thus, the enzyme in its native state is probably a tetramer. Its amino acid composition has been determined, and 64% of its amino acids are polar and 36% nonpolar. It contains two cysteines per monomer. The optimum pH was 6.4. NADH was the specific reductant, since NADPH was ineffective. The K_m for NADH was 1 μM. Broad substrate specificity was observed with 2-hexenoyl-ACP and 2-octenoyl-ACP as the most effective substrate. Crotonyl-CoA was also active. Studies with the avocado reductase showed broad pH optimum between pH 6 and 7.5, a very high K_m for crotonyl-ACP, with an approximate M_r of 62,400 and NADH specificity (Caughey and Kekwick, 1982). In contrast to the spinach enzyme, crotonyl-CoA was inactive as a substrate.

Further comments should be made concerning the specificity for NADH or NADPH with the two spinach reductases and the role of the dehydrase.

The first reduction step in the spinach FAS system catalyzed by β-ketoacyl-ACP reductase was specific for NADPH; similar observations have been made with safflower seeds (Shimakata and Stumpf, 1982a), castor bean (Saito et al., 1980), E. coli (Alberts et al., 1964), yeast (Lynen, 1980), and animal (Stoops and Wakil, 1981). However, we have observed that the PEG (5–15%) fraction from developing castor bean seeds contained β-ketoacyl-ACP reductase activity, which could employ NADH as a replacement for NADPH but with only 15% of the activity of NADPH at saturating concentrations (T. Shimakata, personal communication). One of the avocado reductases also has a NADH-specific reductase. The second reduction step in the spinach system was catalyzed by NADH-dependent enoyl-ACP reductase; in contrast, safflower seed extracts contained both an enoyl-ACP reductase II requiring either NADH or NADPH and a reductase I similar to the spinach reductase. The source of NADH in spinach chloroplasts is probably from electrons flowing through NADPH:ferredoxin reductase, which can also reduce NAD$^+$. In *E. coli*, two distinct enoyl-ACP reductases have been observed, one of which is NADH-specific with broad chain length specificity and the other NADPH-specific with limited chain length specificity (shorter chain specific) (Weeks and Wakil, 1968). Saito *et al.* (1980) reported that castor bean enoyl-ACP reductase was specific for NADPH and that *Chlorella vulgaris* reductase was specific for NADH. The avocado reductase is NADH-specific. The enoyl thioester reductase in multienzyme complexes of yeast (Lynen, 1980) and vertebrates (Stoops and Wakil, 1981) requires NADPH. The spinach β-ketoacyl-ACP reductase and β-hydroxyacyl-ACP dehydrase showed strict stereospecificity (D form) identical to their counterparts in *E. coli*. Although chain length specificity of the spinach β-ketoacyl-ACP reductase was not tested, it could be suggested that, since only a single peak of the activity was observed, albeit only with acetoacetyl-ACP as substrate, presumably the enzyme has a wide specificity. Such is the case with its counterpart in *E. coli* (Birge and Vagelos, 1972). The β-hydroxyacyl-ACP dehydrase and enoyl-ACP reductase purified from spinach leaves were active to 2-enoyl-ACPs having chain lengths from C_4 to C_{16}. Unlike *E. coli* dehydrase, which has its highest activity toward crotonyl-ACP and its lowest activity toward the C_{10} substrate, the spinach dehydrase had the highest activity with 2-octenoyl-ACP. The spinach enoyl-ACP reductase showed higher activity toward 2-hexenoyl-ACP and 2-octenoyl-ACP, as was shown for the NADH reductase in *E. coli*. 3-Decenoyl-ACP was inactive to both the spinach β-hydroxyacyl-ACP dehydrase and enoyl-ACP reductase.

CoA esters were active to the spinach β-ketoacyl-ACP reductase and enoyl-ACP reductase, but inert to β-hydroxyacyl-ACP dehydrase. The NADPH reductase of *E. coli* was absolutely specific for the ACP derivatives, whereas the NADH reductase was active with both ACP and CoA derivatives (Weeks and Wakil, 1968). The three enzymes purified from spinach leaves were sensitive to *p*-CMB but insensitive to NEM and arsenite. This would suggest that the more hydrophobic reagent, *p*-CMB, may penetrate a hydrophobic region of these enzymes where a sulfhydryl group is essential for activity but which cannot be reached by the more hydrophilic reagent, NEM.

In summary, Table I lists the M_r's of the FAS enzyme from a variety of tissues and compares them to the *E. coli* for enzymes. Essentially all the enzymes appear to be quite similar in M_r regardless of their isolation from seed, mesocarp, or chloroplast issue.

As already indicated (Stumpf, 1980), the evidence continues to support the earlier observations that FAS is localized in the chloroplast in leaf tissue and in proplastids in seed and mesocarp tissues. Although the evidence is quite fragmentary, probably most if not all of the enzymes localized in their organelles are nuclear encoded. The best evidence so far is the observations that the ACP gene is nuclear, that transcription and translation to a product somewhat larger than the final ACP molecule are external to the chloroplast, and that the recursor protein with its transit peptide is translocated into the chloroplast and finally processed to the functional ACP (see Chapter 6, Vol. 9 for details). How the seven or more enzymes required for the construction of C_{16} and C_{18} fatty acids are structurally organized in the matrix of the proplastid or in the stroma of the chloroplast, so that a smooth continuous synthesis of the long-chain acids can occur, is still completely unknown.

V. TERMINATION MECHANISMS—LONG- AND MEDIUM-CHAIN FATTY ACID SYNTHESIS

Table II identifies a number of possible mechanisms that determine chain length of fatty acids. The C_2–C_{16} and the C_{16}–C_{18} mechanisms have already been described in Section III.

Recently additional information has been obtained with the systems involved in synthesizing the C_{20} plus series of fatty acids.

A. The Leek System

Cassagne and Lessire (1974) have pioneered in the use of leek leaf tissues to study the synthesis of very long-chain fatty acids (VLCF). It is quite simple to obtain gram quantities of clean epidermal and parenchymal tissue from the leek which then can be used to examine soluble and membrane-bound FAS systems in these specialized cells. Parenchymal cells and the major FAS activity in the leaf (70–90%) and both

TABLE II
Chain Termination Mechanisms

1. $C_2 \rightarrow C_{16}$
   Specificity of β-ketoacyl-ACP synthetase controls this termination step; palmitoyl-ACP is inactive as substrate.
2. $C_{16} \rightarrow C_{18}$
   Palmitoyl-ACP: β-ketostearoyl-ACP synthetase is limited to this reaction.
3. $C_{18} \rightarrow C_{20}$
   Membrane-bound stearoyl-CoA: β-ketoeicosanoyl-CoA synthetase. In some tissues (rapeseed), only oleoyl-CoA: β-ketoeicosenoyl-CoA synthetase functions to guarantee long-chain monoenoic fatty acids.
4. $C_{20} \rightarrow C_n$
   Membrane-bound FAS systems employing acyl-CoAs as substrates plus malonyl-CoA.
5. $C_2 \rightarrow$ Medium-chain fatty acids
   Possible mechanisms:
   a. Specific thioesterase (acyl-CoA or acyl-ACP)
   b. Specific acyltransferases
   c. High levels of acetyl CoA: ACP transacetylase
   d. Separate compartmental synthetases
   e. Controlled β-oxidation
   f. Specific β-ketoacyl synthetase

palmitic and stearic acids were formed via the ACP-dependent FAS pathway in equal amounts. The epidermal FAS system, however, synthesized only palmitic acid with acetyl-ACP as the primer and malonyl-ACP as the condensing unit (Lessire and Stumpf, 1983).

Microsomal preparations of the epidermal cells of leek have, in contrast, an active enzyme system which functions to synthesize only saturated very long-chain fatty acids, which in turn serve as precursors of alkanes. In this system, saturated acyl-CoA's serve as primers, malonyl-CoA's as the condensing units, and ACP is not required. Oleoyl-CoA is an ineffective primer, thereby guaranteeing that the cell synthesizes only saturated acyl thioester derivatives. NADPH was as effective as NADH as a reductant (Agrawal et al., 1984).

There is fragmentary evidence which would suggest two membrane-associated elongation systems, one which elongates stearoyl- or palmitoyl-CoA but not oleoyl-CoA to eicosanoyl-CoA, and the second which elongates eicosanoyl-CoA to the very long-chain homologs. Thus the soluble de novo system converts acetyl-CoA to palmitoyl-CoA via the ACP pathway and the microsomal system then elongates palmitoyl-CoA to the required long-chain homologs.

B. The *Brassica* System

A somewhat similar system has been identified in developing mustard seeds (*Brassica juncea*), but in this tissue oleoyl-CoA is the elongating primer with either NADH or NADPH as the reductant. However, in the conversion of $C_{20:1}$ to $C_{22:1}$ NADPH is the preferred reductant, suggesting that one of the reductases has two isoforms specific for substrates with different chain lengths. Again, ACP does not appear to be involved in the elongation reaction. The condensing unit is malonyl-CoA (Agrawal and Stumpf, 1985).

C. The *Cuphea* System

Compositional studies of seed lipids have from time to time revealed the appearance of unique fatty acids in these tissues. As an example, a large number of *Cuphea* species, an annual that is distributed widely in the Eastern and Western Hemispheres, accumulate in their triacylglycerols medium-chain fatty acids ranging from octanoic to lauric acids in high amounts (Graham et al., 1981).

Therefore, the developing seeds of these *Cuphea* species are excellent models to study termination mechanisms. Unfortunately, because the seeds employed for these studies are exceedingly small, it is difficult to obtain sufficient quantities for purification purposes. Nevertheless, several intriguing papers have recently been published. For example, mature seeds of *Cuphea procumbens* contain over 80% of their fatty acids as decanoic acid (associated with triglycerides). Employing [^{14}C]-acetate, it was shown that seeds in a middle developmental stage almost exclusively synthesize decanoic acid (74 mol %) (Slabas et al., 1982). Palmitic, stearic, and oleic acids together make up an additional 22 mol %. With developing seeds of

Cuphea lutea, [^{14}C]acetate incorporation was similar to that reported earlier (Singh *et al.*, 1984). The possibility of a partial β-oxidation of C_{16} and C_{18} fatty acids to decanoic and lauric acids was explored and, while the evidence is not complete, there may be some contribution of a β-oxidation system to the synthesis of medium-chain fatty acid. *In vitro* results showed that the FAS activity required malonyl-CoA, is ACP-dependent, and is inhibited by cerulenin. With no primer (acetyl-CoA) added, stearic acid was the major product. With high levels of ACP and acetyl-CoA, a significant level of medium-chain fatty acids was formed. These results support the data relating to the control of levels of acetyl-CoA:ACP transacylase mentioned earlier. More recently, acyl-ACP generated *in vivo* were isolated from developing *Cuphea lutea* seeds as well as developing safflower seeds, and rather high levels of octanoic and decanoic acids were associated with the acyl components (Singh *et al.*, 1986). Since octanoic acid is present in very low levels in the mature seeds, it is conceivable that the observation of C_8 ACP accumulation is more the trapping of an intermediate in the chain of events leading to the synthesis of the C_{16} and C_{18} fatty acids. Recently, an ACP-independent, malonyl-CoA-independent FAS system has been reported to occur in mitochondria of a nonphotosynthesizing mutant of *Euglena gracilis* (Inui *et al.*, 1984). The β-oxidation enzymes participate in this synthesis, with the exception that three enoyl-CoA reductases replace the β-oxidation acyl-CoA dehydrogenase. It would be of interest to determine if a system similar to this may be involved in the synthesis of medium-chain fatty acids.

In summary, considerable progress has been made in elucidating the synthesis of fatty acids ranging from C_8 to C_{30} fatty acids. As techniques improve, the next decade should see exciting developments in the control of the type of FAS, the extent of synthesis, and the transfer of these advances into the realm of molecular biology.

REFERENCES

Agrawal, V. P., and Stumpf, P. K. (1985). *Lipids* **20**, 361–366.
Agrawal, V. P., Lessire, R., and Stumpf, P. K. (1984). *Arch. Biochem. Biophys.* **230**, 580–589.
Alberts, A. W., Majerus, P. W., Talamo, B., and Vagelos, P. R. (1964). *Biochemistry* **3**, 1563–1571.
Amrhein, N., and Kionka, C. (1983). *Plant Physiol.* **72**, 5–37.
Birge, C. H., and Vagelos, P. R. (1972). *J. Biol. Chem.* **247**, 4921–4929.
Botha, F. C., and Dennis, D. T. (1986). *Arch. Biochem. Biophys.* **245**, 96–103.
Camp, P. J., and Randall, D. D. (1985). *Plant Physiol.* **77**, 571–577.
Cassagne, C. R., and Lessire, R. (1974). *Physiol. Veg.* **12**, 149–163.
Caughey, I., and Kekwick, R. G. O. (1982). *Eur. J. Biochem.* **12**, 553–561.
Dennis, D. T., and Miernyk, J. A. (1982). *Annu. Rev. Plant Physiol.* **33**, 27–50.
Eastwell, K. C., and Stumpf, P. K. (1983). *Plant Physiol.* **72**, 50–55.
Egin-Bühler, B., and Ebel, J. (1983). *Eur. J. Biochem.* **133**, 335–339.
Finlayson, S. A., and Dennis, D. T. (1983). *Arch. Biochem. Biophys.* **225**, 576–585, 586–595.
Fritsch, H., and Beevers, H. (1979). *Plant. Physiol.* **63**, 687–691.

Givan, C. V. (1983). *Physiol. Plant.* **57**, 311–316.

Givan, C. V., and Hodgson, J. M. (1983). *Plant Sci. Lett.* **32**, 233–242.

Gorham, J. (1980). *In* "Progress in Phytochemistry" (L. Rheinhold, J. B. Harborne, and T. Swaine, eds.), Vol. 6, pp. 203–252. Pergamon, Oxford.

Graham, S. A., Hrisinger, F., and Robbelen, G. (1981). *Am. J. Bot.* **68**, 244–246.

Guerra, D. J., and Ohlrogge, J. B. (1986). *Arch. Biochem. Biophys* **246**, 274–285.

Hahlbrock, K. (1981). *In* "The Biochemistry of Plants: A Comprehensive Treatise" (P. K. Stumpf and E. E. Conn, eds.), Vol. 7, pp. 425–456. Academic Press, New York.

Hilt, K. L. (1984). Ph.D. Thesis, Univ. of California, Davis.

Høj, P. B., and Mikkelsen, J. D. (1982). *Carlsberg Res. Commun.* **47**, 119–141.

Høj, P. B., and Svendsen, I. (1983). *Carlsberg Res. Commun.* **48**, 285–305.

Inui, H., Miyatake, K., Nakano, Y., and Kitaoka, S. (1984). *Eur. J. Biochem.* **142**, 121–126.

Jacobson, B. S., and Stumpf, P. K. (1972). *Arch. Biochem. Biophys.* **153**, 656–663.

Kannangara, C. G. and Stumpf, P. K. (1972). *Arch. Biochem. Biophys.* **152**, 83–91.

Kolattukudy, P. E., Croteau, R., and Buckne, J. S. (1976). *In* "Chemistry and Biochemistry of Natural Waxes" (P. E. Kolattukudy, ed.), pp. 289–347. Elsevier/North-Holland, New York.

Kuhn, D. N., Knauf, M., and Stumpf, P. K. (1981). *Arch. Biochem. Biophys.* **209**, 441–450.

Lessire, R., and Stumpf, P. K. (1983). *Plant Physiol.* **73**, 614–618.

Liedvogel, B. (1985). *Anal. Biochem.* **148**, 182–189.

Liedvogel, B., and Stumpf, P. K. (1982), *Plant Physiol.* **69**, 897–903.

Lynen, F. (1980). *Eur. J. Biochem.* **112**, 431–442.

McLaren, I., Wood, C., Jalil, N. H., Yong, B. C. S., and Thomas, D. R. (1985). *Planta* **163**, 197–200.

Miernyk, J. A., and Dennis, D. I. (1982). *Plant Physiol.* **69**, 825–828.

Miernyk, J. A., and Dennis, D. I. (1983). *J. Exp. Bot.* **34**, 712–718.

Mohan, S. B., and Kekwick, R. G. O. (1980). *Biochem. J.* **187**, 667–676.

Nelson, D. R., and Rinne, R. W. (1977). *Plant Cell Physiol.* **18**, 1021–1027.

Nikolau, B. J., and Hawke, J. C. (1984). *Arch. Biochem. Biophys.* **228**, 86–96.

Nikolau, B. J., Hawke, J. C., and Slack, C. R. (1981). *Arch. Biochem. Biophys.* **211**, 605–612.

Nikolau, B. J., Wurtele, E. S., and Stumpf, P. K. (1984a). *Arch. Biochem. Biophys.* **235**, 555–561.

Nikolau, B. J., Wurtele, E. S., and Stumpf, P. K. (1984b). *Plant Physiol.* **75**, 895–901.

Ohlrogge, J. B., Pollard, M. R., and Stumpf, P. K. (1978). *Lipids* **13**, 203–210.

Packter, N. M. (1980). *In* "The Biochemistry of Plants: A Comprehensive Treatise" (P. K. Stumpf and E. E. Conn, eds.), Vol. 4, pp. 535–540. Academic Press, New York.

Pollard, M. R., and Stumpf, P. K. (1980). *Plant Physiol.* **66**, 649–655.

Saito, K., Kawaguchi, A., Okuda, S., Seyama, Y., Yamakawa, T., Nakumura, Y., and Yamada, M. (1980). *Plant Cell Physiol.* **21**, 9–19.

Schuz, R., Ebel, J., and Hahlbrock, K. (1982). *FEBS Lett.* **104**, 207–209.

Shimakata, T., and Stumpf, P. K. (1982a). *Arch. Biochem. Biophys.* **217**, 144–154.

Shimakata, T., and Stumpf, P. K. (1982b). *Plant Physiol.* **69**, 1257–1262.

Shimakata, T., and Stumpf, P. K. (1982c). *Proc. Natl. Acad. Sci. U.S.A.* **79**, 5808–5812.

Shimakata, T., and Stumpf, P. K. (1982d). *Arch. Biochem. Biophys.* **218**, 77–91.

Shimakata, T., and Stumpf, P. K. (1983a). *J. Biol. Chem.* **258**, 3592–3598.

Shimakata, T., and Stumpf, P. K. (1983b). *Arch. Biochem. Biophys.* **220**, 39–45.

Singh, S. S., Nee, T. Y., and Pollard, M. R. (1984). *In* "Structure, Function and Metabolism of Plant Lipids" (P. A. Siegenthaler and W. Eichenberger, eds.), pp. 161–165. Elsevier, Amsterdam.

Singh, S. S., Nee, T. Y., and Pollard, M. R. (1986). *Lipids* **21**, 143–149.

Slabas, A. R., Roberts, P. A., Ormesher, J., and Hammond, E. W. (1982). *Biochim. Biophys. Acta* **711**, 411–420.

Smirnov, B. P. (1960). *Biokhimiya* **25**, 419–422.

Stitt, M., and Ap Rees, T. (1979). *Phytochemistry* **18**, 1905–1911.

Stoops, J. K., and Wakil, S. J. (1981). *J. Biol. Chem.* **256**, 5128–5133.

Stumpf, P. K. (1980). *In* "The Biochemistry of Plants; A Comprehensive Treatise" (P. K. Stumpf and
    E. E. Conn, eds.), Vol. 4, pp. 177–204. Academic Press, New York.
Stumpf, P. K., and Burris, R. H. (1981). *Plant Physiol.* **68**, 992–995.
Thomas, D. R., and Wood, C. (1982). *Planta* **154**, 145–149.
Trude, H.-J., and Heise, K.-P. (1985). *Z. Naturforsch.* **40C**, 496–502.
Turnham, E., and Northcote, D. H. (1983). *Biochem. J.* **212**, 223–229.
Weeks, G., and Wakil, S. J. (1968). *J. Biol. Chem.* **243**, 1180–1189.

Biochemistry of Plant Acyl Carrier Proteins

6

JOHN B. OHLROGGE

I. INTRODUCTION AND HISTORY

Acyl carrier protein (ACP) is a small, acidic protein which functions as a cofactor for at least a dozen enzymes in plant lipid metabolism. As detailed in this review, the analysis of this relatively simple protein has yielded many important insights into the organization, localization, and regulation of plant fatty acid metabolism. From the study of ACP we have learned that the organization of plant fatty acid synthesis (FAS) is more closely related to bacteria than to fungi or animals, that *de novo* FAS in leaves is localized exclusively in chloroplasts, and that isoforms of fatty acid synthetase proteins are expressed in a tissue-specific fashion. Because ACP has now been cloned, we can anticipate that our first glimpse into the structure and regulation of plant fatty acid synthetase genes will be available soon. Clearly the study of ACP has proved to be, and will continue to be, a most powerful probe into the biology of plant fatty acid metabolism.

Initial evidence for the role of ACP in plants was obtained in 1964 when Overath and Stumpf used ammonium sulfate to fractionate extracts of avocado mesocarp. They observed that FAS could be separated into at least two components which were inactive alone, but when recombined were capable of synthesizing palmitic and stearic acids. One of these components was stable to boiling and dilute acid treatment but was destroyed by proteases. The heat-stable fraction could partially substitute for a similar fraction from *Escherichia coli* in the stimulation of *E. coli* FAS.

These observations provided the first characterization of plant ACPs and the first preliminary evidence for the similarity between plant and bacterial FAS. Since these initial observations, both plant and bacterial ACPs have become the most completely studied proteins in lipid metabolism. Over 200 publications have dealt specifically with ACP. Although most of these concern the *E. coli* protein, in recent years interest in plant ACPs has intensified, leading to a much more complete understanding of this plant protein.

Because *E. coli* ACP has been thoroughly studied and is used extensively in plant lipid research, and because plant ACPs are very similar to their bacterial counterparts, this review will include references to the *E. coli* ACP literature when appropriate.

II. FUNCTIONS OF ACYL CARRIER PROTEINS

A. Fatty Acid Synthesis

The function of ACP is now well understood to be that of a cofactor rather than an enzyme. ACP contains a 4′-phosphopantetheine prosthetic group which is attached to the protein via a phosphodiester linkage to a serine residue. Acyl groups, including the substrates, intermediates, and products of fatty acid biosynthesis, are attached to ACP via a thioester linkage to the terminal sulfhydryl of the prosthetic group.

In the assembly of 16- and 18-carbon fatty acids, ACP serves to carry the acyl chain from one enzyme active site to the next while preserving the high-energy or activated state inherent in a thioester bond. The actual chemistry of each reaction, however, may not always occur with ACP thioesters. It is possible that substrate acyl chains are transferred from ACP to a second sulfhydryl in the enzyme active site and then, following the reaction, the product acyl group is transferred back to ACP. Evidence for such a thioester shuttle is available for *E. coli* (Ruch and Vagelos, 1973) and animals (Wakil *et al.*, 1983), and it presumably occurs also in plants.

The concentration of ACP has a dramatic influence on the products of FAS assays *in vitro*. This effect was first observed in plants by Huang and Stumpf (1971) in extracts of *Solanum tuberosum*, where increasing ACP concentrations led to the production of short-chain fatty acids. Slabas *et al.* (1982) found that increasing the concentrations of ACP 2-fold when assaying rapeseed FAS resulted in a shift from predominantly C_{16} and C_{18} production to predominantly C_8 to C_{14} products. Similar results have been observed by Singh *et al.* (1984) for developing *Cuphea* seeds. These data can be interpreted as a competition by acyl-ACP intermediates for either further condensation or chain termination reactions. Possibly at high ACP levels the chain elongation active sites are saturated and chain termination through hydrolysis or acyl transfer causes the shift to shorter acyl chain products.

B. Desaturation

Nagai and Block (1968) first observed that stearate desaturation in plants utilizes ACP as the cofactor rather than CoA. Jaworski and Stumpf (1974) extended these observations to nonphotosynthetic tissue and confirmed the substrate specificity by using biochemically rather than chemically prepared acyl-ACP substrates. The chemically prepared acyl-ACP substrates are generally less active due to acylation at sites other than the pantetheine sulfhydryl.

The stearoyl-ACP desaturase provides a good example of the importance of kinetic data in evaluating cofactor specificity. The prosthetic group of ACP with its thioester linkage possesses an identical structure to that of acyl-CoA. Although both CoA and ACP have similar acyl carrier and cofactor functions, individual enzymes usually show a distinct selectivity for one cofactor over the other. However, this selectivity may be obscured if examined at nonphysiological substrate concentrations.

When McKeon and Stumpf (1982) compared stearoyl-CoA and stearoyl-ACP as substrates for stearoyl-ACP desaturase, the K_m value for the ACP derivative was 20-fold lower. In contrast, the substrates differed less than 10% in V_{max}. Clearly the enzyme selectivity will be greatly different at low and high substrate concentrations. However, at physiological concentrations of the potential substrates believed to occur in plastids, the desaturase can be expected to favor greatly the ACP substrate. These observations emphasize the necessity to make kinetic comparisons at a range of substrate concentrations. Unfortunately, ACP has often not been available in sufficient quantities for adequate kinetic examinations, and in some cases the difficulty in preparing acyl-ACP substrates has led to their omission from studies altogether. For these reasons it is likely that some plant enzymes reported to act on acyl-CoA actually utilize acyl-ACP *in vivo*.

C. Acyl Transfer

In *E. coli*, acyl transfer to form monoacylglycerol-3-phosphate and diacyl-glycerol-3-phosphate appears to show little preference for acyl-CoA versus acyl-ACP substrates (Rock *et al.*, 1981b). In plants, however, Frentzen *et al.* (1983) have demonstrated that these transfer reactions have a strong selectivity for acyl-ACP. Equal molar ratios of acyl-CoA and acyl-ACP provided to either the soluble glycerol-3-phosphate acyltransferase or the membrane-bound monoacylglycerol-3-phosphate acyltransferase resulted in a 10- to 20-fold preferential transfer of the acyl chain from ACP. Similar results have recently been described for the photosynthetic bacterium *Rhodopseudomonas sphaeroides* (Cooper and Lueking, 1984). The two plastid acyl-ACP transferases also show a strong selectivity for acyl chain structure. In the first acyl transfer, oleoyl-ACP is preferentially used for transfer to the 1-position of glycerol-3-phosphate. In the second acyl transfer, palmitoyl-ACP is selected for esterification at the 2-position on the glycerol backbone. The monoacyl-glycerol-3-phosphate acyltransferase has recently been localized to the inner membrane of the pea chloroplast envelope (Andrews *et al.*, 1985). The action of the two plastid acyltransferases results in a glycerolipid structure that has a saturated fatty acid at position 2 and an unsaturated fatty acid at position 1. Because this arrangement is atypical of most eukaryote glycerolipids but is characteristic of photosynthetic bacteria, these plastid glycerolipids which arise from acyl-ACP reactions are sometimes referred to as "prokaryote-type" lipids. In contrast, the enzymes that utilize acyl-CoA and are presumably localized outside the plastid, generate glycerolipids with saturation at position 1—that is, the typical eukaryote-type structure.

D. Hydrolysis

Fatty acid synthesis can be terminated either by the acyl transfer reactions discussed above or by hydrolysis of the acyl chain from ACP. An acyl-ACP hydrolase (Shine *et al.*, 1976) has been purified 400-fold from avocado mesocarp (Ohlrogge *et al.*, 1978) and 770-fold from safflower (McKeon and Stumpf, 1982), and was shown to be highly specific for oleoyl-ACP thioesters relative to other acyl chains or to CoA esters. Interestingly, *E. coli* does not appear to have a specific acyl-ACP hydrolase, and therefore, acyl transfer may be the sole chain termination step in *E. coli* FAS (Spencer *et al.*, 1978).

In the flow of acyl chains in fatty acid metabolism, a branch point occurs between acyl transfer to glycerol-3-phosphate and hydrolysis of acyl-ACP to release free fatty acid (FFA). It seems likely that this metabolic branch reflects also a control point for the subcellular allocation of acyl chains. Following their release as FFAs, the acyl chains are believed to cross the plastid envelope, become esterified to CoA by the action of the outer-envelope thiokinase (Andrews and Keegstra, 1983), and proceed to the endoplasmic reticulum for further processing by the "eukaryote" pathway (Roughan and Slack, 1984). In contrast, those acyl chains that are transferred to glycerol-3-phosphate remain characteristic of prokaryotic lipids and may never leave the plastid. Therefore, the relative activity of the acyl-ACP transferase and acyl-ACP hydrolase can be expected to play a major part in controlling the proportion of acyl groups exported from the plastid. Examination of the relative activity of these two enzymes in tissues such as leaves and oil-storing seeds which differ greatly in the subcellular location of acyl groups might provide a useful test of this hypothesis.

III. ASSAY OF ACYL CARRIER PROTEINS

Because ACP itself possesses no catalytic activity, most assays have depended either on its cofactor function with other enzymes or on stoichiometric attachment of radiolabeled acyl groups to the pantetheine sulfhydryl. Most early studies used the ability of ACP to participate in the malonyl-CoA–$^{14}CO_2$ exchange reaction. In this assay $H^{14}CO_3$, is incorporated into malonyl-CoA in the presence of malonyl-CoA, acetyl-CoA, and the three enzymes, β-ketoacyl-ACP synthase, malonyl-CoA:ACP transacylase, and acetyl-CoA:ACP transacylase (Majerus et al., 1969). The assay is sensitive but requires standardization with a known quantity of ACP, and the three enzymes involved may be subject to interference by substances present in crude plant extracts.

ACP can also be quantitated by the transfer of [^{14}C]malonate to the ACP sulfhydryl in the presence of malonyl-CoA:ACP transacylase (Ohlrogge et al., 1979; Høj and Svendsen, 1983). Although not as sensitive, this assay is easier to perform than the CO_2 exchange reaction.

Rock and Cronan (1979) developed an assay for E. coli ACP which uses the enzyme acyl-ACP synthetase to transfer [^{14}C]palmitate to the sulfhydryl of ACP. This E. coli enzyme is equally effective with plant ACPs and, with minor modifications, the sensitivity can be increased to permit assay of nanogram quantities of ACP present in crude plant extracts (Kuo and Ohlrogge, 1984a). The E. coli enzyme can also be used for the preparative synthesis of plant ACP derivatives with a variety of acyl chains.

ACPs can also be easily quantitated by radioimmunoassay. Ohlrogge et al. (1979) prepared antibodies to spinach ACP. ACP was then radiolabeled with ^{125}I to establish a sensitive competitive binding assay. This type of assay was further modified by Kuo and Ohlrogge (1984b) by substitution of [^{3}H]palmitoyl-ACP for [^{125}I]ACP as the radiolabeled ligand. Because ACP in crude preparations can be specifically labeled to high specific activity with [^{3}H]palmitate (using E. coli acyl-ACP synthetase), the assay can be applied to a variety of plant species for which pure ACP may not be available. We have found that the radioimmuno-assay is the ACP assay least subject to interference by crude plant tissue extracts.

IV. PURIFICATION OF ACYL CARRIER PROTEINS

ACP is probably one of the most abundant of the plant lipid-biosynthetic proteins due to its interaction at micromolar levels as cofactor with 12 or more enzymes. Nevertheless, the levels of ACP in the plant tissues we have examined generally represent less than 0.1% of the total cell protein. Therefore, to obtain homogeneous ACP requires greater than 1000-fold purification. Høj and Svendsen (1983) reported a 13,000-fold purification for barley ACP. We and others (A. Slabas, personal communication) have observed that ACP is much more easily isolated from leaf tissue than from developing oilseeds. In soybean seeds, the presence of abundant storage lipids and storage proteins with low isoelectric points makes the initial fractionation of the tissue extremely ineffective. Although pure plant ACP is soluble in 50% isopropanol, the extraction method of Rock and Cronan (1980), which is so useful for *E. coli*, results in coprecipitation of ACP with the bulk of plant proteins. Similarly, we have observed major losses of ACP during ammonium sulfate fractionations which we attribute to trapping or coprecipitation with the bulk of protein (J. B. Ohlrogge and T. M. Kuo, unpublished observations).

The first detailed examination of plant ACPs was conducted in a classic and elegant study published by Simoni *et al.* (1967). These workers purified ACP from two bacteria (*E. coli* and *Arthrobacter viscosus*) and two plants (avocado and spinach). The four proteins were found to behave similarly during purification, had similar predominance of acidic amino acids, and ranged in size from 81 to 117 amino acids. Each protein contained a single phosphopantetheine residue. Unlike the other three proteins, the avocado ACP contained two sulfhydryl groups, the second arising from a cysteine residue. Enzymatic cross-reactivity of the four ACPs was evaluated in the malonyl-CoA:CO_2 exchange reaction and in FAS. The bacterial ACPs were 4- to 10-fold more active than the plant ACPs when added to the *E. coli* enzyme systems. Surprisingly, the bacterial ACPs were also 2- to 4-fold more active with the plant enzymes! This result might have been attributable to a greater lability of the plant ACPs during purification; however, a similar preference of plant FAS for *E. coli* ACP has recently been observed for barley (Høj and Svendsen, 1983).

Most purification protocols for plant ACP have relied extensively on anion-exchange chromatography. The very low isoelectric point of all ACPs examined (3.9–4.2) results in their tight binding to anion exchangers such as DEAE–cellulose, and effective purification is achieved with salt gradient elution. Several other column purification materials have been tested for purification of plant ACPs. *Escherichia coli* ACP is reported to bind calcium and other divalent metals (Schulz, 1977), suggesting that hydroxyapatite might provide effective purification. However, we observed essentially irreversible binding of ACP to this matrix (J. B. Ohlrogge, unpublished observations). Barley ACP binds weakly to octyl-Sepharose (Høj and Svendsen, 1983), and we have observed similar results with spinach ACP. ACPs can be covalently bound to sulfhydryl columns and eluted with reducing agents (Ohlrogge *et al.*, 1978). Chromatofocusing was used by Høj and Svendsen

(1983) to purify barley ACP and is also effective in separating spinach ACP isoforms (D. J. Guerra and J. B. Ohlrogge, unpublished observations).

Ernst-Fonberg *et al.* (1977) developed an immunoaffinity purification procedure for ACPs. Antibodies raised to *E. coli* ACP were linked to CNBr-activated Sepharose. This column was capable of single-step purification of ACP from two species of *Euglena*. We have followed a similar protocol to prepare an immunoaffinity column for plant ACPs using antibodies to spinach ACP (Ohlrogge and Kuo, 1985). Although limited in capacity, these columns have great value in rapid purification of small quantities of cross-reactive ACP from crude preparations.

V. STRUCTURE OF ACYL CARRIER PROTEINS

The structure of *E. coli* ACP has been extensively studied (reviewed in Rock and Cronan, 1982). In addition to the amino acid sequence published in 1968 (Vanaman *et al.*, 1968), additional studies have included CD/ORD (Schulz, 1977), NMR (Gally *et al.*, 1978; Rock *et al.*, 1981a; Mayo *et al.*, 1983), UV (Rock, 1983), and various electrophoretic and chromatographic techniques (Rock and Cronan, 1979). Recently, crystals of *E. coli* ACP were obtained and preliminary X-ray diffraction results are now available (McRee *et al.*, 1985).

Escherichia coli ACP is now understood to possess a markedly asymmetric shape. In gel filtration experiments it elutes with a Stokes radius equivalent to a globular protein of MW 20,000. The high content of acidic amino acids results in an isoelectric point of 4.1. The low content of hydrophobic amino acids results in a reduced mobility in SDS–PAGE analysis such that *E. coli* ACP migrates with a relative MW of 20,400 (Rock and Cronan, 1979), whereas the actual molecular mass based on the amino acid sequence is 8847. Plant ACPs have similar properties: for example, spinach ACP has an isoelectric point of approximately 4.2 (Kuo and Ohlrogge, 1984c), and barley ACP elutes from a chromatofocusing column at pH 3.75 (Høj and Svendsen, 1983). In SDS–PAGE analysis, barley ACP comigrates with *E. coli* ACP (Høj and Svendsen, 1983), whereas the spinach ACP mobility corresponds to a standard protein of MW 14,500.

Rock and Cronan (1979) used the predictive algorithm of Chou and Fasman (1978) to predict the secondary structure of *E. coli* ACP. Their model, combined with additional NMR and nuclear Overhauser data, has been extended by Mayo *et al.* (1983) to yield the three-dimensional prediction of ACP structure shown in Fig. 1. More recent two-dimensional proton NMR studies by T. A. Holak and J. H. Prestegard (personal communications) do not detect the second short helix proposed from P. Y. Chou and G. D. Fassman calculations. Additional nuclear Overhauser experiments or perhaps X-ray diffraction data should lead to refinement of the Fig. 1 model.

Immunological and enzymological cross-reactivity and the amino acid composition and sequence data summarized in Fig. 2 and Table I indicate that ACP is a highly conserved protein. Therefore, it is likely that many of the structural features

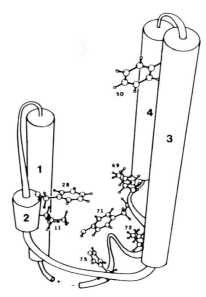

Fig. 1. Tertiary structure of *E. coli* ACP as proposed by Mayo *et al.* (1983). Four α-helical regions are depicted by cylinders; β-turns and random coils are shown as solid lines. Hydrophobic residues (numbered with position in amino acid sequence) are shown in ball and stick form and are positioned based on nuclear Overhauser experiments.

described for *E. coli* ACP apply also to plant ACP. Furthermore, application of the Chou and Fasman algorithm to the spinach ACP sequence results in the prediction of β-turns in positions analogous to those in *E. coli* ACP (Kuo and Ohlrogge, 1984c). Therefore, if the major features of the model structure shown in Fig. 1 are valid for *E. coli*, they probably are also valid for plant ACPs.

The amino acid compositions for *E. coli* and several plant ACPs are summarized in Table I. Data for *Euglena gracilis* (DiNello and Ernst-Fonberg, 1973), *Mycobacterium smegmatis* (Matsumura *et al.*, 1970), *Arthrobacter viscosus* (Simoni *et al.*, 1967), *Clostridium butyricum* (Ailhaud *et al.*, 1967), and *Rhodopseudomonas sphaeroides* (Cooper *et al.*, 1985) have also been published. All ACP compositions so far examined are characterized by a high content of acidic amino acids and a low content of aromatic and hydrophobic residues. Walker and Ernst-Fonberg (1982) used amino acid composition data to examine the relatedness of ACPs. A phylogenetic tree constructed from their analysis suggested a close relationship between plant and *E. coli* ACPs but a more distant association of plants with *Clostridium*, *Arthrobacter*, or *Euglena* ACPs. A variety of structure–function studies on chemically modified forms of *E. coli* ACP have been conducted (for review see Høj and Svendsen, 1983).

The complete amino acid sequence of *E. coli* ACP was published in 1968. In the same year Matsumura and Stumpf published a partial sequence of 17 amino acids from the prosthetic-group region of spinach ACP. For the next 15 years no additional

Table I
Amino Acid Composition of Acyl Carrier Proteins from Plants and *E. coli*[a]

Amino acid	Spinach leaf (1)		Barley leaf (2)		Avocado mesocarp (3)	Castor oilseed endosperm (4)	*E. coli* (5)
	I	II	I	II			
Ala	9	11	11	8	11	12	7
Asx	12	7	9	9	12	5	9
Glx	13	17	15	12	22	20	18
Phe	2	2	2	2	3	2	2
Gly	4	4	6	5	7	5	4
Ile	7	6	4	4	5	4	7
Lys	9	8	7	5	10	7	4
Leu	7	6	6	7	9	7	5
Met	1	3	3	1	1	1	1
Pro	0	1	3	1	3	2	1
Ser	5	4	4	8	10	4	3
Thr	4	6	8	5	7	5	6
Val	9	8	8	8	10	9	7
Tyr	0	0	0	0	1	0	1
His	0	0	0	0	1	0	1
Arg	0	0	1	0	1	0	1
Cys	0	nd[b]	1	nd	1	1	0
Trp	0	nd	0	nd	nd	nd	0
Total	82	83	87	75	114	84	77

[a] References: 1. Ohlrogge and Kuo (1985); 2. Høj and Svendsen (1984); 3. Simoni *et al.* (1967); 4. Drennan (1971); 5. Vanaman *et al.* (1968).

[b] nd, Not determined.

extensive ACP sequence data were published. This picture has now changed substantially, as ACP sequence data for rabbit (McCarthy *et al.*, 1983), barley (Høj and Svendsen, 1983), spinach (Kuo and Ohlrogge, 1984c), goose (Poulose *et al.*, 1984), *Rhodopseudomonas* (Cooper *et al.*, 1985), and rapeseed (A. Slabas, personal communication) have recently become available. Figure 2 summarizes the amino acid sequence data available for several species.

The *E. coli* sequence shown was determined for strain E-15. Recently, ACP from the more common *E. coli* K-12 has been sequenced and found to differ in having Asn instead of Asp at position 24 and Ile substituted for Val at position 43 (C. Rock, personal communication).

The spinach and *E. coli* sequences exhibit approximately 40% homology, whereas spinach and barley are 70% homologous. This degree of homology between diverse species indicates that ACP is a highly conserved protein. A high degree of structure conservation might be predicted for a cofactor protein which must interact with 10 or more other proteins (Kaplan, 1968). Structural mutations in such cofactors are less likely to be tolerated because they would require

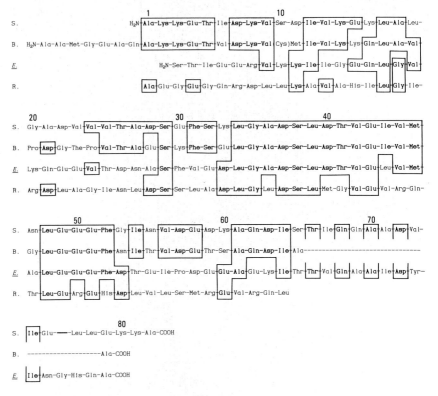

Fig. 2. Amino acid sequences of spinach I (S), barley I (B), *E. coli* (*E*), and rabbit (R) ACP. The sequences are aligned to maximize homology using Ser[38] of the spinach sequence, which is the phosphopantetheine attachment site, as a reference. Amino acid residues in the boxed areas are identical for all ACPs. A portion of the barley sequence which has not been determined, is shown by a broken line.

accommodation by the large number of other proteins which must bind to the cofactor.

McCarthy *et al.* (1983) have suggested that the homology between animal and *E. coli* ACP sequences indicates that these structures evolved from a common ancestor. The data in Fig. 2 indicate that this hypothesis can also be extended to the plant ACPs. Interestingly, the eukaryote plant ACPs have less homology with animal ACP (25%) than with bacterial ACP (40%). This comparison emphasizes the intriguing question of the evolution of plant FAS systems. Eukaryotic plant cells appear to have retained an organellar compartmentation and a type II nonassociated organization of their fatty acid-biosynthetic proteins similar to that observed in most prokaryotes. In contrast, fungi and animals have cytoplasmic FAS that is catalyzed by a multi-functional polypeptide (type I organization). Thus, cells of higher plants are similar to bacteria and differ fundamentally from all other eukaryotes in the cellular and molecular organization of FAS. The amino acid sequence data in Fig. 2 confirm this relationship. The plant ACP structure is clearly more closely related to that of the prokaryotic *E. coli* than to the eukaryotic rabbit sequence. This result contrasts with the spinach chloroplast-coded ATP synthase proton-translocating subunit, which had similar degrees of homology with its *E. coli* and bovine mitochondria counterparts (Howe *et al.*, 1982), and also contrasts with cytochrome *c,* which shows closer homology between plants and animals than between plants and bacteria (Dayhoff *et al.*, 1972).

For practical reasons, almost all studies of those plant enzymes which require ACP as cofactor have used *E. coli* ACP rather than plant ACP. This compromise has been adopted primarily because 50–100 mg of *E. coli* ACP can be readily purified in several days from 1 kg of frozen cells (Rock and Cronan, 1980), whereas purification of an equivalent quantity of plant ACP may require several months and 100 kg of starting material (Simoni *et al.*, 1967; Ohlrogge and Kuo, 1984a). The validity of using *E. coli* ACP to study plant enzymes is based primarily on the observations of Simoni *et al.* (1967) that bacterial ACP was an effective cofactor for spinach FAS and yielded products similar to reactions using plant ACP. However, in the same study it was shown that plant ACP added to bacterial fatty acid synthetase yielded lower rates of synthesis and an abnormal product distribution. Just as some enzymes are more selective than others in ACP–CoA comparisons, it is likely that some enzymes will be sensitive to species differences in ACP structure. It is hoped that cloning of plant ACP and its expression in *E. coli* will provide an abundant supply of plant ACP for future biochemical studies.(Beremand *et al.*, 1987).

VI. IMMUNOLOGICAL CHARACTERIZATION OF ACYL CARRIER PROTEINS

The enzymatic cross-reactivity of plant and bacterial ACPs observed by Simoni *et al.* (1967) suggested that antibodies raised against one ACP species would cross-react with ACP from another species. Ernst-Fonberg *et al.* (1977) found this to be the case when antibodies raised against *E. coli* ACP were shown to inhibit the activity of *Euglena* ACP in the malonyl-CoA:CO_2 exchange reaction. A precipitin line could not be obtained in double-diffusion analysis, but an immunoaffinity column of immobilized anti-*E. coli*-ACP immunoglobulins was capable of purifying ACP from two Euglena species.

In 1979 Ohlrogge *et al.* prepared antibodies to spinach ACP. The immunological cross-reactivity of ACP from various other species was assessed by their ability to compete with ^{125}I-spinach ACP for binding to the antibodies to spinach ACP. By this measure of cross-reactivity *E. coli* ACP was observed to compete with only approximately 20% of the binding sites for spinach ACP. Soybean and sunflower ACPs were able to block approximately 30 and 50% of the binding sites, respectively. These rather limited extents of cross-reactivity were somewhat surprising in light of the interchangeability of ACPs from various species in several enzymatic reactions. They can be interpreted either by suggesting that the antibodies have a greater discrimination of ACP structural features than the enzymes or that the regions of ACP recognized by the antibodies are less highly conserved than those regions recognized by the enzymes.

VII. SUBCELLULAR LOCALIZATION OF ACYL CARRIER PROTEINS

An early study of *E. coli* ACP using autoradiography of cells labeled with [β-³H]alanine indicated that ACP was preferentially associated with the cell membrane (van den Bosch *et al.*, 1970). However, a recent reevaluation of this topic using immunoelectron microscopy could detect no such membrane association (Jackowski *et al.*, 1985). Additionally, in this study it was shown that ACP and acyl-ACP did not bind significantly *in vitro* to *E. coli* membranes.

In animals and fungi, FAS is catalyzed by a multifunctional polypeptide localized in the cytoplasm. Early attempts to localize FAS in plants led to reports of *de novo* synthesis by a wide variety of subcellular fractions including the cytoplasm (reviewed in Harwood, 1975). Many of these early studies preceded the development of gentle techniques for organelle isolation and the application of the marker enzyme concept for evaluation of organelle purity and integrity. Nevertheless, it was clear that at least a portion of plant FAS could be considered "particulate" or organelle-associated. For examples, a 10,000-*g* pellet from avocado mesocarp (Stumpf and Barber, 1957), isolated and washed chloroplasts from spinach (Mudd and McManus, 1962) and sucrose gradient-purified plastids from developing castor oilseeds (Zilkey and Canvin, 1969) were all observed to be active in [¹⁴C]acetate incorporation into fatty acids.

It was not clear, however, if additional sites of FAS might exist in other organelles or, as with animals and fungi, in the cytoplasm. Evaluation of this question was difficult for two major reasons. First, the plant cell wall made it difficult to isolate organelles without encountering major breakage and mixing of the organelle contents with the cytoplasmic fraction. Second, the assay for *de novo* FAS requires the coordinate action of at least six enzymes. Because of the abundance in plants of proteases and enzyme-inactivating factors such as phenolics, the observed absence of FAS activity in a subcellular fraction must be interpreted with great caution.

In 1979, Ohlrogge *et al.* reevaluated the localization of plant FAS. The problem of organelle breakage was minimized by application of cell wall-digesting enzymes to prepare spinach leaf protoplasts which could then be very gently ruptured to release intact organelles. The problem of FAS enzyme inactivation was solved by the selection of ACP as a marker for the FAS pathway. This small, stable protein was expected to be much less susceptible to inactivation than the complete FAS pathway. Furthermore, the ACP assay was accomplished using antibodies which are capable of recognizing both active and inactive proteins.

The antibodies raised to spinach ACP were shown to inhibit *all* of the *de novo* FAS by crude spinach leaf homogenates. Therefore, the spinach ACP could be considered a valid marker for the cell's complete FAS activity. Separation of the organelles of spinach leaf protoplasts on sucrose density gradients revealed that essentially all of the cell's ACP could be attributed to the chloroplasts. From these results it became clear that plants are fundamentally different from other eukaryotes by the absence of a *de novo* FAS system in the cytoplasm. In addition, localization of ACP leads to the conclusion that plastids must be involved in the export of fatty acids. Therefore, the synthesis of endoplasmic reticulum, mitochondrial, tonoplast, plasma, and other plant cell membranes must be dependent on the plastid FAS as a source of acyl chains. At present, very little is understood about how the needs for fatty acids for organelle biogenesis are communicated to the plastid FAS. Therefore, study of the subcellular transport of acyl chains and glycerolipids should be a fruitful area for future research.

The radioimmunoassay of ACP in purified spinach chloroplasts allowed the concentration of ACP in the stroma to be estimated as 8 μM (Ohlrogge *et al.*, 1979). This value is considerably below the ACP level in *E. coli*, which can be estimated to be 100–200 μM. However, 8 μM is well above the estimated K_m values for acyl-ACP hydrolase (Ohlrogge *et al.*, 1978) and stearoyl-ACP desaturase (McKeon and Stumpf, 1982).

The ACP localization discussed above was carried out with spinach mesophyll cells, which are characterized by a high proportion of their membranes and associated fatty acids located in the chloroplast. Extrapolation of these results to other cell types should be made with caution. In data from the doctoral thesis of Drennan (1971) the ACP of developing castor oilseeds was shown to be present in sucrose gradient-purified plastids. Vick and Beevers (1978) reported that plastids of germinating castor oilseeds were the sole site of fatty acid-biosynthetic activity. Because of organelle breakage or incomplete accounting or recovery of all activity, these and other related studies have in general failed to rule out the cytoplasm as a site of FAS in non-photosynthetic tissue. Although it is clear that the plastid in all plant tissues so far examined is the major site of *de novo* FAS, the possibility of additional subcellular locations can at this time not be discounted.

VIII. ISOFORMS OF ACYL CARRIER PROTEINS

While obtaining a partial amino acid sequence for spinach ACP, Matsumura and Stumpf (1968) observed heterogeneity in the amino acid composition of some peptide fragments. From these data they suggested that spinach ACP might exist in two forms. This preliminary observation has now been confirmed and extended for barley (Høj and Svendsen, 1984), spinach (Kuo and Ohlrogge, 1984c), castor oilseed and soybean (Ohlrogge and Kuo, 1985). In spinach the two ACP isoforms are very similar in structure, making their separate characterization difficult without HPLC. In barley and castor oil plants the isoforms appear to differ more significantly. Purification of the spinach and barley isoforms to homogeneity has allowed their amino acid compositions and N-terminal amino acid sequences to be determined. As shown in Table I and Fig. 3, these data suggest that the ACP isoforms are products of separate genes rather than being generated through posttranslational protein modifications.

Spinach ACP-II and barley ACP-I each contain three methionine residues, whereas spinach ACP-I and barley ACP-II each have one methionine residue. Similarly, for many of the other amino acids a closer relationship exists between the spinach ACP-I and barley ACP-II compositions than between the two spinach species or two barley isoforms. Furthermore the N-terminal amino acid sequences of barley ACP-II and spinach ACP-I are as homologous as spinach ACP-I and spinach ACP-II (Fig. 3). Together these comparisons suggest that the divergence of the ACP isoform structures occurred early in plant evolution, before the divergence of monocots and dicots.

Antibodies raised to spinach ACP-I cross-react with ACP-II, but in a competitive binding assay ACP-II is able to compete with only 40% of the ACP-I-binding sites (Ohlrogge and Kuo, 1985). A number of other observations on the expression of ACP isoforms have been made using immunoblot procedures. As indicated in Fig. 4, the expression of ACP-I and ACP-II in spinach and castor oil plant is tissue-specific. ACP-I is the major species present in spinach leaves but is absent or barely detectable in developing seeds (Ohlrogge and Kuo, 1985). A very similar situation appears to exist with castor oilseed (Fig. 4B). Additionally, in both barley (Høj and Svendsen, 1984) and spinach (Ohlrogge and Kuo, 1985) the relative proportions of ACP-I and II in leaves can be altered by growth of plants in the dark. Both forms of ACP are associated with the chloroplasts as indicated by lane 5, Fig. 4A.

Høj and Svendsen (1984) reported the possible presence of a third form of barley leaf ACP which was present in much lower concentrations and lost activity upon purification. We have also noticed minor peaks of spinach ACP activity during protein purifications but have not been able to obtain enough material for characterization. It is not yet clear if additional ACP forms occur *in vivo* or if these variants are generated during the purification process.

The functional significance of multiple isoforms of ACP is at this time unknown and needs further evaluation. However, the following possible explanations can be considered.

Fig. 3. Comparison of N-terminal amino acid sequences of isoforms of spinach and barley ACP. Amino acids which are not conserved are highlighted.

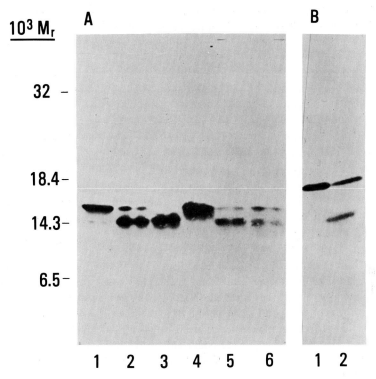

Fig. 4. Immunoblot analysis of ACPs in spinach and castor bean extracts. Proteins electrophoresed in 15% SDS gels were transferred to 0.2 μm nitrocellulose membranes and probed with antiserum to spinach ACP-I. (A) lane 1, spinach seed extract; lane 2, spinach leaf extract; lane 3, purified ACP-I (3 ng); lane 4, purified ACP-II (3 ng); lane 5, extract of spinach chloroplasts; lane 6, extract of dark-grown spinach leaves. B: lane 1, castor oil plant seed extract; lane 2, castor oil plant leaf extracts.

1. The isoforms may be localized in different subcellular compartments. Plant metabolism is often characterized by the existence of separate isozymes in different locations. For example, the glycolytic enzymes exist in both plastids and cytoplasm and, in most cases, a form of each enzyme specific to each compartment can be isolated (Dennis and Miernyk, 1982). This does not appear to be an explanation for ACP because, as reported by Høj and Svendsen (1983) for barley and as shown in lane 5 of Fig. 4A, both isoforms of ACP are associated with isolated chloroplasts. At this time, the possibility of different locations within the plastid or within different subpopulations of plastids cannot be ruled out.

2. A second possible function for the ACP isoforms might relate to the different reactions in which ACP must participate. As discussed above, ACP not only acts as a cofactor for the six reactions in FAS but also is required for desaturation of stearate, release of oleate by acyl-ACP hydrolase, and acyl transfer to glycerol-3-phosphate and monoacylglycerol-3-phosphate. Possibly ACP-I and ACP-II vary

in reactivity in these reactions, and thereby control could be exerted over the distribution of acyl chains by regulating the ratio of ACP-I and ACP-II. Recently, we obtained data in support of this possibility (Guerra *et al.*, 1986). We tested purified preparations of spinach leaf ACP-I and ACP-II in several *in vitro* reactions of fatty acid metabolism. Total *de novo* FAS and malonyl-CoA:ACP transacylase do not appear to discriminate between ACP isoforms. In contrast, the K_m value of oleoyl-ACP thioesterase is 10-fold higher for ACP-II than for ACP-I, whereas acyl-ACP glycerol-3-phosphate acyltransferase exhibits a 5-fold higher K_m for oleoyl-ACP-I than for oleoyl-ACP-II. Because both the acyltransferase and the thioesterase compete for the same substrate within the plastid, the apportionment of oleate between esterification to glycerol in the plastid or release by hydrolysis for export to the cytoplasm may be regulated by expression or acylation of ACP isoforms. Therefore, these data suggest that differential expression of ACP isoforms in plastids may partially determine the metabolic fate of oleoyl-ACP.

3. The presence of multiple genes for ACP may provide a mechanism for regulation of ACP levels (and FAS) under a greater variety of cellular demands. FAS serves a "housekeeping function" needed in all cells for membrane biosynthesis; in addition, fatty acids are synthesized in developing seeds for the tissue-specific, developmentally regulated process of oil storage. ACP levels increase markedly in maturing soybean seeds during the period of maximum oil synthesis (Ohlrogge and Kuo, 1984a). Therefore, the existence of two genes for ACP may be associated with promoters and regulatory sequences under different developmental and environmental controls, so that the regulation of these two diverse functions of fatty acids can be kept distinct.

IX. REGULATION OF ACYL CARRIER PROTEINS

In bacteria there is some evidence that ACP levels are responsive to cellular demand for FAS (Sabaitis and Powell, 1976). However, depletion of *E. coli* ACP pools by 50% in a pantothenate auxotroph did not result in decreased phospholipid/protein ratios, suggesting that ACP concentration is not rate-limiting for *E. coli* membrane lipid synthesis (Jackowski and Rock, 1983). In plants, ACP concentrations are at least 10-fold lower than in *E. coli* (Ohlrogge *et al.*, 1979), and it is possible that these lower concentrations are rate-limiting for plant FAS.

In order to evaluate the relationship between rates of FAS and levels of fatty acid-biosynthetic proteins in plants, Ohlrogge and Kuo (1984a) examined the developing soybean seed. This tissue greatly increases its rate of lipid synthesis over a 15- to 30-day period after flowering. This increased capacity for oil synthesis could be the result of a variety of mechanisms including increased photosynthate supply, hormonal activation of existing enzymes, or *de novo* production of FAS proteins. Measurement of ACP levels by both enzymatic and immunochemical assays allowed a partial resolution of these alternatives. A close correlation was found between rates of FAS and ACP content. ACP levels in the seed increased even after the period of cell division ceased. As the seeds became mature and lipid synthesis rates declined, a decrease in ACP levels also occurred. Together these results suggest that *de novo* synthesis of fatty acid-biosynthetic proteins may be a rate-determining component of the seed's overall lipid-biosynthetic capacity. Although other factors such as substrate and cofactor supply may also limit fatty acid production, the results with ACP provide encouragement that molecular genetic modification of FAS protein levels may provide a means to influence oilseed metabolism.

Although the prosthetic group in plant ACP is almost certainly derived from CoA, nothing is known about the plant enzymes involved in this process. In *E. coli*, holo-[ACP] synthase catalyzes the transfer of the 4'-phosphopantetheine moiety of CoA to apo-ACP (Elovson and Vagelos, 1968). Prosthetic-group turnover occurs through the action of [ACP]-phosphodiesterase, which cleaves the prosthetic group from ACP (Vagelos and Larrabee, 1967). More research is needed to evaluate the existence and subcellular location of these enzymes in plants.

The possibility that control over FAS might be exerted through modifications in the ratio of holo-ACP to apo-ACP has been examined. Jackowski and Rock (1983) could not detect apo-ACP *in vivo* in *E. coli*. In plants, Ohlrogge and Kuo (1984a) measured similar levels of ACP in developing soybeans, using an enzymatic assay which measures only holo-ACP and an immunological assay which detects both apo- and holo-ACP. Høj and Svendsen (1983) concluded from their amino acid analysis that purified barley ACP contains insignificant levels of apo-ACP. From these results it seems unlikely that a holo-ACP/apo-ACP cycle is important in the regulation of FAS.

Soll and Roughan (1982) estimated that 60% of the ACP pool was acylated during FAS by isolated spinach chloroplasts. This value may be higher than the *in vivo* situation where complex lipid synthesis removes acyl chains from the acyl-ACP pool. In *E. coli*, acyl-ACP is estimated to constitute approximately 10% of the total ACP *in vivo* (Rock and Jackowski, 1982).

Escherichia coli extracts contain a membrane-bound enzyme activity that can ligate long-chain fatty acids to ACP (Ray and Cronan, 1976). For some time it was believed that plant cells did not possess this ability, because addition of exogenous palmitate or stearate to a variety of tissues did not result in their desaturation, whereas incorporation into complex lipids proceeded readily. These results suggested that exogenous long-chain fatty acids could be activated to CoA thioesters but not to ACP. However, in 1979 Lessire and Cassagne reported that a microsomal fraction from leek epidermal cells synthesized stearoyl-ACP from stearate, ACP, ATP, and Mg^{2+}. This rate of ACP esterification was approximately 1% of the rate of CoA esterification to stearate (Lessire *et al.*, 1982).

X. MOLECULAR BIOLOGY OF ACYL CARRIER PROTEINS

A. *In Vitro* Translation of Acyl Carrier Proteins mRNA

Most of the proteins present in the plastid are not coded by the plastid genome; instead, genes for these proteins are in the nucleus. These plastid proteins are synthesized on cytoplasmic ribosomes and then transported across the plastid envelope membranes. In most cases so far examined, the nuclear-coded plastid proteins are synthesized as higher molecular weight precursors with additional amino acids at the N-terminus, referred to as a transit sequence. Transit sequences are believed to guide these proteins across the envelope membranes. Subsequently, a specific plastid protease removes the transit peptide, yielding a mature protein (Robinson and Ellis, 1984).

The presence of a transit sequence can be detected by examination of *in vitro* mRNA translation products. The presence of a transit sequence is implied if a plastid protein synthesized *in vitro* is 2,000–15,000 larger in size than the corresponding *in vivo* protein. This, in turn, suggests that the protein is coded by the nuclear genome. We have obtained evidence that this situation applies to ACP (Ohlrogge and Kuo, 1984b).

PolyA$^+$ RNA isolated from spinach leaves was translated in a wheat germ system with [^{35}S]methionine. Immunoprecipitation of the translation products with antibodies to ACP revealed two radioactive proteins with apparent molecular weights of approximately 20,000. This size is approximately 5,000–6,000 larger than the mobility of native ACP in the same gel system. Incubation of the immunoprecipitated proteins with a crude chloroplast extract resulted in proteolytic processing to a lower molecular weight product (P. D. Beremand and J. B. Ohlrogge, unpublished observations). These results are consistent with the hypothesis that ACP is a nuclear-coded protein that is synthesized in the cytoplasm as a precursor and that an ACP transit sequence is removed during or after uptake into the plastid. Dorne *et al.* (1982) also concluded that the genes for plant lipid-biosynthetic enzymes are nuclear encoded, based on observations of a plastid ribosome-deficient mutant of barley.

B. Cloning of Acyl Carrier Proteins

Interest in the molecular genetics of plant fatty acid biosynthesis has accelerated in large part due to the hope that genetic engineering techniques will eventually allow design of improved oilseed crops. The availability of amino acid sequence data for ACP provides a strategy for selecting and confirming clones of the ACP genes. Recently various laboratories have attempted to clone ACP from barley, spinach, rapeseed, and sunflower, and preliminary information indicates success has been achieved. In addition, a synthetic spinach ACP-I gene has now been constructed, cloned, and expressed in *E. coli* (Beremand *et al.*, 1987). Although clones of ACP may not be directly useful for oilseed improvement goals, the clones should be of great benefit in evaluating the structure and regulation of genes in the plant lipid-biosynthetic pathway. The regulation of fatty acid biosynthesis is of particular interest and significance, because fatty acid biosynthesis can be considered as *both* a housekeeping function and a developmentally regulated, organ-specific function. Because FAS is essential for membrane biogenesis in all plant cells, the FAS genes must be expressed constitutively as a housekeeping function. In addition, during seed development fatty acids serve not only in their structural role in membranes but also as a major form of carbon storage. The accumulation of this storage lipid requires expression of the FAS system at substantially higher levels than in leaf or root tissue. Because FAS proteins and their coding sequences must function in both housekeeping and oil production, their study offers a potentially fruitful method to examine how plants regulate genes that are needed for two divergent functions.

REFERENCES

Ailhaud, G. P., Vagelos, P. R., and Goldfine, H. (1967). *J. Biol. Chem.* **242**, 4459–4465.

Andrews, J., and Keegstra, K. (1983). *Plant Physiol.* **72**, 35–740.

Andrews, J., Ohlrogge, J. B., and Keegstra, K. (1985). *Plant Physiol.* **78**, 459–465.

Beremand, P. D., Hannapel, D. J., Guerra, D. J., Kuhn, D. N., and Ohlrogge, J. B. (1987). *Arch. Biochem. Biophys.* (in press).

Chou, P. Y., and Fasman, G. D. (1978). *Annu. Rev. Biochem.* **47**, 251–276.

Cooper, C. L., and Lueking, D. R. (1984). *J. Lipid Res.* **25**, 1222–1232.

Cooper, C. L., Boyce, S. G., and Lueking, D. R. (1985). *Proc. Annu. Meet. Fed. Am. Soc. Exp. Biol.*, 69th. **44**, 1412.

Dayhoff, M. O., Park, C. M., and McLaughlin, P. J. (1972). *In* "Atlas of Protein Sequence and Structure" (M. O. Dayhoff, ed.), Vol. 5, pp. 7–16. Natl. Biomed. Res. Found., Silver Spring, Maryland.

Dennis, D. T., and Miernyk, J. A. (1982). *Annu. Rev. Plant Physiol.* **33**, 27–50.

DiNello, R. K., and Ernst-Fonberg, M. L. (1973). *J. Biol. Chem.* **248**, 1707–1711.

Dorne, A. J., Corde, J. P., Joyard, J., Borner, T., and Douce, R. (1982). *Plant Physiol.* **69**, 1467–1470.

Drennan, C. H. (1971). Ph.D. Thesis, Queens Univ., Ontario, Canada.

Elovson, J., and Vagelos, P. R. (1968). *J. Biol. Chem.* **243**, 3603–3611.

Ernst-Fonberg, M. L., Schongalla, A. W., and Walker, T. A. (1977). *Arch. Biochem. Biophys.* **178** 166–173.

Frentzen, M., Heinz, E., McKeon, T. A., and Stumpf, P. K. (1983). *Eur. J. Biochem.* **129**, 629–636.

Gally, H. U., Spencer, A. K., Armitage, I. M., Prestegard, J. H., and Cronan, J. E., Jr. (1978). *Biochemistry* **17**, 5377–5382.

Guerra, D. J., Ohlrogge, J. B., and Frentzen, M. (1986). *Plant Physiol.,* **82**, 448–453.

Harwood, J. L. (1975). *In* "Recent Advances in the Chemistry and Biochemistry of Plant Lipids" (T. Galliard and E. I. Mercer, eds.), pp. 44–93. Academic Press, London.

Høj, P. B., and Svendsen, I. (1983). *Carlsberg Res. Commun.* **48**, 285–306.

Høj, P. B., and Svendsen, I. (1984). *Carlsberg Res. Commun.* **49**, 483–492.

Howe, C. J., Auffret, A. D., Doherty, A., Bowman, C. M., Dyer, T. A., and Gray, J. C. (1982). *Proc. Natl. Acad. Sci. U.S.A.* **79**, 6903–6907.

Huang, K. P., and Stumpf, P. K. (1971). *Arch. Biochem. Biophys.* **143**, 412–427.

Jackowski, S., and Rock, C. O. (1983). *J. Biol. Chem.* **258**, 16186–16191.

Jackowski, S., Edwards, H. H., Davis, D., and Rock, C. D. (1985). *J. Bacteriol.* **162**, 5–8.

Jaworski, J. G., and Stumpf, P. K. (1974). *Arch. Biochem. Biophys.* **162**, 158–165.

Kaplan, N. O. (1968). *In* "Homologous Enzymes and Biochemical Evolution" (N. Van Thoai and J. Roche, eds.), pp. 405–432. Gordon & Breach, New York.

Kuo, T. M., and Ohlrogge, J. B. (1984a). *Arch. Biochem. Biophys.* **230**, 110–116.

Kuo, T. M., and Ohlrogge, J. B. (1984b). *Anal. Biochem.* **136**, 497–502.

Kuo, T. M., and Ohlrogge, J. B. (1984c). *Arch. Biochem. Biophys.* **234**, 290–296.

Lessire, R., and Cassagne, C. (1979). *Plant Sci. Lett.* **14**, 43–48.

Lessire, R., Moureau, P., and Cassagne, C. (1982). *Physiol. Veg.* **20**, 691–702.

McCarthy, A. D., Aitken, A., and Hardie, D. G. (1983). *Eur. J. Biochem.* **136**, 501–508.

McKeon, T. A., and Stumpf, P. K. (1982). *J. Biol. Chem.* **257**, 12141–12147.

McRee, D. E., Richarson, J. S., and Richardson, D. C. (1985). *J. Mol. Biol.* **183**, 467–468.

Majerus, P. W., Alberts, A. W., and Vagelos, P. R. (1969). *In* "Methods in Enzymology (J. M. Lowenstein, ed.), Vol. 14, pp. 43–50. Academic Press, New York.

Matsumura, S., and Stumpf, P. K. (1968). *Arch. Biochem. Biophys.* **125**, 932–941.

Matsumura, S., Brindley, D. N., and Bloch, K. (1970). *Biochem. Biophys. Res. Commun.* **38**, 369–377.

Mayo, K. H., Tyrell, P. M., and Prestegard, J. H. (1983). *Biochemistry* **22**, 4485–4493.

Mudd, J. B., and McManus, T. T. (1962). *J. Biol. Chem.* **237**, 2057–2063.

Nagai, J., and Bloch, K. (1968). *J. Biol. Chem.* **243**, 4626–4633.

Ohlrogge, J. B., and Kuo, T. M. (1984a). *Plant Physiol.* **74**, 622–625.

Ohlrogge, J. B., and Kuo, T. M. (1984b). *In* "Structure, Function, and Metabolism of Plant Lipids" (P. A. Siegenthaler and W. Eichenberger, eds.), pp. 63–67. Elsevier, Amsterdam.

Ohlrogge, J. B., and Kuo, T. M. (1985). *J. Biol. Chem.* **260**, 8032–8037.

Ohlrogge, J. B., Shine, W. E., and Stumpf, P. K. (1978). *Arch. Biochem. Biophys.* **189**, 382–391.

Ohlrogge, J. B., Kuhn, D. N., and Stumpf, P. K. (1979). *Proc. Natl. Acad. Sci. U.S.A.* **76**, 1194–1198.

Overath, P., and Stumpf, P. K. (1964). *J. Biol. Chem.* **239**, 4103–4110.

Poulose, A. J., Bonsall, R. F., and Kolattukudy, P. E. (1984). *Arch. Biochem. Biophys.* **230**, 117–128.

Ray, T. K., and Cronan, J. E., Jr. (1976). *Proc. Natl. Acad. Sci. U.S.A.* **73**, 4374–4378.

Robinson, C., and Ellis, R. J. (1984). *Eur. J. Biochem.* **142**, 337–342.

Rock, C. O. (1983). *Arch. Biochem. Biophys.* **225**, 122–129.

Rock, C. O., and Cronan, J. E., Jr. (1979). *J. Biol. Chem.* **254**, 9778–9785.

Rock, C. O., and Cronan, J. E., Jr. (1980). *Anal. Biochem.* **102**, 362–364.

Rock, C. O., and Cronan, J. E., Jr. (1982). *In* "Membranes and Transport" (A. N. Martonosi, ed.), pp. 333–337. Plenum, New York.

Rock, C. O., Cronan, J. E., Jr., and Armitage, I. M. (1981a). *J. Biol. Chem.* **256**, 2669–2674.

Rock, C. O., Goalz, S. E., and Cronan, J. E., Jr. (1981b). *J. Biol. Chem.* **256**, 736–742.

Rock, C. O., and Jackowski, S. (1982). *J. Biol Chem.* **257**, 10759–10765.

Roughan, G., and Slack, R. (1984). *Trends Biochem. Sci.* **9**, 383–386.

Ruch, F. E., and Vagelos, P. R. (1973). *J. Biol. Chem.* **248**, 8095–8106.

Sabaitis, J. E., Jr., and Powell, G. L. (1976). *J. Biol. Chem.* **251**, 4706–4712.

Schulz, H. (1977). *FEBS Lett.* **78**, 303–306.

Shine, W. E., Mancha, M., and Stumpf, P. K. (1976). *Arch. Biochem. Biophys.* **172**, 110–116.

Simoni, R. D., Criddle, R. S., and Stumpf, P. K. (1967). *J. Biol. Chem.* **242**, 573–581.

Singh, S. S., Nee, T., and Pollard, M. R. (1984). *In* "Structure, Function, and Metabolism of Plant Lipids" (P. A. Siegenthaler and W. Eichenberger, eds.), pp. 161–165. Elsevier, Amsterdam.

Slabas, A., Roberts, P., and Ormesher, J. (1982). *In* "Biochemistry and Metabolism of Plant Lipids" (J. F. G. M. Wintermans and P. J. C. Kuiper, eds.), Elsevier, Amsterdam.

Soll, J., and Roughan, G. (1982). *FEBS Lett.* **146**, 189–193.

Spencer, A. K., Greenspan, A. D., and Cronan, J. E., Jr. (1978). *J. Biol. Chem.* **253**, 5922–5926.

Stumpf, P. K., and Barber, G. A. (1957). *J. Biol. Chem.* **227**, 407–417.

Vagelos, P. R., and Larrabee, A. R. (1967). *J. Biol. Chem.* **242**, 1776–1781.

Vanaman, T. C., Wakil, S. J., and Hill, R. L. (1968). *J. Biol. Chem.* **243,** 6420–6431.

Van Den Bosch, H., Williamson, J. R., and Vagelos, P. R. (1970). *Nature (London)* **228,** 338–341.

Vick, B., and Beevers, H. (1978). *Plant Physiol.* **62,** 173–178.

Wakil, S. J., Stoops, J. K., and Joshi, V. C. (1983). *Annu. Rev. Biochem.* **52,** 537–539.

Walker, T. A., and Ernst-Fonberg, M. L. (1982). *Int. J. Biochem.* **14,** 879–882.

Zilkey, B. F., and Canvin, D. T. (1969). *Biochem. Biophys. Res. Commun.* **34,** 646–653.

Biosynthesis of Monoenoic and Polyenoic Fatty Acids

7

I. Introduction
II. Oleic Acid Biosynthesis
    A. Stearoyl-ACP Desaturase
    B. Oleic Acid Synthesis in Cyanobacteria
III. Polyunsaturated Fatty Acid Biosynthesis
    A. Introduction
    B. 16:3 Plants versus 18:3 Plants
    C. 18:2 Biosynthesis
    D. 18:3 Biosynthesis
IV. Summary
    References

I. INTRODUCTION

Polyunsaturated fatty acids are found throughout the plant kingdom. They are a principal component of the photosynthetic membranes of leaves and make up the bulk of commercially important vegetable oils. Because of their obvious biochemical and commercial importance, numerous studies have attempted to determine the pathway of their biosynthesis. In spite of these efforts, this pathway is not completely elucidated.

The broad aspects of the biosynthesis of these fatty acids, including the sequential desaturation sequence suggested by Harris and James (1964) (Fig. 1), have been known since the mid-1960s. However, many of the details of how polyunsaturated fatty acids are synthesized are still missing. Among the principal reasons for our limited information are that (1) only one of the desaturases (18:0 → 18:1) has been studied in sufficient detail to allow reasonably substantial understanding of this reaction, (2) the other two desaturase activities (18:1 → 18:2 and 18:2 → 18:3)

The Biochemistry of Plants, Vol. 9
Copyright © 1987 by Academic Press, Inc.
All rights of reproduction in any form reserved.

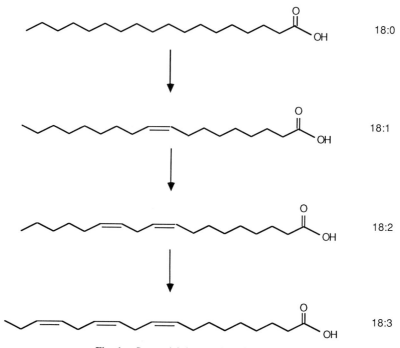

Fig. 1. Sequential desaturation of fatty acids.

have defied extensive *in vitro* study, and (3) the biosynthesis of this group of acids apparently occurs in more than one cellular compartment and is intimately associated with lipid metabolism. As a consequence, most of the studies have been *in vivo* radiolabeling experiments, which are inherently limited in the scope of information they can provide.

This chapter will present recent developments in the biosynthesis of mono- and polyunsaturated fatty acids, and will be limited to the biosynthesis of oleic (18:1), linoleic (18:2), and linolenic (18:3) acids. The principal emphasis will be on work presented since 1980. While earlier work will be referenced, choices will be selective rather than exhaustive and have the purpose of putting current work in perspective. For a more complete discussion of the earlier work, readers are also referred to other reviews on this topic by Roughan and Slack (1982), and Stumpf (1980).

II. OLEIC ACID BIOSYNTHESIS

A. Stearoyl-ACP Desaturase (Acyl-[acyl-carrier-protein] Desaturase)

The biosynthesis of oleic acid in plants is catalyzed by stearoyl-acyl carrier protein (stearoyl-ACP) desaturase (Fig. 2). It is the only plant desaturase which has

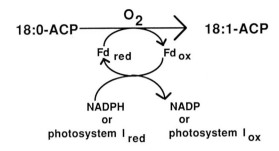

Fig. 2. Aerobic biosynthesis of oleic acid catalyzed by stearoyl-ACP desaturase.

been studied extensively *in vitro*, including its purification, and as a result, is the most completely understood of these desaturases. This reaction was first studied in photoauxotrophic *Euglena gracilis* and spinach by Nagai and Bloch (1966, 1968). The desaturase was subsequently studied in developing safflower seeds (Jaworski and Stumpf, 1974) and ultimately purified from this source (McKeon and Stumpf, 1982).

1. Localization

In green leaves, the stearoyl-ACP desaturase has been obtained from isolated spinach chloroplast in very active preparations (Nagai and Bloch, 1966; Jacobson *et al.*, 1974). Since this enzyme is specific for stearoyl-ACP as its substrate, and all the ACP is localized in the chloroplast of green leaves (Ohlrogge *et al.*, 1979), it is assumed that the chloroplast is the only site of stearoyl-ACP desaturase in green leaves. While the evidence is much less firm in nonphotosynthetic tissue, the desaturase in this tissue is presumed to be localized in the proplastids. In three separate tissues—castor bean (Nakamura and Yamada, 1974), avocado mesocarp, and cauliflower (Weaire and Kekwick, 1975)—the incorporation of acetate into long-chain fatty acids including 18:1 was localized in the proplastid. Thus in all plant tissue thus far examined, the stearoyl-ACP desaturase is localized in the same organelle as the fatty acid synthetase. Among the aerobic desaturases, the stearoyl-ACP desaturase is unique in that it is completely soluble and thus located in the stroma of the chloroplast. All the other aerobic desaturases which have been studied *in vitro* are localized in membranes.

2. Purification

The soluble nature of the stearoyl-ACP desaturase offered an opportunity to study a purified aerobic desaturase without first having to release the enzyme from a membrane environment and then later reconstitute the membrane system. However, due to the instability of the stearoyl-ACP desaturases, it initially proved difficult to carry out a significant purification. The purification of the stearoyl-ACP desaturase from safflower was eventually accomplished with the introduction of a key affinity chromatography step utilizing an ACP affinity column (McKeon and Stumpf,

1982). This allowed a simple three-step purification involving an ammonium sulfate precipitation, passing the enzyme through a DEAE–cellulose column, and, finally, affinity chromatography. This resulted in a near-homogeneous enzyme preparation with a 200-fold increase in specific activity. However, even this rapid purification procedure resulted in loss of 80% of the starting desaturase activity. Storage at 4°C or dialysis procedures resulted in major losses in activity.

3. Characterization

The molecular weight of the safflower stearoyl-ACP desaturase was determined utilizing two techniques (McKeon and Stumpf, 1982). Gel filtration yielded a molecular weight of 68,000, while SDS–polyacrylamide gel electrophoresis indicated a MW of 36,000. Thus, the stearoyl-ACP desaturase appears to be a 68,000 MW dimer consisting of two similarly sized subunits. Analysis to determine if the subunits were identical was not undertaken.

The safflower desaturase is highly specific for stearoyl-ACP. The desaturase activity for palmitoyl-ACP and stearoyl-CoA were only 1% and 5%, respectively, compared to stearoyl-ACP. Kinetic analysis of the desaturase activity with these three substrates indicates the substrate requirements of this enzyme lie in both the acyl group and the acyl carrier (Table I). The high K_m value for the stearoyl-CoA compared to the palmitoyl- and stearoyl-ACP indicates that the ACP is necessary for binding, while the low V_{max} value for the palmitoyl-ACP compared to the other two substrates indicates that the acyl chain length determines desaturase activity. The stearoyl-ACP desaturase in the crude preparations from *Euglena gracilis* showed a substrate specificity similar to the safflower desaturase. However, when partially purified, the *Euglena gracilis* desaturase has equal activity with stearoyl-ACP and stearoyl-CoA (Nagai and Block, 1968). This may indicate a difference between the

TABLE I
Summary of Properties of Plant Stearoyl-ACP Desaturase[a]

Localization
In seed tissue, proplastid;
    in leaf tissue, chloroplast
Size
MW 68,000; structure, dimer

Substrate specificity	K_m (nM)	V_{max} (nmol/min/mg)
Palmitoyl-ACP	511	0.96
Stearoyl-ACP	380	106
Stearoyl-CoA	8300	106
Inhibitors		
Cyanide, hydrogen peroxide		

[a] Data tabulated from Mckeon and Stumpf (1982) and Jaworski and Stumpf (1974).

safflower and *Euglena* systems, or it may be an artifact of purification of the unstable desaturase.

The desaturase reaction requires, in addition to a stearoyl-ACP, a source of electrons and molecular oxygen. Reduced ferredoxin has been repeatedly implicated as the immediate donor of electrons to the desaturase (Nagai and Block, 1968; Jaworski and Stumpf, 1974; McKeon and Stumpf, 1982). The stearoyl-ACP desaturase is capable of distinguishing ferredoxins from different sources (Nagai and Block, 1968). Flavodoxin could partially replace ferredoxin as an electron donor in a crude system, but only when endogenous ferredoxin had been removed (Jaworski and Stumpf, 1974). The system responsible for the reduction of the ferredoxin may depend on the tissue. Ferredoxin reduction by NADPH is catalyzed by ferredoxin-$NADP^+$ oxidoreductase, and this system has been effective in all stearoyl-ACP desaturases studied thus far. In nonphotosynthetic tissue, NADPH is the most likely source of electrons, although this has yet to be conclusively demonstrated. In photosynthetic tissue in which the photosystems, ferredoxin, and the desaturase all are found in the chloroplast, the photosynthetic electron transport chain may be the normal source of electrons. Stearoyl-ACP desaturase from spinach chloroplasts was more active when the NADPH–ferredoxin-$NADP^+$ oxidoreductase system was replaced by isolated lamella membranes in the presence of dichlorophenolindo-phenol-ascorbic acid and light (Jacobson *et al.* 1974). This system was also more active when used with the safflower desaturase, even though the photosystems are obviously not normally functional in this seed tissue (Jaworski and Stumpf, 1974).

The K_m value for oxygen is 56 μM for the safflower desaturase (McKeon and Stumpf, 1982). This concentration is well below the oxygen concentration in water at room temperature and indicates that normal variations in temperature would not affect the oxygen concentration sufficiently to alter significantly the *in vivo* desaturase activity. Thus, while temperature may affect the level of desaturation in plants by altering the activity of desaturases, this apparently is not due simply to changes in oxygen concentration.

Hydrogen peroxide has been shown to be an inhibitor of the stearoyl-ACP desaturase (McKeon and Stumpf, 1982). In the presence of catalase, a peroxide scavenger, the desaturase activity was stimulated 4- to 5-fold in crude enzyme preparations and 3-fold in purified preparations. While addition of hydrogen peroxide resulted in loss of desaturase activity, the addition had no effect in the presence of added catalase. The effect of the peroxide was not irreversible, since after inhibition by peroxide addition of catalase restored most of the desaturase activity. The nature of hydrogen peroxide inhibition is not known, and thus the mechanistic implications of peroxide inhibition cannot be assessed.

The relationship between stearoyl-ACP desaturase and other desaturases or electron carrier systems is still unclear. There is no evidence to suggest cytochrome-b_5, or cytochrome-b_5 reductase are components of the system. These two components are known to be part of the stearoyl-CoA desaturase system found in animals (Enoch *et al.*, 1976). Spectral analysis of dilute solutions of the desaturase revealed no obvious peaks or clues to the identity of prosthetic groups in the active site

(McKeon and Stumpf, 1982). While the desaturase reaction is inhibited by cyanide (but unaffected by carbon monoxide) (Jaworski and Stumpf, 1974), the site of cyanide inhibition has not been established.

B. Oleic Acid Synthesis in Cyanobacteria

The similarity in the lipid composition between plants and cyanobacteria (Mudd, 1982), as well as possible evolutionary relationships (Gray and Doolittle, 1982), suggests possible overlap in their biosynthetic pathways. Recently, it has been established that the basic reactions leading to stearoyl-ACP are very similar in the cyanobacteria *Anabaena variabilis* and plants (Lem and Stumpf, 1984; Stapleton and Jaworski, 1984a,b). While there is overlap in the early reactions of fatty acid biosynthesis, current evidence suggests that there is a considerable difference in the biosynthesis of oleic acid in plants and cyanobacteria.

Pulse–chase experiments using $^{14}CO_2$ and *A. variabilis* indicate that newly synthesized 18:0 is transferred to monoglucosyldiacylglycerol (MGlcDG) before desaturation occurs. The 18:0-MGlcDG is desaturated to 18:1-MGlcDG, and this molecule may either be further desaturated or the glucose moiety can undergo epimerization to galactose to form the corresponding monogalactosyldiacylglycerol (MGDG) (Sato and Murata, 1982a,b) (Fig. 3). Since plants appear to synthesize

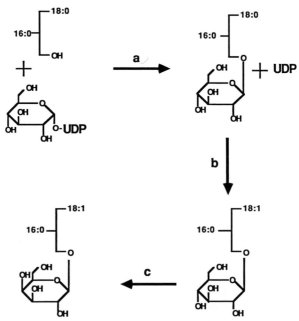

Fig. 3. Oleic acid biosynthesis in *Anabaena variabilis*. Reaction sequence includes: a, glucosyl transfer to diacylglycerol; b, desaturation of stearoyl moiety; and c, epimerization of glucose to yield MGDG.

18:1 exclusively by the desaturation of 18:0-ACP, the *Anabaena* system is a significant metabolic departure for 18:1 biosynthesis. Lending support to this mode of 18:1 synthesis is the failure to detect any stearoyl-ACP desaturase activity in homogenates of *A. variabilis* (Lem and Stumpf, 1984; Stapleton and Jaworski, 1984b).

III. POLYUNSATURATED FATTY ACID BIOSYNTHESIS

A.   Introduction

Compared to our knowledge of the biosynthesis of oleic acid, the corresponding body of knowledge for polyunsaturated fatty acid biosynthesis is substantially incomplete. Questions to which answers are incomplete or unavailable include the following:

1. To which carrier(s) or lipid(s) is the acyl group attached during desaturation?
2. In which cellular compartments do the desaturation reactions occur?
3. What are the characteristics of the 18:1 and 18:2 desaturases?

While there is substantial agreement on the answers to all of these questions for the 18:0-ACP desaturase, there are few definitive answers for these questions for 18:2 and 18:3 biosynthesis. The problem is clearly more complex with respect to the synthesis of the latter fatty acids. The membranes of both the chloroplast and endoplasmic reticulum are probable sites of desaturation. More than one lipid may act as an acyl carrier for a particular desaturase reaction. In addition, 16:3 plants (plants containing both 16:3 and 18:3, such as spinach) may have additional desaturation pathways when compared to those found in 18:3 plants (plants lacking 16:3, such as barley and soybean). Further affecting progress in this area has been the consistent observation that rates of desaturation are greatly affected by the age and condition of the tissue. The development of *in vitro* systems which can be readily studied has also been an elusive problem. Thus the 18:1 and 18:2 desaturases have proved to be difficult to study and progress has been slow.

B.   16:3 Plants versus 18:3 Plants

The pathway of biosynthesis of polyunsaturated fatty acids may be greatly dependent on the type of plant under study. A significant consideration appears to be the differences in 16:3 plants and 18:3 plants. The difference between these plants is apparent in the molecular species of their MGDG (Douce and Joyard, 1980). In the 16:3 plants, the MGDG is predominantly a mixture of 18:3/18:3-MGDG and 18:3/16:3-MGDG. In the case of the latter species, the 16:3 is esterified almost exclusively to the C-2 of the glycerol. In 18:3 plants, there is essentially no 16:3 present, and the predominant molecular species is 18:3/18:3-MGDG.

Since the glycolipids are the principal lipids containing 18:3, this variation in molecular species is probably related to the biosynthesis of these polyunsaturated fatty acids. Examination of the molecular species of MGDG found in prokaryotes such as cyanobacteria reveals that C_{18} fatty acids are found at C-1, and C_{16} fatty acids (but not C_{18} fatty acids) are found at C-2 of the glycerol moiety (Heinz *et al.*, 1979). Thus cyanobacteria and 16:3 plants have in common molecular species of MGDG, indicating that the biosynthesis of the diglycerides used in MGDG synthesis in 16:3 plants may be a metabolic remnant related to the origin of chloroplasts. For this reason, the desaturation reactions which lead to 18:3/16:3-MGDG are labeled the "prokaryotic" pathway of plant MGDG biosynthesis. In contrast, 18:3 plants lack these molecular species of MGDG, indicating the pathway for C_{18}/C_{16} diglyceride synthesis has been lost and a second "eukaryotic" pathway has evolved which is responsible for the synthesis of C_{18}/C_{18} diglycerides. Since the C_{18}/C_{18} MGDG species are also found in 16:3 plants, the eukaryotic pathway would also be present in these as well. Thus, 16:3 plants may have multiple desaturation pathways, and interpretation of *in vivo* experiments must take this into account. Furthermore, direct comparison of data between 16:3 and 18:3 plants may not always be valid.

C. 18:2 Biosynthesis

1. 18:1 Substrate for Desaturase

During the early studies of 18:2 biosynthesis, a major question concerned the nature of the substrate for the 18:1 desaturase, that is, which acyl carrier or donor acted as the substrate. The early suggestion, based on work with *Chlorella vulgaris* (Gurr *et al.*, 1969; Gurr and Brawn, 1970) and pumpkin leaves (Roughan, 1970), was that phosphatidylcholine (PC) was the substrate. This conclusion was drawn from *in vivo*-labeling studies using either $^{14}CO_2$ or [^{14}C]acetate. The lipid labeled earliest was the PC, and its specific radioactivity was the highest of all the major lipids. Further, changes in the fatty acid labeling pattern followed a classic precursor–product relationship, leading to the proposal that the desaturase reaction involved 18:1-PC → 18:2-PC.

Subsequently, the first major *in vitro* study of an 18:1 desaturase from plants concluded 18:1-CoA was the substrate (Vijay and Stumpf, 1971). Using isolated microsomes from developing safflower seeds, they concluded that 18:1-CoA was directly desaturated and that previously observed labeling patterns of PC were a result of the combined activities of 18:1-CoA desaturase and an acyltransferase which rapidly transferred 18:1 or 18:2 from CoA to PC. The low levels of 18:2-CoA which were consistently observed were presumably due to an 18:1-CoA desaturase, since a reversible acyltransferase was thought to be unlikely.

Following the *in vitro* work with safflower, numerous other reports called into question the nature of the substrate for the 18:1 desaturase. Nichols and Safford (1973) demonstrated that the method used by Vijay and Stumpf for analysis of acyl-CoA's also detected PC. Thus the amount of 18:2-CoA may have been overestimated (although the extent of overestimation is not known). Stymne and Appelqvist (1978) reexamined the safflower microsomal system and demonstrated that after the 18:1-CoA had disappeared, 18:2 biosynthesis continued. Preincubation of 18:1-CoA and safflower microsomes in the absence of NADH prevented 18:2 biosynthesis and resulted in extensive incorporation of 18:1 in PC and other lipids. Subsequent addition of NADH to the reaction mixture resulted in 18:1 desaturation with the formation of principally 18:2-PC. Furthermore, this preincubation without NADH did not diminish the quantity of the 18:2 produced, indicating no need for 18:1-CoA for desaturation. In this study, the acyl-CoA's were quantitated by measuring the amount of water-soluble acyl ester following a Bligh and Dyer extraction and assuming radioactivity in this fraction would be almost exclusively in acyl-CoA's.

A second reexamination of the safflower system (Slack et al., 1979) also measured the water-soluble acyl esters to check for the appearance of 18:2-CoA. Under normal desaturase assay conditions, the appearance of 18:2-CoA could not be confirmed, while the 18:2-PC was easily detected. Under conditions where acyl-CoA concentrations were maintained by supplementing the reaction with ATP and CoA, however, 18:2-CoA was detected, though in substantially smaller quantities than 18:2-PC. Thus, this study was inconclusive with respect to a possible role for 18:1-CoA in 18:2 synthesis.

An improved analysis was accomplished using a reverse-phase C_{18} column, on which acyl-CoA, CoA, and lipids could be quantitatively separated (Stymne and Glad, 1981). When oleoyl-CoA was the substrate for a desaturase in a soybean microsomes, 18:2-PC appeared substantially before 18:2-CoA. After a 30-min incubation, 19% of the ^{14}C from oleoyl-CoA appeared in PC and 12% was converted to 18:2-PC. In that time, no 18:2-CoA was detected. After 60 min, small quantities of 18:2-CoA appeared and could be accounted for as products of the acyl exchange reaction.

The possible involvement of an acyl exchange between PC and acyl-CoA has been demonstrated (Stymne and Glad, 1981; Stobart et al., 1983; Stymne and Stobart, 1984). In homogenates of both safflower and soybean cotyledons, oleoyl-CoA underwent exchange with PC to produce 18:2- and 18:3-CoA. While bovine serum albumin was essential for exchange, the ATP-dependent acyl-CoA synthetase was not required (Stymne and Glad, 1981). The nature of the acyl exchange reaction appears to involve lysophosphatidylcholine (lyso-PC) and the reversibility of the acyl-CoA:lyso-PC acyltransferase and is thus specific for C-2 of the glycerol (Stymne and Stobart, 1984). While the reverse of this reaction is not normally favored, the binding of acyl-CoA by the serum albumin apparently lowers the concentration of acyl-CoA sufficiently to shift the equilibrium. Overall, reversal of this acyltransferase results in acyl exchange:

$$
\begin{aligned}
\text{Forward:} \quad & \text{R-CoA + lyso-PC} \rightarrow \text{CoA + R-PC} \\
\text{Reverse:} \quad & \text{R'-PC + CoA} \quad \rightarrow \text{lyso-PC + R'-CoA} \\
\hline
\text{Overall:} \quad & \text{R-CoA + R'-PC} \rightarrow \text{R-PC + R'-CozA}
\end{aligned}
$$

The physiological significance of this acyl exchange as yet has not been demonstrated. Furthermore, it is not clear if it relates to the observed 18:2-CoA by Vijay and Stumpf (1971). Neither CoA nor serum albumin were added to these assays. On the other hand, Slack *et al.* (1979) observed 18:2-CoA only when the CoA concentration was 1.0 m*M*, which may have been a sufficiently high concentration to shift the equilibrium of the transferase enough, even in the absence of serum albumin, to observe exchange (Stymne and Stobart, 1984).

Each of the systems cited above were from seed tissue. Leaf tissue can also desaturate 18:1, and study of pea leaf homogenates indicated that the 18:1 desaturase system had the same properties as those in seed (Slack *et al.*, 1976a). Using 18:1-CoA as the substrate, 18:1-PC was rapidly formed and subsequently 18:2-PC was formed. Even after the 18:1-CoA had disappeared, 18:2 biosynthesis continued, indicating once again that 18:1-PC was the substrate for the desaturase.

Consistent with the *in vitro* studies have been the data from numerous *in vivo* studies. Leaf tissue in which PC has been demonstrated to be the major site of 18:1 desaturation include maize (Slack and Roughan, 1975; Hawke and Stumpf, 1980), barley (Stobart *et al.*, 1980), and broad bean (Williams, 1980). Similarly, *in vivo* studies of oilseeds are consistent with 18:1-PC desaturation (Slack *et al.*, 1978)

Are there other substrates for 18:1 desaturation? When safflower microsomes were incubated with 18:1-CoA and lysophosphatidylethanolamine, there was increased synthesis of 18:1-phosphatidylethanolamine (PE) and 18:2-PE, suggesting that 18:1-PE may also be a substrate for the 18:1 desaturase (Sanchez and Stumpf, 1984). In addition, isolated spinach chloroplasts have some capacity to synthesize 18:2 (see, e.g., Kannangara and Stumpf, 1972; Slack *et al.*, 1976b; Heinz and Roughan, 1983). This 18:2 is found in phosphatidylglycerol (PG) and MGDG, and, at least in 16:3 plants, chloroplastic 18:2 synthesis probably does not involve 18:1-PC (Heinz and Roughan, 1983; Roughan and Slack, 1984). The substrates in these latter cases are believed to be the 18:1-PG and 18:1-MGDG, although there is no direct evidence to support this. Tremolieres *et al.* (1980a) have also measured 18:2 biosynthesis from 18:1-CoA using a plastid preparation from pea leaves.

2. *Localization of 18:1 Desaturation*

Essentially all the studies of the 18:1-PC desaturase indicate that this enzyme is localized in the microsomal fraction, presumably the endoplasmic reticulum. In oilseeds, including safflower (Vijay and Stumpf, 1971) and soybean (Stymne and Glad, 1981), and in potato slices (Ben Abdelkader *et al.*, 1973), the 18:1 desaturase was in the 105,000 *g* pellet. Similarly, in leaf tissue from maize (Slack *et al.*, 1976a) and pea (Trémolières and Mazliak, 1974; Trémolières *et al.*, 1980a; Murphy *et al.*, 1984), the endoplasmic reticulum is the site of 18:1-PC desaturation.

As indicated in the above section, the chloroplasts of 16:3 plants represent a second site of 18:2 synthesis, although very little is known about the chloroplast 18:1 desaturase.

3. Properties of the 18:1 Desaturase

The initial characterization of the 18:1 desaturase by Vijay and Stumpf (1972) described the major properties of this enzyme system. As in most studies of the 18:1 desaturases, 18:1-CoA was the substrate used, and a crude microsomal preparation from developing safflower seeds was the enzyme source. In addition to the 18:1 substrate, there was an absolute requirement for molecular oxygen and a source of electrons, usually in the form of a reduced pyridine nucleotide. NADH was preferred over NADPH as an electron donor (K_m values for NADH and NADPH were 0.18 and 2.5 mM, respectively). Photochemically reduced ferredoxin could also replace NADH in this system, but unlike the 18:0-ACP desaturase (Jaworski and Stumpf, 1974), it was a poorer electron donor than NADH. While CO had no effect on the desaturase, CN at a concentration of 1.0 mM caused 58% inhibition. Metal-chelating reagents such as p-OH-quinoline and thiol inhibitors such as N-ethylmaleimide were potent inhibitors of desaturation. In this crude system, however, it was unclear if the site inhibition was the desaturase itself, or an acyl transferase.

Attempts to define further the components of the 18:1 desaturase from safflower were recently carried out (Gennity and Stumpf, 1985). They concluded that in addition to the desaturase, an acyl-CoA lysophospholipid acyltransferase was associated with the system. Attempts to define the electron carrier components of the system were unsuccessful. The cytochrome-b_5 and cytochrome-b_5 reductase which are associated with the 18:1-CoA desaturase in rat liver (Enoch et al., 1976) could not be confirmed as components of the safflower system. Attempts to solubilize the desaturase were also unsuccessful.

Similar studies using pea leaf microsomes indicate essentially no differences with the properties of the safflower system (Slack et al., 1976a; Murphy et al., 1983, 1984). The association of an acyl-CoA:lyso-PC acyltransferase was demonstrated, and detergent solubilization of this enzyme was described (Murphy et al., 1983). Attempts to solubilize the pea 18:1 desaturase resulted in complete loss of activity (Murphy et al., 1983).

The association of the acyltransferase and the 18:1 desaturase appears to be very close. Newly synthesized 18:1-PC was preferentially desaturated by pea microsomes (Murphy et al., 1984), indicating this lipid pool was mixing very little with endogenous 18:1-PC. This would explain the consistent observation that 18:1-CoA is very rapidly transferred and desaturated to 18:2-PC. The close functional and, presumably, physical association of the two enzymes eliminates the need for extended diffusion of newly formed 18:1-PC through the membrane to the 18:1 desaturase. This would also explain the observation that exogenous 18:1-PC was a much poorer substrate than 18:1-PC generated in situ from 18:1-CoA (Slack et al., 1979; Trémolières et al., 1980b).

D. 18:3 Biosynthesis

1. In Vivo *Studies*

The early proposal (Harris and James, 1964) that 18:3 is the product of a sequential desaturation involving $18:1 \rightarrow 18:2 \rightarrow 18:3$ is generally accepted as the major route of 18:3 biosynthesis. However, the characterization of the final desaturation step has remained very elusive. The 18:3 resides primarily in the MGDG located in the chloroplast membranes, and, as discussed earlier, the different molecular species of MGDG are what distinguish 18:3 plants from 16:3 plants. If PC is the major site of 18:2 biosynthesis, the connection between it and MGDG must also be established to complete our understanding of this pathway. Thus the study of 18:3 biosynthesis has in many cases been a study of MGDG biosynthesis, and the nature of the plant (16:3 versus 18:3) has been critical to our understanding of the data.

In several 18:3 plants, including maize (Slack *et al.*, 1977), *Vicia faba* (broad bean) (Heinz and Harwood, 1977), barley, wheat, pea (Wharfe and Harwood, 1978), and *Avena sativa* (oat) (Ohnishi and Yamada, 1980a,b), *in vivo* studies of the incorporation of a fatty acid precursor, such as CO_2 or acetate, into 18:3 indicated that 18:2 desaturation occurred to 18:2-MGDG rather than 18:2-PC. This was based on the general absence of labeled 18:3-PC early in the incubation. A similar conclusion was drawn by Hawke and Stumpf (1980) when following the incorporation of $[^{14}C]18:1$ into 18:2 and 18:3 in maize leaves.

In disagreement with the above studies was the synthesis of 18:3-PC from 18:1 by young barley leaves (Stobart *et al.*, 1980). The small but significant synthesis of 18:3-PC always preceded the appearance of 18:3 in any other lipid. In another study using *Vicia faba*, Williams (1980) determined the molecular species of MGDG which first contained $[^{14}C]$galactose labeled from $^{14}CO_2$. Assuming a minimal desaturation of fatty acids would have occurred at the earliest incubation times, he concluded that highly unsaturated fatty acids, including 18:3, were being donated with the diacylglycerol, presumably from PC, for the synthesis of MGDG. The data also suggested that a significant amount of desaturation occurred after incorporation of the acyl groups into MGDG.

In 16:3 plants, incorporation of label into 18:3-MGDG follows a pattern very different from 18:3 plants. With leaves of *Brassica napus* (tower plant) (Williams and Khan, 1982; Williams *et al.*, 1983), *Anthriscus cerefolium* (chervil), *Chenopodium album* (goosefoot), and spinach (Siebertz and Heinz, 1977), the incorporation of label from acetate and CO_2 was approximately equal into both PC and MGDG. Unlike the situation in 18:3 plants, the fatty acids initially labeled in MGDG were 16:0 and 18:1 and the changing of the labeled fatty acids was consistent with a sequential desaturation of 16:1-MGDG and 18:1-MGDG to yield 16:3-MGDG and 18:3-MGDG, respectively. Siebertz and Heinz (1977) did not detect the synthesis of 18:3-PC in the 16:3 plants they studied. This was in contrast to the study of *Brassicus napus* (Williams and Khan, 1982), in which 18:3-PC was labeled by CO_2. In the latter case, however, there appeared to be little or no movement of the 18:3 from PC to MGDG. The 18:3 on PC is found at both C-1 and C-2 of the glycerol, whereas none of the considerable amount of labeled 18:3 in the MGDG was in the C-2 of the MGDG. This would indicate that the newly synthesized 18:3 found in MGDG is in the 18:3/16:3-MGDG and that movement of acyl groups from PC to MGDG is slow in this type of tissue.

2. In Vitro *Studies*

While there have been numerous *in vivo* studies of 18:3 biosynthesis, there have been relatively few comparable studies which have successfully exploited cell-free systems. Chloroplasts isolated from spinach, pea, and lettuce were capable of synthesizing 18:3 from acetate (Roughan *et al.*, 1979) or 18:2 (Jones and Harwood, 1980). In each case, 18:3 synthesis was dependent on using chloroplasts isolated by the method of Nakatani and Barber (1977), as well as incubating under conditions which promoted MGDG biosynthesis, with glycerol 3-phosphate and UDPgalactose. The 18:3 was found almost exclusively associated with the MGDG, and the data were entirely consistent with 18:2-MGDG acting as the substrate for the final desaturation step.

In contrast, homogenates of developing soybean cotyledons were capable of producing 18:3 when incubated with 18:1-CoA (Stymne and Appelqvist, 1980). The 18:1-CoA was predominantly incorporated into PC and the initial 18:3 synthesized appeared in the PC. Further, while as much as 8% of the 18:1 was incorporated into the 18:3, the maximum incorporation into 18:3-MGDG was only 0.5%. Thus, 18:3 biosynthesis by isolated chloroplasts from leaves is consistent with 18:2-MGDG desaturation, whereas whole homogenates from a developing seed apparently carry out the desaturation of 18:2-PC.

3. *Cooperation between Phospholipid and Glycolipid Metabolism*

The major repository of 18:3 is the MGDG of the chloroplast. The *in vivo* and *in vitro* studies of 16:3 plants are consistent with the biosynthesis of 18:3/16:3-MGDG occurring in complete association with the chloroplast (Roughan *et al.* 1979; Jones and Harwood, 1980; Heinz and Roughan, 1983). This pathway has been labeled the "prokaryotic pathway" of 18:3 biosynthesis due to the similarity of this molecular species to those found in prokaryotes such as cyanobacteria (Roughan and Slack, 1984).

On the other hand, biosynthesis of 18:3/18:3-MGDG appears to be synthesized by a different pathway, labeled the "eukaryotic pathway" because of its departure from the prokaryotic-type molecular species. The principal molecular species of PC are 18/18-PC at varying levels of unsaturation. The evidence is consistent with a pathway in which the biosynthesis of PC and desaturation of acyl groups on the endoplasmic reticulum is intermediate in the biosynthesis of 18:3/18:3-MGDG. Incubations with whole leaves support transfer of polyunsaturated fatty acids (Slack and Roughan, 1975; Williams, 1980) or the entire diglyceride moiety (Slack *et al.*, 1977) from PC to MGDG. The *in vitro* data also support cooperation of the chloroplasts and endoplasmic reticulum in the biosynthesis of PC (Trémolières *et al*., 1980b; Drapier *et al*., 1982; Roughan *et al*., 1980). While discussion of phospholipid and galactolipid biosynthesis is beyond the scope of this chapter, it is clear that a thorough understanding of their synthesis is essential for a complete elucidation of the pathway for 18:3 biosynthesis.

IV. SUMMARY

Based on our current state of knowledge, the scheme in Fig. 4 presents probable routes of 18:2 and 18:3 biosynthesis. The desaturation of 18:1-PC and 18:2-PC on

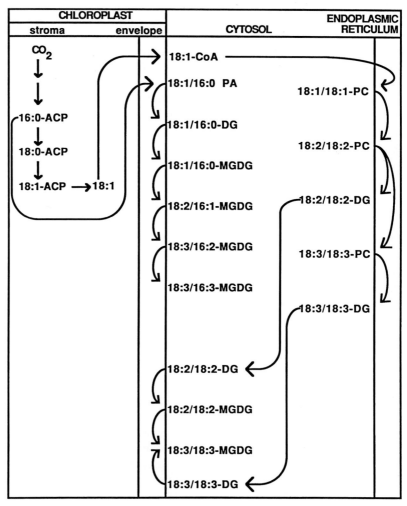

Fig. 4. Routes of biosynthesis of polyunsaturated fatty acids.

the endoplasmic reticulum and 18:1-MGDG and 18:2-MGDG on the chloroplast envelope, however, represent only two of the most probable routes of desaturation. The scheme does not attempt to take into account the synthesis of polyunsaturated fatty acids found in other lipids such as phosphatidylglycerol or sulfolipids, since little is known about their synthesis. For similar reasons, the biosynthesis of 16:3 was not addressed. While a large amount of progress has been made since the mid-1970s toward our understanding of the biosynthesis of polyunsaturated fatty acids, there are clearly major questions which are unresolved. With the exception of the stearoyl-ACP desaturase, the bulk of our knowledge is derived from extensive *in vivo* studies which are open to some misinterpretation.

Beyond the simple elucidation of the pathway of desaturation of fatty acids, there are many additional questions to te resolved. For example, how many different fatty acid desaturases are there in a cell? Even limiting our attention to the scheme presented in Fig. 3, there is a proposal of at least two different desaturases for the synthesis of each polyunsaturated fatty acid. What is the mechanism of desaturation? No detailed mechanism for a single desaturase has yet been elucidated. Obviously, the answers to many of these questions await the isolation of active desaturase systems.

REFERENCES

Ben Abdelkader, A., Cherif, C., Demandre, C., and Mazliak, P. (1973). *Eur. J. Biochem.* **32,** 155–165.
Douce, R., and Joyard, J. (1980). *In* "Biochemistry of Plants" (P. K. Stumpf and E. E. Conn, eds.), Vol. 4, pp 321–362. Academic Press, New York.
Drapier, D., Dubacq, J. P., Trémolières , A., and Mazliak, P. (1982). *Plant Cell Physiol.* **23,** 125–135.
Enoch, H. G., Catala, A., and Strittmatter, P. (1976). *J. Biol. Chem.* **251,** 5095–5103.
Gennity, J. M., and Stumpf, P. K. (1985). *Arch. Biochem. Biophys.* **239,** 444–454.
Gray, M., and Doolittle, W. F. (1982). *Microbiol. Rev.* **46,** 1–42.
Gurr, M. I., and Brawn, P. (1970). *Eur. J. Biochem.* **17,** 19–22.
Gurr, M. I., Robinson, M. P., and James, A. T. (1969). *Eur. J. Biochem.* **9,** 70–78.
Harris, R. V., and James, A. T. (1964). *Biochim. Biophys. Acta* **106,** 456–464.
Hawke, J. C., and Stumpf, P. K. (1980). *Arch. Biochem. Biophys.* **203,** 296–306.
Heinz, E., and Harwood, J. L. (1977). *Hoppe-Seyler's Z. Physiol. Chem.* **358,** 897–908.
Heinz, E., and Roughan, P. G. (1983). *Plant Physiol.* **72,** 273–279.
Heinz, E., Siebertz, H. P., Linscheid, M., Joyard, J., and Douce, R. (1979). *In* "Advances in the Biochemistry and Physiology of Plant Lipids" (L. A. Appelqvist and C. Liljenberg, eds.), pp. 99–120. Elsevier, Amsterdam.
Jacobson, B. S., Jaworski, J. G., and Stumpf, P. K. (1974). *Plant Physiol.* **54,** 484–486.
Jaworski, J. G., and Stumpf, P. K. (1974). *Arch. Biochem. Biophys.* **162,** 158–165.
Jones, A. V. M., and Harwood, J. L. (1980). *Biochem. J.* **190,** 851–854.
Kannangara, G., and Stumpf, P. K. (1972). *Arch. Biochem. Biophys.* **148,** 414–424.
Lem, N. W., and Stumpf, P. K. (1984). *Plant Physiol.* **74,** 134–138.
McKeon, T. A., and Stumpf, P. K. (1982). *J. Biol. Chem.* **257,** 12141–12147.
Mudd, J. B. (1982). *In* "On the Origins of the Chloroplast" (J. A. Schiff, ed.), pp. 131–148, Elsevier/North-Holland, New York.
Murphy, D. J., Woodrow, I. E., Latzko, E., and Mukherjee, K. D. (1983). *FEBS Lett.* **162,** 442–446.
Murphy, D. J., Mukherjee, K. D., and Woodrow, I. E. (1984) *Eur. J. Biochem.* **139,** 373–379.

Nagai, J., and Bloch, K. (1966). *J. Biol. Chem.* **241**, 1925–1927.
Nagai, J., and Bloch, K. (1968). *J. Biol. Chem.* **243**, 4626–4633.
Nakamura, Y. and Yamada, M. (1974). *Plant Cell Physiol.* **15**, 37–48.
Nakatani, H. Y., and Barber, J. (1977). *Biochim. Biophys. Acta* **461**, 510–512.
Nichols, B. W., and Safford, R. (1973). *Chem. Phys. Lipids* **11**, 222–227.
Ohlrogge, J. B., Kuhn, D. N., and Stumpf, P. K. (1979). *Proc. Natl. Acad. Sci. U.S.A.* **76**, 1194–1198.
Ohnishi J., and Yamada, M. (1980a). *Plant Cell Physiol.* **21**, 1595–1606.
Ohnishi J., and Yamada, M. (1980b). *Plant Cell Physiol.* **21**, 1607–1618.
Roughan, P. G. (1970). *Biochem. J.* **117**, 1–8.
Roughan, P. G., and Slack, C. R. (1982). *Annu. Rev. Plant Physiol.* **33**, 97–132.
Roughan, P. G., and Slack, C. R. (1984). *Trends Biochem. Sci.* **9**, 383–386.
Roughan, P. G., Mudd, J. B., McManus, T. T., and Slack, C. R. (1979). *Biochem. J.* **184**, 571–574.
Roughan, P. G., Holland, R., and Slack, C. R. (1980). *Biochem. J.* **188**, 17–24.
Sanchez, J., and Stumpf, P. K. (1984). *Arch. Biochem. Biophys.* **228**, 185–196.
Sato, N., and Murata, N. (1982a). *Biochim. Biophys. Acta* **710**, 271–278.
Sato, N., and Murata, N. (1982b). *Biochim. Biophys. Acta* **710**, 279–289.
Siebertz, H. P., and Heinz, E. (1977). *Z. Naturforsch., C* **32C**, 193–205.
Slack, C. R., and Roughan, P. G. (1975). *Biochem. J.* **152**, 217–228.
Slack, C. R., Roughan, P. G., and Terpstra, J. (1976a). *Biochem. J.* **155**, 71–80.
Slack, C. R., Roughan, P. G., and Holland, R. (1976b). *Biochem. J.* **158**, 593–601.
Slack, C. R., Roughan, P. G., and Balasingham, N. (1977). *Biochem. J.* **162**, 289–296.
Slack, C. R., Roughan, P. G., and Balasingham, N. (1978). *Biochem. J.* **170**, 421–433.
Slack, C. R., Roughan, P. G., and Browse, J. (1979). *Biochem. J.* **179**, 649–656.
Stapleton, S. R., and Jaworski, J. G. (1984a). *Biochim. Biophys. Acta* **794**, 240–248.
Stapleton, S. R., and Jaworski, J. G. (1984b). *Biochim. Biophys. Acta* **794**, 249–255.
Stobart, A. K., Stymne, S., and Appelqvist, L.-A. (1980). *Phytochemistry* **19**, 1397–1402.
Stobart, A. K., Stymne, S., and Glad, G. (1983). *Biochim. Biophys. Acta* **754**, 292–297.
Stumpf, P. K. (1980). *In* "Biochemistry of Plants" (P. K. Stumpf and E. E. Conn, eds.), Vol. 4, pp. 177–204. Academic Press, New York.
Stymne, S., and Appelqvist, L.-Å. (1978). *Eur. J. Biochem.* **90**, 223–229.
Stymne, S., and Appelqvist, L.-Å. (1980). *Plant Sci. Lett.* **17**, 287–294.
Stymne, S., and Glad, G. (1981). *Lipids* **16**, 298–305.
Stymne, S., and Stobart, A. K. (1984). *Biochem. J.* **223**, 305–314.
Trémolières , A., and Mazliak, P. (1974). *Plant Sci. Lett.* **2**, 193–201.
Trémolières , A., Drapier, D., Dubacq, J. P., and Mazliak, P. (1980a). *Plant Sci. Lett.* **18**, 257–269.
Trémolières , A., Dubacq, J. P., Drapier, D., Muller, M., and Mazliak, P. (1980b). *FEBS Lett.* **114**, 135–138.
Vijay, I. K., and Stumpf, P. K. (1971). *J. Biol. Chem.* **246**, 2910–2917.
Vijay, I. K., and Stumpf, P. K. (1972). *J. Biol. Chem.* **247**, 360–366.
Weaire, P. J., and Kekwick, R. G. O. (1975). *Biochem. J.* **146**, 439–445.
Wharfe, J., and Harwood, J. L. (1978). *Biochem. J.* **174**, 163–169.
Williams, J. P. (1980). *Biochim. Biophys. Acta* **618**, 461–472.
Williams, J. P., and Khan, M. U. (1982). *Biochim. Biophys. Acta* **713**, 177–184.
Williams, J. P., Khan, M. U., and Mitchell, K. (1983). *In* "Biosynthesis and Function of Plant Lipids" (W. W. Thomson, J. B. Mudd, and M. Gibbs, eds.), pp. 28–39. Am. Soc. Plant Physiol., Rockville, Maryland.

Triacylglycerol Biosynthesis | 8

STEN STYMNE
ALLAN KEITH STOBART

I. INTRODUCTION

The triacylglycerol constituents of vegetable oils are one of the world's most important plant commodities, with enormous potential for further nutritional and industrial exploitation. The world production of the major plant oils has in general shown a continuous increase since the mid-1970s (Table I), and this is expected to

TABLE I
Production Projections for Individual Edible Vegetable Oils (in 10^3 tonnes)[a]

	Actual: 1972–1974 avereage	Projected: 1985 adjusted	Growth rates: projected 1972–1974 to 1985, basic (% per annum)
Edible vegetable oils	29,380	45,380	4.5
Soft oils	23,820	35,000	3.3
Cottonseed oil	2,920	3,870	2.4
Groundnut oil	3,210	4,600	3.0
Maize oil	470	690	3.3
Olive oil	1,580	1,840	1.3
Rapeseed oil	2,550	3,800	3.4
Safflower seed oil	240	340	2.9
Sesame seed oil	640	810	2.0
Soybean oil	8,270	13,090	3.9
Sunflower seed oil	3,940	5,960	3.5
Lauric oils	3,090	4,770	3.7
Coconut oil	2,470	3,760	3.6
Palm kernel oil	620	1,010	4.2
Other edible/soap fats and oils			
Palm oil	2,470	6,060	7.8

[a]Source: FAO, Rome. From Bozzini (1982). Reproduced by permission of the author and publisher, International Atomic Energy Agency, Vienna.

be sustained. Although the major use of the plant oils is for human consumption, they have important industrial application, particularly in the production of paints, detergents, and specialized lubricants. The anticipated upward trend in the utilization of vegetable oil will require a tremendous increase not only in the area of land under cultivation and crop yield but also in the ability to produce new varieties and strains which will synthesize triacylglycerols with specific fatty acid compositions. A thorough knowledge, therefore, of the regulatory mechanisms which govern oil quality, particularly in seeds, is an important prerequisite in the development of varieties and genetically engineered systems.

In light of the important role that the plant triacylglycerols play in human welfare, it is surprising to find that little information concerning the detailed biochemical events leading to the formation of vegetable oils has hitherto been available. M. I. Gurr, in a chapter in a previous volume in this series (Gurr, 1980), suggested several reasons for this paucity in knowledge. He considered that it was perhaps due to the difficulties experienced in obtaining experimental material at the necessary precise stages of seed development and the fact that plant lipid biosynthesis was not a fashionable subject. Much of the slow progress in our understanding of triacylglycerol biosynthesis certainly can be attributed to the noncritical approach adopted by many workers in obtaining experimental material at synchronized and correct stages of development for biochemical studies. It should also be noted that, while research with animal systems has contributed markedly to plant studies, it has led many plant biochemists and physiologists to relate findings in the animal field erroneously directly to the synthesis of plant triacylglycerols. This has led in some cases to the incorrect interpretation and understanding of how triacylglycerols are synthesized and accumulated in plants. During the last few years, however, new concepts on the biosynthesis of triacylglycerols have emerged which now make it possible to describe with more certainty the mechanisms which regulate the fatty acid quality of the plant oil.

II. SEED OIL COMPOSITION AND DEPOSITION

A. Triacylglycerol Structure

Energy reserves in the form of fatty acids are in nearly all plant cells contained in a triacylglycerol molecule. The only known exception to this is the jojoba (*Simmondsia chinensis* L.) seed, in which the fatty acids are esterified to long-chain alcohols to form waxes (Yermanos, 1975). The structure of a triacylglycerol molecule is depicted in Fig. 1. Three fatty acids are esterified, one to each of the hydroxyl groups of a glycerol molecule, to form a triacylglycerol. Since the glycerol molecule does not have rotational symmetry, all the carbon atoms are readily distinguished from each other. They are therefore classified *sn* (stereochemical numbering)-1, *sn*-2, and *sn*-3 (Fig. 1) according to the recommendation of the IUPAC-IUB Commission of Biochemical Nomenclature. The plant triacylglycerols can possess a great number of different fatty acids, although eight particular ones account for some 97% of those present in commercial vegetable oils (Table II). The seed triacylglycerols are usually characterized by the predominance of C_{18}-unsaturated and polyunsaturated fatty acids, and this distinguishes them from animal fats, which are generally of a more saturated nature. The C_{18}-unsaturated fatty acids (oleic, linoleic, and linolenic) are particularly important and govern, to a large degree, the physical properties of the oil and hence its use and commercial value.

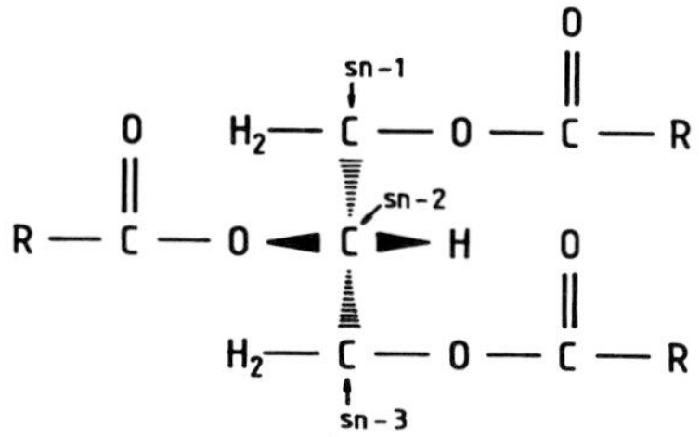

Fig. 1. Triacylglycerol structure. R, Fatty acid acyl chains.

TABLE II
The Major Fatty Acids in Commercial Plant Oils[a,b]

Trivial name	Systematic name	Symbol	%
Lauric	Dodecanoic	12:0	4
Myristic	Tetradecanoic	14:0	2
Palmitic	Hexadecanoic	16:0	11
Stearic	Octadecanoic	18:0	4
Oleic	Octadec-9-enoic	18:1	34
Linoleic	Octadec-9,12-dienoic	18:2	34
α-Linolenic	Octadec-9,12,15-trienoic	18:3	5
Erucic	Docos-13-enoic	22:1	3

[a]Adapted from Gunstone and Norris (1983), with permission of Pergamon Press,Oxford.
[b]Estimates are based on 1964–1970 production of commercial oilseeds.

Linoleic acid is also an essential fatty acid (vitamin F) in the human diet, where it functions as a precursor of arachidonic acid and many of the prostaglandins.

The particular fatty acids in the plant triacylglycerols are not distributed randomly between the different sn-carbon atoms. It is a general rule that saturated species of fatty acid are confined to positions sn-1 and sn-3 with some enrichment in the first position, and that the polyunsaturated C_{18} fatty acids are located mainly at position sn-2 (Gunstone and Ilyas-Qureshi, 1965; Gunstone et al., 1965; Hilditch and Williams, 1964). The distribution of the fatty acids in the triacylglycerols of some common oilseeds is given in Table III. A detailed description of the methods used

TABLE III
Stereochemical Analysis of the Fatty Acids in the Triacylglycerols of Oil-Rich Seeds

Species	sn Position	16:0	18:0	18:1	18:2	18:3	20:1	22:1	References
Safflower	1	9	3	8	81	—	—	—	Ichihara and
	2	0	0	8	92	—	—	—	Noda (1980)
	3	4	1	7	88	—	—	—	
Soybean	1	14	6	23	48	9	—	—	Gunstone
	2	1	Trace	22	70	7	—	—	(1979)
	3	13	6	28	45	8	—	—	
Rape	1	4	2	23	11	6	16	35	Brockerhoff
(high erucate)	2	1	0	37	36	20	2	4	(1971)
	3	4	3	17	4	3	17	51	
Rape	1	6	2	65	16	7	2	1	Töregard
(low-erucate,	2	Trace	Trace	53	31	16	Trace	Trace	and Podhala
var. Oro)	3	8	2	71	10	5	3	1	(1974)

in the positional analysis of the fatty acids in triacylglycerols is not the subject of the present work, and the reader is referred to an earlier volume in this series (Gurr, 1980). The variety of fatty acids that are available for oil synthesis can result in a heterogeneous mixture of triacylglycerols in a particular seed oil (Gurr and James, 1980). Generally only a few species of triacylglycerol predominate, and these can be elucidated by their degree of unsaturation upon argentation chromatography (Gurr et al., 1972; Shewry et al., 1972; Fatemi and Hammond, 1977).

B. Oil Deposition

Although triacylglycerols are present in nearly all the organs of a plant, they generally only accumulate to any extent in the seed or fruit (Table IV). In most economically important oil-producing plants the storage lipid is located in the cotyledons of the seed (safflower, sunflower, soy, rape, linseed), the mesocarp of certain fruits (avocado, oil palm), and endosperm tissue (castor bean, coconut). Our present knowledge, however, on the synthesis of triacylglycerols is largely based on biochemical studies with the developing cotyledons of seeds, and hence much of this chapter is devoted to work with this particular tissue.

The accumulation of the triacylglycerols in a maturing oilseed is well illustrated by safflower (Fig. 2). Oil deposition in the developing cotyledons of the seed generally commences within 10–15 days after flowering (Ichihara and Noda, 1980) and is normally completed within a further 15 days. Accumulation, however, is at its most rapid within a 4- to 5- day period (13–18 days after flowering), when some 70% of the final oil is produced (Slack et al., 1985; Fig. 2). The precise and predicted time in the commencement of oil deposition has been commented on by many workers. Such synchronization probably indicates the involvement of some

TABLE IV
Oil Content Of Some Common Oil Sources[a]

Source	Oil Content (%)
Copra	66
Corn	5
Cottonseed	19
Palm, fruit	48
Palm, kernel	48
Peanut	48
Rapeseed	43
Safflower	33
Soybean	19
Sunflower	40

[a]Adapted from Gunstone and Norris (1983), with permission of Pergamon Press, Oxford.

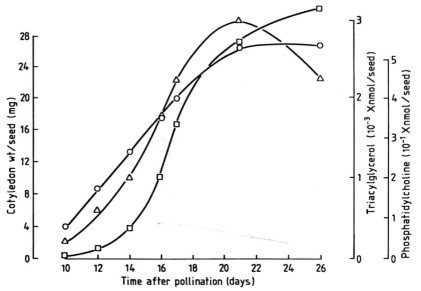

Fig. 2. Triacylglycerol (□) and phosphatidylcholine (△) accumulation in the developing cotyledons (○) of safflower. [From Slack *et al.* (1985). Reproduced with permission from Elsevier Science Publishers.]

hormonal regulation in the development of the seed. The fact that active oil deposition largely occurs only over a short period in the development of the seed is important for biochemical studies on the synthesis of triacylglycerols, and it requires good timing to obtain tissue with adequate enzyme activity. With some oilseeds, such as soybean and linseed, it is also beneficial to select the seed material at a stage which is as early as possible in the phase of oil deposition, since these seeds rapidly accumulate some unknown but potent inhibitors of the enzymes involved in lipid synthesis.

C. Oil Body Formation

1. Introduction

The most notable feature of cells from seeds rich in triacylglycerols is the appearance of characteristic oil storage structures. Generally these bodies are spherical in shape in early and intermediate stages of seed development but become tightly compressed together in the oil-rich seed as maturity is reached and dehydration occurs. The biogenesis of the oil body has been, and still is, a contentious area of research, as evidenced by the diverse nomenclature that has developed in the literature to describe "oil-rich" structures in plant cells (e.g., spherosomes, oleosomes, lipid bodies, oil bodies, lipid droplets, lipid vesicles). Associated with this has been a number of suggestions as to where in the cell the oil bodies originate as well as conflicting reports on their fundamental structure. The term oil body appears to be synonymous with the rest of the nomenclature and should be used to describe all the "oil-rich" inclusions found particularly in oilseeds. There still appears to be some confusion, however, over the distinction between oil bodies and spherosomes. In the early work on the origin of oil bodies in the cytoplasm of rape (*Brassica napus*) and mustard (*Sinapsis alba*) cotyledon cells, Frey-Wyssling *et al.* (1963) suggested that spherosomes originated as vesicles from the endoplasmic reticulum and that, as they accumulated neutral lipids, they developed into oil bodies. Yatsu and co-workers (1971; Yatsu and Jacks, 1972) concluded that spherosomes, however, were the same as "oil droplets." In the previous volume to this, Gurr (1980) stated that "spherosomes and oil bodies are separate entities with distinct functions and one does not develop into the other." He also suggested biochemical and functional differences between the two. The distinctions made by Gurr, however, do not appear to be tenable at the moment, and the subsequent work of Wanner and co-workers (Wanner and Theimer, 1978; Wanner *et al.*, 1981) implies that they perhaps should be considered as one and the same.

2. Structure and Ontogeny

During seed development the oil bodies differ little in size, and various reports based on a number of plant species give them a mean diameter of slightly greater than 1 μm (Slack *et al.*, 1980). The spherical oil bodies in the cell do not coalesce, and their surfaces appear to be stable. This has led to the belief that the oil body is delimited by a discrete boundary between the triacylglycerol lipid phase and the cytosol. At various times it has been claimed that this boundary represents a typical unit membrane (Frey-Wyssling *et al.*, 1963; Mollenhauer and Totten, 1971), a half-unit membrane (Yatsu and Jacks, 1972; Wanner and Theimer, 1978; Wanner *et al.*, 1981), or no membrane at all (Harwood *et al.*, 1971; Rest and Vaughan, 1972). Bergfeld *et al.* (1978), in an interesting study with developing mustard cotyledons, concluded that "nascent" oil bodies first arose in the cytoplasm close to plastids and that they lacked, at this stage, any detectable boundary structures. Later in development they became encircled by a cisterna of rough endoplasmic reticulum and an osmiophyllic coat of about ~3 nm thickness became detectable at the lipid–water boundary. Kleinig *et al.*(1978) also concluded, in studies on oil body development in suspension cultures of carrot (*Daucus carota*), that the boundary between the lipid phase and the cytosol was not built up by a true phospholipid membrane and that the phospholipid content of the oil body was insufficient to account for even a half-unit membrane. Freeze-etched electron microscopy studies (Slack *et al.*, 1980) have demonstrated that isolated oil bodies from developing safflower (*Carthamus tinctorius*) cotyledons are bounded by "some form of membrane" with a particle-free outer surface. It is therefore well established that some form of interfacial structure is present at the lipid–cytoplasm interphase. Its homology, however, with a unit membrane or part of a unit membrane is still problematical. In this context the endoplasmic reticulum half-unit membrane hypothesis has recently been revived by Wanner and co-workers (Wanner and Theimer, 1978; Wanner et al., 1981), and they have produced attractive models for the development of oil bodies in a number of species (Fig. 3), which go a long way to explain many of the ontological events perhaps being described by other authors. In general they consider that the uncharged triacylglycerols are not expelled from the lipid-producing membranes but form an accumulation between the phospholipid layers which grows into a lipid body that is finally covered by a half-unit membrane derived from the endoplasmic reticulum.

Analyses of isolated oil bodies from a number of plant sources have shown that they appear to contain, in addition to triacylglycerol, phospholipids and protein (Yatsu et al., 1971; Gurr et al., 1974). Carefully purified preparations of the oil bodies from linseed and safflower (Slack et al., 1980) have an essentially constant composition of 2.3% protein, 0.7% polar lipid, and 97% neutral lipid. The polar lipid and protein content of the oil body would be sufficient, therefore, to constitute at least a mono molecular, half-unit membrane. Analyses of the proteins in the oil body by SDS electrophoresis indicated the presence of at least four peptides, including two characteristic major ones with an apparent molecular weight of 17,500 and 15,700 (Bergfeld et al., 1978; Slack et al., 1980). The spectrum of

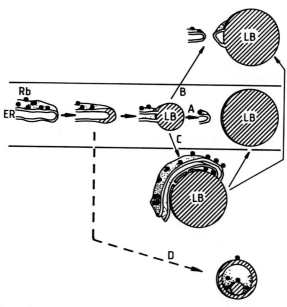

Fig. 3. Hypotheses for the development of lipid bodies (LB) from the endoplasmic reticulum (ER). (A) Lipid bodies pinch off the ER without a remaining piece of ER membrane. (B) A membranous appendix derived from the ER is left with the lipid body. (C) The ER strand develops multiple contacts with the nascent lipid body. (D) Vesicles of lipid-rich membranes develop from ER vesicles by continued lipid incorporation into the membrane lipid layer. [From Wanner et al. (1981). Reproduced with permission from Springer-Verlag.]

polypeptides from oil body preparations of mustard (*S. alba*) differed markedly from that of the peptides in the endoplasmic reticulum (Bergfeld *et al.*, 1978), and this observation was taken as evidence to suggest that the oil body becomes bounded by particular and specific proteins in an event which occurs soon after oil deposition. The fact that the peptides of the oil body and the endoplasmic reticulum are different has been explained (Wanner *et al.*, 1981) on the basis of endomembrane differentiation. It should be taken into account, however, that endomembrane differentiation would presumably require processing by the Golgi complex (Tartakoff, 1983), and in the theories on oil body development proposed by Wanner *et al.* (1981) the oil body coat is derived directly from, and at an early stage has continuity with, the membranes of the endoplasmic reticulum. Further work is obviously required on the origins of the boundary proteins in order to interpret more fully the biogenesis of the oil body in developing cotyledons.

3.   *Oil Body Formation* In Vitro

We have developed a novel system for the study of oil body formation *in vitro* and its relationship to the biosynthesis of triacylglycerols (Stobart *et al.*, 1986). Microsomal preparations, prepared from the developing cotyledons of safflower, are, under optimum conditions, capable of high rates of triacylglycerol formation. We have observed initial rates of triacylglycerol synthesis in the order of 45 nmol/mg protein per hour (Stobart and Stymne, 1985a) when microsomal preparations of safflower were incubated with acyl-CoA and glycerol phosphate. Triacylglycerol synthesis in *in vitro* membrane preparations from seeds has been found by other workers and Rochester and Bishop (1984) reported rates of ~0.8 nmol/mg protein per hour for microsomes from developing sunflower. Sunflower preparations are, however, a particularly active system for the synthesis of triacylglycerol (Stymne and Stobart, 1984a) and the observations of Rochester and Bishop (1984) are probably due to limiting substrate in the reaction mixtures.

When microsomal preparations from safflower are incubated with glycerol phosphate, Mg^{2+}, and supplied with linoleoyl-CoA, in low concentrations and at regular intervals to reduce membrane disruption and the accumulation of phosphatidic acid, they are capable of sustained and high rates of triacylglycerol synthesis (Fig. 4). The rates of triacylglycerol formation observed *in vitro* (42 nmol/100 nmol phosphatidylcholine per hour) in these experiments approach those measured *in vivo* at the most active phase of oil deposition (65 nmol/100 nmol phosphatidylcholine per hour) as depicted in Fig. 2 (Slack *et al.*, 1985). Negative staining and electron microscopy of the microsomal preparations show the presence of membrane vesicles and fragments and very few, if any, oil bodies (Fig. 5A). Within 15 min incubation with glycerol phosphate and acyl-CoA the membranes began to accumulate small lipid droplets (Fig 5B). With further incubation a large population of oil droplets is present in the reaction mixture (Fig. 5C). The oil droplets which are synthesized *in vitro* have a diameter of 20–40 nm and are considerably smaller than those observed *in vivo* in oil synthesizing cells. It is interesting to note that the reaction mixtures began to appear "milky" within 1 h of

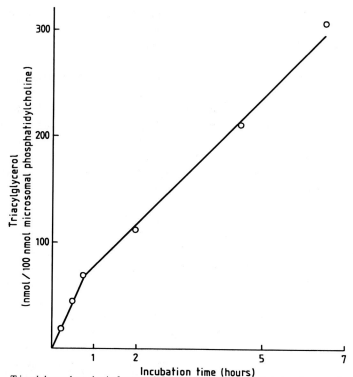

Fig. 4.   Triacylglycerol synthesis from acyl-CoA and glycerol phosphate in microsomal membranes from the developing cotyledons of safflower. [From Stobart *et al.* (1986). Reproduced with permission from Springer-Verlag.]

incubation and with prolonged incubation took on a thick white souplike appearance (Fig. 6A). On centrifugation the majority of the synthezised neutral lipids accumulated as a white, floating fat pad at the surface (Fig. 6B).

It is of interest to compare oil body formation *in vitro* with observations on *in vivo* systems. The electron micrographs of the oil bodies formed *in vitro* show no clear evidence for a delimiting membrane, half-unit or otherwise. In fact, the oil bodies which are formed, either on the surface and/or in the membrane of the microsomal vesicles, appear to be released as "naked" droplets. There is no coalescing of the oil droplets in the incubation mixture, and it is probable that they are stabilized relatively quickly after their release from the membrane, perhaps by the availability of the bovine serum albumin which is present in the reaction mixture. In the context of these observations *in vitro*, the findings of Bergfeld *et al.* (1978) on the initial formation of nascent oil bodies *in vivo* take on some significance. If one can extrapolate our observations with those of *in vivo* studies, with all the reservations necessary in such a comparison, we would consider that the triacylglycerol-rich bodies are synthesized on or in the membranes of the endoplasmic reticulum and that they are released as almost "naked" entities. The release may occur through the

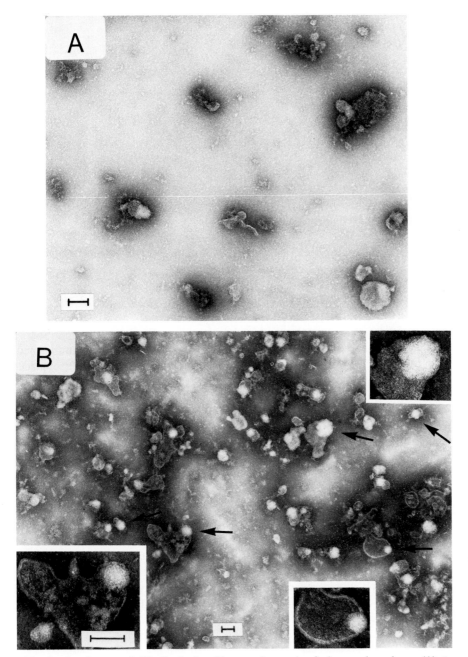

Fig. 5. Development of oil bodies *in vitro*. Electron micrographs of microsomal membranes (A) at the start of incubation with acyl-CoA and glycerol phosphate. ×56,250; (B) after 15 min incubation. ×37,500; insets ×93,750; (C) after 2 h incubation. ×37,500; bar = 100 nm. [From Stobart *et al.* (1986). Reproduced with permission from Springer-Verlag.]

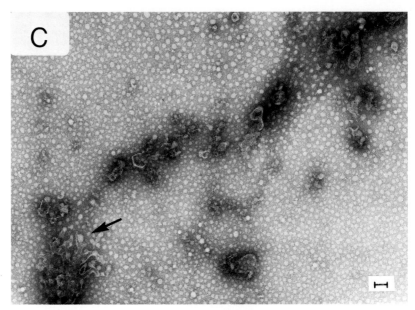

Fig. 5. *(Continued)*

outer membrane by this layer being put under strain, stretching and parting, and, after the oil body has passed through, perhaps reuniting (see also Fig. 3, sequence A). The protein boundary layer of the oil body would then have to arise as a postendoplasmic membrane event, perhaps as described by Bergfeld *et al.*, (1978). In the *in vitro* system described, oil droplet formation continues unabated for many hours with no, or very little, loss in synthesizing activity. If the microsomal membrane was giving rise to a boundary layer around the vesicle, one might anticipate that the rate of oil deposition would slow considerably as more and more membrane lipid–protein is utilized. That this does not occur strongly implies that the outer membrane of the endoplasmic reticulum is not involved, certainly during development *in vitro* , in forming a delimiting boundary layer around the oil body.

In vitro model systems are now available, therefore, in which all the enzymes for the biosynthesis of triacylglycerol, from glycerol phosphate and acyl-CoA, are present in one membrane, and in which they have specific activities that are compatible with the rates of oil accumulation observed in the developing seed cotyledon. Such an experimental model is not far divorced from the reconstituted systems which were recently speculated upon by Slack and Browse (1984) as being valuable for future studies on triacylglycerol biosynthesis.

4. *Biosynthetic Activity*

The activity of several enzymes associated with lipid metabolism have been attributed to the oil bodies present in developing seeds. In earlier work, the fat

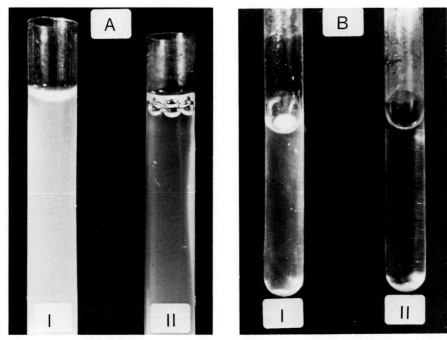

Fig. 6. Oil accumulation in incubations with safflower microsomes. (A) I, 4 h incubation; II, control (B) I, 4 h incubation; II, control, after centrifugation. [From Stobart *et al.* (1986). Reproduced with permission from Springer-Verlag.]

fraction, from maturing castor bean endosperm, was found to synthesize fatty acids from acetate and malonyl-CoA (Harwood *et al.*, 1971) and was capable of synthesizing triacylglycerol from acyl-CoA and glycerol phosphate (Harwood and Stumpf, 1972). Since it is now generally accepted that the sole site of *de novo* fatty acid synthesis in the plant cell is the plastid present in seeds (Yamada and Usami, 1975) and other plant tissues (Weaire and Kekwick, 1975; Ohlrogge *et al.*, 1979), it would appear that the oil body preparations used in the earlier studies were severely contaminated with other cell membranes. Oil bodies purified from carrot suspension cultures (Kleinig *et al.*, 1978), safflower, and linseed (Slack *et al.*, 1980) were particularly deficient in enzymes associated with the synthesis of fatty acids and triacylglycerols. On the other hand, oil bodies isolated from seeds of *Crambe abyssinica* (Gurr *et al.*, 1974; Appleby *et al.*, 1974) were considered to be the site of the elongation of oleoyl-CoA to erucate. Unfortunately it has proved impossible, to date, to obtain active microsomal fractions from the developing seeds of plants such as rape, and hence the relationship of triacylglycerol synthesis in the endoplasmic reticulum to any modifying reactions in the oil body awaits further clarification.

III. BIOSYNTHESIS

A. *De Novo* Biosynthesis of Fatty Acids

The two-carbon source for the biosynthesis of fatty acids *de novo*, as well as the biosynthesis of the fatty acids which are utilized in triacylglycerol production, is dealt with in detail elsewhere in this volume. The ultimate carbon source from which the whole of the triacylglycerol molecule is derived is the sucrose, which is synthesized by the photosynthetic apparatus of the plant and which is then transported from the leaves, through the phloem, to the seed tissues in which oil deposition is to occur. In developing castor bean (*Ricinus communis*) endosperm the plastids possess all the enzymes of a glycolytic sequence, and these exhibit specific activities which are sufficient to account for the observed rates of *de novo* fatty acid synthesis (Simcox *et al.*, 1977; Dennis and Miernyk, 1982). The plastids also appear to be self-sufficient in the production of reducing equivalents and ATP for the synthesis of fatty acids to the C_{16} and C_{18} level (Slack and Browse, 1984). Recent evidence on the intracellular location of the fatty acid synthetase complex in plant cells strongly indicates that it is confined to the chloroplast of photosynthetic tissues and to the proplastids present in the nonphotosynthetic parts of the plant (Weaire and Kekwick, 1975; Vick and Beevers, 1978; Ohlrogge *et al.*, 1979; Zilkey and Canvin, 1972; Yamada and Usami, 1975; Browse and Slack, 1985). The end product of the *de novo* synthesis is, in most plants, palmitate, which can then be elongated further to stearate. The stearate, in its turn, is rapidly desaturated to yield oleate. The elongation of palmitate and the desaturation of stearate in the plastid requires the fatty acid to be esterified to acyl-carrier-protein (ACP). In oil storage tissues the palmitate, stearate, and oleate are transferred out from the plastid to the cytoplasm, where they are made available for further metabolic events as their CoA esters. The sequence of reactions and the transport mechanism involved in the transfer of a fatty acid from its ACP ester in the plastid to the CoA ester in an extraplastidic compartment has not been completely resolved. However, an active acyl-ACP hydrolase, with a broad substrate specificity, is present in the plastid (Ohlrogge *et al.*, 1978) and may participate in liberating the free fatty acid for movement through the membranes of the plastid envelope. The ligation of CoA to the free fatty acid may be catalyzed by the action of an acyl-CoA synthetase which is located in or on the outer membrane of the plastid (Roughan and Slack, 1977; Joyard and Douce, 1977). The mode of transport of the fatty acids through the plastid membranes, however, is unknown. It is of interest, therefore, that carnitine and carnitine long-chain acyltransferase enzymes are present in the plastid (Thomas *et al.*, 1983).

B. Origin of the Glycerol Backbone

The initial steps in the synthesis of triacylglycerol in plants involve a sequential acylation of *sn*-glycerol 3-phosphate (Section III,C). In animals the synthesis of *sn*-glycerol 3-phosphate for triacylglycerol formation is accomplished by a glycerol-3-phosphate dehydrogenase NAD$^+$ (E.C. 1.1.1.8), which catalyzes the reduction of dihydroxyacetone phosphate (Lin, 1977). It is only recently that this particular enzyme has received attention in plants. Previously, an alternative route for glycerol phosphate formation was thought to occur which involved the direct phosphorylation of glycerol through the action of a glycerol kinase (Gurr, 1980). The glycerol kinase, however, probably plays only a minor role in the synthesis of glycerol phosphate, since the activity of this enzyme is particularly low in oil-synthesizing plant tissues (Gurr *et al.*, 1974; Barron and Stumpf, 1962). It is more likely that the major contribution to glycerol phosphate is through, as in animals, the dihydroxyacetone oxidoreductase. The activity and cytosolic location of this enzyme in oil-rich tissue can account for the observed rates of triacylglycerol deposition *in vivo* (Finlayson and Dennis, 1980). The dihydroxyacetone phosphate that is necessary for glycerol phosphate formation must be derived through the activity of glycolytic enzymes which operate in the cytoplasm (Dennis and Miernyk, 1982). The plastid, as mentioned previously, appears to contain all the enzymes of a glycolytic sequence in sufficient quantity to account for the supply of carbon for *de novo* fatty acid synthesis. This would imply that some form of regulation must exist for the channeling of the sugar moieties to the plastids for the synthesis of the acyl components that are required for triacylglycerol and for the supply of substrate in the cytoplasm for the subsequent formation of the glycerol backbone (Fig. 7).

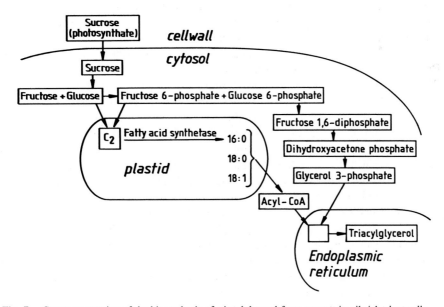

Fig. 7. Compartmentation of the biosynthesis of triacylglycerol from sucrose in oil-rich plant cells.

C.   Formation of Phosphatidic Acid

sn-Glycerol 3-phosphate + acyl-CoA → sn-1 acyl-lysophosphatidic acid
sn-1 acyl-lysophosphatidic acid + acyl-CoA → phosphatidic acid

The glycerol phosphate pathway, or the so-called Kennedy pathway (Kennedy, 1961), was found to operate in the synthesis of triacylglycerols in plants as early as 1962 (Barron and Stumpf, 1962). The individual steps in that pathway are depicted in Fig. 11. The formation of phosphatidic acid (phosphatidate; sn-1,2 diacylglycerol 3-phosphate) from glycerol phosphate is a two-step event and requires the participation of two distinct enzymes (Yamashita et al., 1972; Tamai and Lands, 1974). The first enzyme, a glycerophosphate acyltransferase (E.C. 2.3.1.15), brings about the acylation of sn-glycerol 3-phosphate at the sn-1 position to form a lysophosphatidic acid. The lysophosphatidic acid is the substrate for the second enzyme, 1-acylglycerol-3-phosphate acyltransferase (E.C. 2.3.1.51), which completes the acylation at position sn-2 with the formation of phosphatidic acid. The acylation of glycerol phosphate with acyl groups from acyl-CoA to form phosphatidic acid occurs in cell-free preparations from oil-rich plant tissues (Barron and Stumpf, 1962; Vick and Beevers, 1977; Porra, 1979) and in microsomal membrane fractions prepared from developing cotyledons of safflower (Stobart et al., 1983; Stymne et al., 1983; Ichihara, 1984), sunflower (Stymne and Stobart, 1984a), and linseed (Stymne and Stobart, 1985a),. The activity of the enzymes which synthesize phosphatidic acid from glycerol phosphate and acyl-CoA in vitro is sufficient to account fully for the rate of triacylglycerol accumulation which is observed in vivo.

The acyl selectivity in the formation of lysophosphatidic acid in safflower microsomes shows a particular preference for palmitate (Ichihara, 1984; Griffiths et al., 1985). When an equimolar mixture of acyl-CoA species was incubated with safflower microsomes and glycerol phosphate, the selectivity in the acylation of the sn-1 position was in the order palmitate > oleate = linoleate > stearate (Ichihara, 1984). On the other hand, Griffiths et al. (1985) observed that linoleate was incorporated into the sn-1 position of sn-glycerol 3-phosphate at half the rate of oleate and somewhat lower than that of stearate. The apparent discrepancy in these acyl-selectivity studies is due to the fact that Ichihara (1984) analyzed the incorporation of [14C]glycerol phosphate into different molecular species of lysophosphatidic acid in safflower microsomes which were provided with unlabeled acyl-CoA. Griffiths et al. (1985), however, measured the radioactive fatty acids incorporated into position sn-1 of sn-phosphatidic acid from [14C]acyl-CoA in the presence of glycerol phosphate. In safflower microsomes, as well as in preparations from other linoleate-rich oilseeds, there is a rapid exchange of acyl groups between acyl-CoA and position sn-2 of sn-phosphatidylcholine (Section III,F,2). The acyl exchange results in the rapid dilution of radioactive linoleate in the acyl-CoA pool with unlabeled linoleate derived from phosphatidylcholine. Hence, the observed incorporation of [14C]linoleate from labeled acyl-CoA into phosphatidic acid will always be an underestimate, as pointed out by Griffiths et al. (1985). The relative selectivity for linoleate, therefore, in the acylation of the sn-1 position of sn-glycerol

3-phosphate, as reported by Ichihara (1984), will be the more correct. These observations illustrate the importance of being aware of any competing reactions which effectively alter the specific radioactivity of a substrate before reaching definitive conclusions regarding the properties of a particular reaction. The positional acyl selectivity in the acylation of glycerol phosphate for saturated and unsaturated fatty acids has now been demonstrated in *in vivo* experiments with developing cotyledons of safflower (Griffiths, 1986).

Ichihara (1984) reported that lysophosphatidic acid was the major product in incubations of membrane fractions from safflower with acyl-CoA and glycerol phosphate. We find, however, that the lysophosphatidic acid only accumulates in incubations when saturated acyl-CoAs are presented alone to the microsomes in the presence of glycerol phosphate. If the microsomes, on the other hand, are provided with linoleoyl-CoA and glycerol phosphate, then both *sn*-1 and *sn*-2 positions of *sn*-glycerol 3-phosphate are acylated almost simultaneously to yield phosphatidic acid (Griffiths *et al.*, 1985).

The acylation of position *sn*-2 in *sn*-glycerol 3-phosphate shows a strong selectivity for linoleate over oleate and the total exclusion of the saturated fatty acids, palmitate and stearate (Griffiths *et al.*, 1985).If microsomes are presented with saturated [^{14}C]acyl-CoAs, then the formation of phosphatidic acid is almost totally dependent on the acylation of the *sn*-2 position with the unlabeled linoleate which arises through acyl exchange between the saturated fatty acids in acyl-CoA and the predominant fatty acid (linoleate) in position *sn*-2 of *sn*-phosphatidylcholine (Griffiths *et al.*, 1985).

The preference for palmitate in the acylation of the *sn*-1 position of *sn*-glycerol 3-phosphate and the total exclusion of saturated fatty acids from position *sn*-2 can fully account, therefore, for the nonrandom distribution of saturated and unsaturated fatty acids in positions *sn*-1 and *sn*-2 of the *sn*-triacylglycerols in safflower (Table III).

Microsomal preparations from the developing cotyledons of linseed also readily synthesize phosphatidic acid from acyl-CoA and glycerol phosphate (Stymne and Stobart, 1985a). The acyl selectivity was in the order linoleate > oleate > linolenate. The linolenate was utilized in the acylation of glycerol phosphate at only half the rate of linoleate. This observation appears at first rather unexpected, since the ratio of linolenate to linoleate in the linseed triacylglycerols was 3.5. These results, however, may indicate that most of the linolenate in linseed is made available for triacylglycerol synthesis after the formation of phosphatidic acid. A similar explanation was also suggested in *in vivo* studies with developing cotyledons of linseed (Slack *et al.*, 1983).

Little is known about the properties of the glycerol phosphate acyltransferase system in plants, although the enzyme(s) from castor bean endosperm (Vick and Beevers, 1977) was reported to have an absolute requirement for divalent cations. No such requirement, however, is necessary for the enzymes in microsomes from the cotyledons of safflower (Ichihara, 1984; Griffiths *et al.*, 1985; Stobart and Stymne, 1985b). The subcellular location of the glycerol phosphate acylating enzymes in oil-rich seed tissue has not been fully established, although it is clearly extraplastidic, with a particularly high specific activity residing in a membrane fraction which sediments after high-speed centrifugation (Vick and Beevers, 1977; Ichihara, 1984; Griffiths *et al.*, 1985). It is therefore a reasonable assumption that the enzymes are localized in the endoplasmic reticulum. Extraplastidic glycerol phosphate-acylating enzymes have been found in the endoplasmic reticulum membrane fraction isolated from leaves of spinach (Frentzen *et al.*, 1984). The enzymes present in the endoplasmic reticulum of the leaf appear to be similar in their acylating properties to those found in oil-rich tissues and direct saturated and unsaturated fatty acids to positions *sn*-1 and *sn*-2 of *sn*-glycerol 3-phosphate, respectively. The positional specificity for acyl-CoA of the extraplastidic glycerol phosphate-acylating enzymes is distinctly different from those enzymes present in the chloroplast, where saturated fatty acids are utilized, particularly in the acylation of the *sn*-2 position (Bertrams and Heinz, 1981; Frentzen *et al.*, 1983).

D. Diacylglycerol Production

$$\text{Phosphatidic acid} \rightarrow \text{diacylglycerol} + P_i$$

The enzyme phosphatidase (phosphatidate phosphatase, E.C. 3.1.3.4) catalyzes the cleavage of phosphatidic acid to yield diacylglycerol and inorganic phosphate. The enzyme is active in cell-free particulate fractions from plant tissues which are synthesizing triacylglycerols (Barron and Stumpf, 1962; Stymne *et al.*, 1983; Ichihara, 1984). The properties of the enzyme have only recently been studied (Griffiths *et al.*, 1985). The activity of the enzyme in microsomal preparations from developing safflower cotyledons was enhanced by Mg^{2+} and substantially inhibited by EDTA. The inhibition by EDTA, however, was completely alleviated by an excess of Mg^{2+}. In this respect, the phosphatidase in oilseeds resembles the enzyme in animal tissues (Call and Williams, 1973; Sturton and Brindley, 1980) and is different from that described for the enzyme which is located in the inner membrane of the chloroplast (Block *et al.*, 1983).

The phosphatidase in oilseeds and avocado mesocarp is an extraplastidic membrane-bound enzyme and is located in the same microsomal fraction that contains all the other enzymes involved in the synthesis of triacylglycerol from glycerol phosphate and acyl-CoA (Barron and Stumpf, 1962; Stymne *et al.*, 1983; Ichihara, 1984). In particulate preparations from castor bean endosperm the major activity of the phosphatidase was coincident with marker enzymes for the endoplasmic reticulum (Moore *et al.*, 1973).

The phosphatidase step in triacylglycerol synthesis in animal tissues is considered to be rate-limiting (Brindley, 1978). In microsomal preparations from the developing cotyledons of safflower and sunflower the limiting step in the formation of triacylglycerol from glycerol phosphate and acyl-CoA was the production of diacylglycerol from phosphatidic acid (S. Stymne and A. K. Stobart, unpublished observations). This indicates that, as in animals, the phosphatidase enzyme may play an important regulatory role and control the flow of intermediates toward triacylglycerol.

E. Acylation of Diacylglycerol

Diacylglycerol + acyl-CoA → triacylglycerol

The ultimate step in the biosynthesis of the triacylglycerols is the acylation of the 1,2-diacylglycerol at position sn-3 and this is catalyzed by a diacylglycerol acyltransferase (E.C. 2.3.1.20). Although the enzyme is present in the microsomal fractions prepared from plant tisssue (Barron and Stumpf, 1962; Shine et al., 1976; Stymne et al., 1983), its properties have not been studied in any detail until recently (Ichihara and Noda, 1982; Cao and Huang, 1986). The enzyme also appears to be present in the envelope of chloroplasts prepared from the leaves of spinach (Martin and Wilson, 1984). The available biochemical evidence, however, does not indicate any significant involvement of plastids in the assembly of triacylglycerol in oil-rich seeds (Cao and Huang, 1986).

The stereospecific distribution of the fatty acids in the triacylglycerol from commercially important oil plants (Table III) indicates that the acyl selectivity in the acylation of the sn-3 position of the glycerol backbone is similar to that of position sn-1 and is usually characterized by the presence of substantial saturated fatty acid. Ichihara and Noda (1982) found, in particulate fractions from the developing cotyledons of safflower, that the acylation of added emulsified diacylglycerol occurred with no preference for different acyl species of diacylglycerol or acyl-CoA. It should be noted, however, that the activity of the enzyme investigated by Ichihara and Noda (1982) was almost a magnitude lower than that found later in safflower microsomes, which would catalyze the whole sequence of reactions from glycerol phosphate to triacylglycerol (Stobart and Stymne, 1985a).

The acylation of diacylglycerol is the only reaction which is unique to the triacylglycerol-biosynthetic pathway and it is responsible for a third of the fatty acids present in the plant oil. Its importance, therefore, in regulating the acyl quality of the triacylglycerol should not be underrated and it requires further investigation in oilseeds.

F. Involvement of Phosphatidylcholine in Triacylglycerol Synthesis

1. Synthesis of C_{18} Polyunsaturated Fatty Acids

The C_{18} mono-unsaturated and polyunsaturated fatty acid are generally the most prevalent in seed triacylglycerols, where they form by far the most important nutritional constituents. Much research, therefore, has been directed toward understanding the biosynthesis of C_{18} polyunsaturated fatty acids and its relationship to triacylglycerol production with the view to complement plant breeding programs and enable the development of new varieties and strains which will produce oils of a desired acyl quality.

In 1969 Gurr and co-workers presented evidence to show that, in a particulate fraction from *Chlorella,* the oleate esterified to phosphatidylcholine could undergo desaturation to linoleate (Gurr *et al.,* 1969). *In vivo* radiolabeling studies with developing cotyledons of oilseeds (Dybing and Craig, 1970; Slack *et al.,* 1978; Wilson *et al.,* 1980) were also consistent with the view that oleate entered phosphatidylcholine and was there desaturated to linoleate. The linoleate in the phosphatidylcholine was then, by some unknown mechanism, incorporated into diacyl- and triacylglycerols. The desaturation of oleate in cell-free particulate fractions from safflower seed was first demonstrated by McMahon and Stumpf (1964). Some properties of the phosphatidylcholine desaturase system were also studied by Vijay and Stumpf (1971, 1972). These authors, however, used an assay method for determining acyl-CoA and oxygen esters which was not sufficiently discriminating and incorrectly concluded that acyl-CoA was the immediate substrate for the desaturase enzyme. At a later date the desaturase in safflower was reinvestigated, and convincing evidence was obtained to show that oleoyl-phosphatidylcholine was the true substrate for the desaturation of oleate to linoleate (Stymne and Appelqvist, 1978; Slack *et al.,* 1979). In these experiments it was found that oleate from oleoyl-CoA was first transferred to position *sn*-2 of *sn*-phosphatidylcholine and there desaturated to linoleate. These observations are particularly well demonstrated in an experiment with microsomal membranes from the developing cotyledons of soybean (Fig. 8). Microsomes were incubated with [^{14}C]oleoyl-CoA under desaturating conditions (plus NADH), and at intervals the radioactivity present in the oleate and linoleate in the acyl-CoA and the complex lipids was determined. The results (Fig. 8) clearly show a substrate–product relationship in the formation of linoleoyl-phosphatidylcholine from oleoyl-phosphatidylcholine. The desaturation of oleate in phosphatidylcholine continues after essentially all the oleoyl-CoA in the reaction mixture has been utilized.

Exogenous oleoyl-phosphatidylcholine was also desaturated to some extent by oilseed microsomes, and here the synthesis of linoleate was found to occur at both *sn*-1 and *sn*-2 positions of *sn*-phosphatidylcholine (Slack *et al.,* 1979). The desaturation of oleate at both the *sn*-1 and *sn*-2 positions of *sn*-phosphatidylcholine was also demonstrated in safflower microsomal phosphatidylcholine that had been *in situ* labeled with [^{14}C]oleate from glycerol phosphate and oleoyl-CoA through the glycerol phosphate pathway (Stobart and Stymne, 1985a).

The desaturation of oleate in microsomes from oilseeds shows a complete dependence on the addition of NADH or other reducing equivalents (Vijay and Stumpf, 1972; Stymne and Appelqvist, 1978), and this enables the conversion of oleoyl-phosphatidylcholine to linoleoyl-phosphatidylcholine to be readily demonstrated. Microsomes from developing cotyledons of safflower can be incubated with [^{14}C]oleoyl-CoA under nondesaturating conditions (i.e., minus reductant) until all the [^{14}C]oleoyl-CoA is metabolized. A major part of the labeled oleate is then found in position *sn*-2 of *sn*-phosphatidylcholine. Upon the subsequent addition of NADH the desaturation of the [^{14}C]oleate to [^{14}C]linoleate in the phosphatidylcholine can be followed. A radioactive gas–liquid chromatogram which illustrates such an

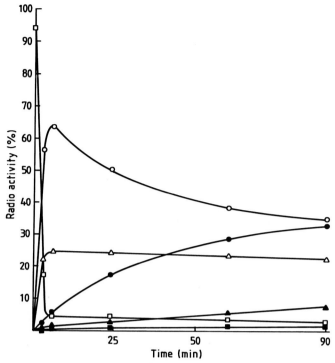

Fig. 8. Radioactive oleate and linoleate in acyl-CoA and complex lipids in soy microsomes incubated for various times, under desaturating conditions, with [14C]oleoyl-CoA. Symbols used: (○, ●) oleate and linoleate in phosphatidylcholine, respectively; (□, ■) oleate and linoleate in acyl-CoA, respectively; (△, ▲) oleate and linoleate in other lipids, respectively (Stymne, 1980).

experiment is given in Fig. 9. The chromatogram in Fig. 9A is the mass and radioactive distribution of 14C in the fatty acids in phosphatidylcholine just prior to the addition of NADH. In Fig. 9B the chromatogram gives the mass and radioactivity in the same lipid 8 min after the addition of the reductant. No significant change has taken place in the total radioactivity in phosphatidylcholine during the 8-min. incubation period. Note also the decrease in mass of oleate together with the efficient conversion of [14C]oleate to [14C]linoleate. The de-saturation of oleoyl-phosphatidylcholine also occurs in microsomal preparations from developing cotyledons of soy and sunflower (Stymne, 1980; Rochester and Bishop, 1982; Stymne and Stobart, 1984a).

Kinetic studies on the desaturation of oleate in phosphatidylcholine show that the complete conversion of oleate to linoleate is never achieved in the microsomal preparations. The rate of desaturation is dependent on the concentration of oleate in phosphatidylcholine and rapidly ceases when the oleate approaches amounts that are equivalent to those found in the endogenous phosphatidylcholine in the oilseed (Stymne *et al.*, 1983; Stymne and Stobart, 1986a). This may be due to the affinity

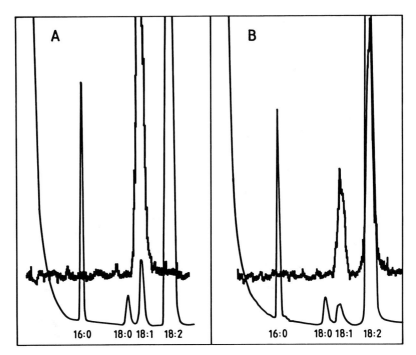

Fig. 9.   Desaturation of [^{14}C]oleoyl-phosphatidylcholine in microsomes of safflower.   Radio active
gas–liquid chromatograms of the fatty acids in phosphatidylcholine before (A) and 8 min after (B) the
addition of NADH. Upper and lower traces represent the radioactivity and the mass of the fatty acids,
respectively.

of the oleoyl desaturase for its substrate and/or a product inhibition by the linoleate
in phosphatidylcholine. In either case, these properties of the oleate desaturase
could govern the quantity of unsaturated fatty acids found in phosphatidylcholine in
oilseeds of different species and thus be one of the key factors in regulating the ratio
of oleate to linoleate in the final triacylglycerol. This suggestion is also supported
by the view that virtually all the phosphatidylcholine in the tissue of a developing
oilseed is probably present as one active pool and is therefore all available to
participate in the desaturation of oleate and the synthesis of linoleate (Slack, 1983).

   In vivo radio labeling experiments also indicate that phosphatidylcholine is
involved in the further desaturation of linoleate to α-linolenate in linolenate-rich seeds
(Slack *et al.*, 1978). Little success, however, has been achieved so far in obtaining
cell-free fractions from such seeds, which will consistently synthesize linolenate.
The data that are available for homogenates of the developing cotyledons of soy
(Stymne and Appelqvist, 1980) and microsomal preparations from linseed (Browse
and Slack, 1981) indicate that linoleate, as in the case of oleate, is desaturated while
esterified to phosphatidylcholine. The linoleate desaturase enzyme is extremely
labile in cell-free preparations of soy and differs in this respect from the oleate

desaturase, which is readily demonstrated in microsomal fractions from that tissue (Stymne, 1980; Stymne and Glad, 1981). The linoleoyl desaturase in a microsomal preparation from linseed was stabilized, to some degree, by the presence of catalase in the extraction and assay media (Browse and Slack, 1981), although it still exhibited a highly temperamental behavior which has prohibited any detailed investigation (J. A. Browse, personal communication).

The oleate and linoleate desaturase enzymes appear to have many of the properties of other eukaryotic fatty acid desaturases and require oxygen, NADH, or NADPH, and are inhibited by CN^- but not CO (Vijay and Stumpf, 1972; Stymne, 1980). The oleoyl-phosphatidylcholine desaturase activity in a particulate fraction from oilseeds closely correlates to the activity of enzyme markers for the endoplasmic reticulum (Stymne, 1980), whereas the subcellular site of the linoleate desaturase, although activity was present in microsomal fractions of linseed (Browse and Slack, 1981), is still debatable (Stymne, 1980).

Although the desaturation of oleate and the incorporation of linoleate into the triacylglycerols in oilseeds are now reasonably well understood, the synthesis and metabolism of linolenate require much more intensive study. While it is reasonable to assume, on the basis of the present fragmentary data, that the site of linolenate synthesis and its relationship to triacylglycerol formation are similar to those described for linoleate, this still awaits confirmation.

2. Exchange of Fatty Acids between Acyl-CoA and Phosphatidylcholine

The concept of the synthesis of polyunsaturated C_{18} fatty acids from substrates esterified to the complex lipid, phosphatidylcholine, in oilseeds (Stymne and Appelqvist, 1978; Slack et al., 1979) immediately raised the problems of how did the oleate enter phosphatidylcholine and how were the polyunsaturated fatty acids products made available for triacylglycerol synthesis. In incubations of microsomes from developing oilseeds, with [^{14}C]oleoyl-CoA, the radioactive oleate was transferred to position sn-2 of sn-phosphatidylcholine (Slack et al., 1979, Stobart et al., 1983). It was assumed that this reaction was catalyzed by the activity of the well-documented enzyme, acyl-CoA:lysophosphatidylcholine acyltransferase (E.C. 2.3.1.23) (Stymne and Appelqvist, 1978). If this route, however, was significant for the influx of oleate to phosphatidylcholine, there must be a mechanism which regenerates the acyl acceptor, lysophosphatidylcholine. In animal systems this is generally thought to occur by the action of an intracellular phospholipase A_2 enzyme which specifically removes a fatty acid from position sn-2 of sn-phospholipids (Van den Bosch et al., 1972). To date, however, no such phospholipase has been demonstrated in plant tissues. It was observed that microsomes from developing soybean catalyzed the transfer of [^{14}C]oleate from double-labeled [^{14}C]oleoyl-[^{3}H]CoA to phosphatidylcholine in amounts that were substantially greater than could be accounted for by the preexisting pool of lysophosphatidylcholine (Stymne and Glad, 1981). During the incorporation of [^{14}C]oleate into phosphatidylcholine there was the simultaneous appearance of linoleate and linolenate in the acyl-CoA together with a decrease in the ratio of ^{14}C:^{3}H in this fraction. The results suggested that there was an exchange of acyl groups between the acyl-CoA and the microsomal phosphatidylcholine. The acyl exchange was later confirmed in a number of investigations with safflower (Stymne *et al.*, 1983), sunflower (Stymne and Stobart, 1984a), and linseed (Stymne and Stobart, 1985a).

The acyl exchange in microsomal preparations is illustrated in Fig. 10. In this particular example we have used membranes from developing cotyledons of the High-Oleate variety of safflower. The High-Oleate variety has ~65% oleate and 35% linoleate in position *sn*-2 of microsomal *sn*-phosphatidylcholine. Acyl exchange in such a membrane will also readily utilize linolenoyl-CoA as substrate, a fatty acid that in this case is absent in the endogenous phosphatidylcholine. Acyl exchange, therefore, between linolenoyl-CoA and the oleate and linoleate in

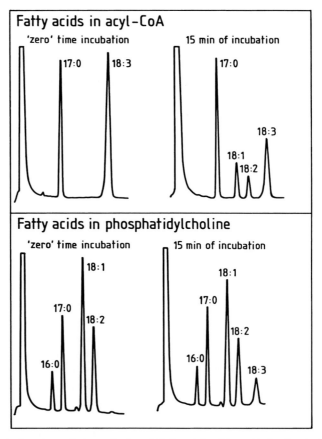

Fig. 10. Acyl exchange between linolenoyl-CoA and the fatty acids in phosphatidylcholine in microsomes of safflower (var. High-Oleate). The figure shows GLC traces of the fatty acids in acyl-CoA and microsomal phosphatidylcholine at zero time (~5 sec) and 15 min after the addition of linolenoyl-CoA. Heptadecanoic acid (17:0) was added as an internal standard in the analytical procedure.

position sn-2 of sn-phosphatidylcholine is readily monitored. Figure 10 shows a typical gas–liquid chromatogram of the fatty acids in acyl-CoA and phosphatidylcholine in the High-Oleate microsomal membranes initially and after 15 min incubation with linolenoyl-CoA. The fatty acid methyl esters, derived from acyl-CoA and phosphatidylcholine, were measured relative to heptadecanoic acid (C17:0), 100 nmol of which was added as an internal standard in the analytical procedure. The chromatograms show clearly that at zero time the linolenate was present only in the acyl-CoA fraction and that phosphatidylcholine was composed largely of oleate and linoleate. After 15 min incubation, however, the acyl-CoA fraction contained appreciable quantities of oleate and linoleate. Linolenate was now present in the phosphatidylcholine. A quantitative analysis of these results shows that some 60 nmol linolenate had entered phosphatidylcholine at the expense of oleate and linoleate. The amount of oleate and linoleate which has appeared in the acyl-CoA pool closely corresponds to the amount of linolenate transferred to phosphatidylcholine. Note that no change in mass of the phosphatidylcholine occurred during the incorporation of linolenate.

The mechanism of acyl exchange was investigated in some detail in microsomal membranes from developing seeds of safflower by Stymne and Stobart (1984b), and the evidence indicates that it is catalyzed by the reverse and forward reactions of the acyl-CoA:lysophosphatidylcholine acyltransferase working in concert. These conclusions were based on the following observations:

1. The addition of free CoA to the incubation mixtures resulted in an 8-fold stimulation of acyl exchange.
2. [^{3}H]CoA was acylated with nonradioactive fatty acids derived from position sn-2 of sn-phosphatidylcholine.
3. Acylation of phosphatidylcholine via acyl exchange takes place with the same acyl specificity as the acylation of exogenously supplied palmitoyl-lysophosphatidylcholine.
4. Bovine serum albumin in the reaction mixture enhanced quite considerably the acyl exchange.

The initial step in the acyl exchange process is considered to be the transesterification of fatty acids from position sn-2 of sn-phosphatidylcholine [X in Eq. (1)] to CoA by the reverse reaction of the acyltransferase:

$$\text{X-phosphatidylcholine } + \text{ CoA} \rightarrow \text{lysophosphatidylcholine } + \text{ X-CoA} \qquad (1)$$

The next stage in the exchange is the reacylation of the lysophosphatidylcholine from a preexisting pool of acyl-CoA [Y in Eq. (2)] by the forward reaction of the same enzyme:

$$\text{Y-CoA } + \text{ lysophosphatidylcholine} \rightarrow \text{Y-phosphatidylcholine } + \text{ CoA} \qquad (2)$$

The combined forward and reverse reactions of the acyltransferase will then result in acyl exchange between acyl-CoA and phosphatidylcholine:

$$\text{X-phosphatidylcholine } + \text{ Y-CoA} \rightarrow \text{X-CoA } + \text{ Y-phosphatidylcholine} \qquad (3)$$

The equilibrium of the reaction is far to the right toward the synthesis of phosphatidylcholine, since it involves the cleavage of a high-energy thioester bond and the formation of an oxygen ester. The rate-limiting step in the exchange, therefore, is the removal of acyl groups from phosphatidylcholine, and thus any conditions which drive the equilibrium of the reaction in that direction will enhance acyl exchange. It is obvious that increased concentrations of free CoA and low amounts of lysophosphatidylcholine and acyl-CoA will act in this way. As mentioned previously, free CoA and bovine serum albumin both have a profound stimulatory effect on the acyl exchange. Since bovine serum albumin was found to inhibit the acylation of exogenously supplied lysophosphatidylcholine (Stymne and Stobart, 1984b), it was suggested that the stimulation of the exchange was due to the reversible binding of the acyl-CoA to the albumin. The binding of acyl-CoA to the protein will effectively lower the apparent concentration of the free acyl-CoA which is available to the acyltransferase, since the enzyme does not appear to utilize the protein-bound thioester. No other phospholipids in the microsomal membrane, other than phosphatidylcholine, participate in acyl exchange to any real extent, and phosphatidylcholine is the only phospholipid that responds to increased acylation upon the addition of CoA and serum albumin (Stymne and Stobart, 1984b).

The rates of the acylation of lysophosphatidylcholine and acyl exchange in microsomal preparations of safflower are impressive (Stymne and Stobart, 1984b). Under optimum conditions for the two reactions, the acylation of added lyso-phosphatidylcholine with oleoyl-CoA and acyl exchange were in the order of 300 and 18 nmol min^{-1} mg $protein^{-1}$, respectively. Acyl exchange provides an elegant method for manipulating the fatty acid composition of microsomal phos-phatidylcholine without changing other parameters. The exchange has therefore been used to alter substantially the oleate content of phosphatidylcholine in microsomal membranes from safflower from levels of 2% up to 30% (equivalent to 60% of all the fatty acids in position sn-2 of sn-phosphatidylcholine undergoing exchange) (Stymne and Stobart, 1986b).

Acyl exchange appears to be a common feature of, perhaps, all acyl-CoA: lysophosphatidylcholine acyltransferase enzymes. Microsomal preparations from rat liver (Stymne and Stobart, 1984b) and rat lung (Stymne and Stobart, 1985b) can carry out, albeit at a lower rate than the oilseed microsomes, acyl exchange. Other workers have also found that the reverse reaction of the acyltransferase can occur in a variety of animal tissues (Irvine and Dawson, 1979; Colard et al., 1984; Flesh et al., 1984). Rochester and Bishop (1984) showed that free CoA enhanced the incorporation of exogenously added radioactive lysophosphatidylcholine into phos-phatidylcholine in microsomal preparation from developing cotyledons of sun-flower. The most probable explanation for the observations of Rochester and Bishop (1984) is that a trans-esterification is occurring between phosphatidylcholine and lysophosphatidylcholine and that this is catalyzed by the reverse and forward reactions of the acyltransferase. The observation that acyl groups from position 2 of sn-phosphatidylcholine can enter the acyl-CoA pool does not rely solely on the analysis of the acyl-CoA fraction. It is also evident from specific radioactivity data

(Stobart *et al.*, 1983) and in pulse–chase experiments (Griffiths *et al.*, 1985) that there is a flow of linoleate from phosphatidylcholine through the acyl-CoA pool to phosphatidic acid.

The acyl selectivity of the acyl exchange in the two oilseeds investigated so far, safflower (Stymne *et al.*, 1983; Griffiths *et al.*, 1985) and linseed (Stymne and Stobart, 1985a) both exhibited a similar preference for species of acyl-CoA. Oleate and linoleate were acylated to phosphatidylcholine at a similar rate and linolenate at about half that rate. The saturated fatty acids, palmitate and stearate, were almost completely excluded in the acylation to phosphatidylcholine when presented in a acyl-CoA mixture together with the unsaturated C_{18} fatty acids. On the other hand, if palmitoyl- or stearoyl-CoA were presented as a single substrate to the microsomes, then acyl exchange was observed but at a considerable lower rate than for the unsaturated fatty acids. The acyl selectivity of the reverse reaction of the acyltransferase (the acylation of CoA with fatty acids from position *sn*-2 of *sn*-phosphatidylcholine) appeared to have a similar selectivity as the forward reaction (acylation of lysophosphatidylcholine). It is of interest, however, that the acyl selectivity of both the reactions of the acyltransferase are not necessarily similar. The enzyme in microsomes from rat lung prefers saturated species of acyl-CoA in the acylation to phosphatidylcholine and polyunsaturated fatty acids, notably arachidonate, in the reacylation of CoA with acyl groups from position *sn*-2 of *sn*-phosphatidylcholine (Stymne and Stobart, 1985b). The properties of the acyltransferase in rat lung might be explained by a difference in the affinity of the enzyme for saturated and unsaturated acyl groups (Stymne and Stobart, 1985b). At low concentration of acyl donor (acyl-CoA and phosphatidylcholine) the enzyme selects for saturated fatty acids, whereas at higher concentrations it prefers unsaturated species. If the free acyl-CoA concentration is kept low it selects for saturated fatty acids in the acylation to phosphatidylcholine. In the reverse reaction, the acyltransferase, which moves within the phospholipid bilayer of the microsomal membrane, is in an environment of high acyl donor concentration (phosphatidylcholine) and will then select for the unsaturated fatty acids in position *sn*-2 of *sn*-phosphatidylcholine. In pneumocyte cells the differential in K_m values for the different acyl substrates will, in effect, control the movement of saturated fatty acids into phosphatidylcholine and the selective removal of unsaturated fatty acids from that lipid. This would result in species of phosphatidylcholine which are specifically enriched with palmitate. Our observations with systems from animals and plants that are specialized in producing specific lipids lead us to the opinion that particular acyl-CoA:lysophosphatidylcholine acyltransferase enzymes play an important role in controlling the acyl quality of the acyl-CoA pool and phosphatidylcholine.

In vivo experiments with detached cotyledons from developing safflower and sunflower seeds strongly support the proposal that an extensive recycling of fatty acids in position *sn*-2 of *sn*-phosphatidylcholine plays a major role in supplying oleate for desaturation and linoleate for triacylglycerol production (Griffiths, 1986). In these experiments the developing cotyledons were incubated with radioactive oleate or acetate. In the shorter incubation periods some 90% of the label in unsaturated fatty acids in *sn*-phosphatidylcholine was confined to position *sn*-2 (Table V). With prolonged incubation the label in position *sn*-2 decreased to $\sim 80\%$. The results indicate that the major proportion of oleate in the cotyledons is channeled through position *sn*-2 of *sn*-phosphatidylcholine for desaturation while only a minor part enters phosphatidylcholine via the glycerol phosphate pathway (Section III,F,3).

It is pertinent to speculate on the conditions for acyl exchange *in vivo* in developing oilseed. There is no information available on the size of the acyl-CoA pool *in vivo*. It might be anticipated that the acyl-CoA exists only in a small and highly active metabolic pool. The availability of unbound acyl-CoA would be even lower in the cell due to its binding to cytoplasmic proteins. It is possible that some of the cytoplasmic proteins may have a specific binding function for acyl-CoA (Rickers and Spener, 1984). It is possible, therefore, that *in vivo*, the limiting step in the reactions catalyzed by the acyltransferase is not the removal of fatty acids from phosphatidylcholine but rather, because of the low concentration of free acyl-CoA, merely the acylation of lysophosphatidylcholine.

3. Interconversion of Diacylglycerol with Phosphatidylcholine

In vivo experiments with detached cotyledons of developing oilseeds (Slack *et al.*, 1978) demonstrated that [³H]glycerol in phosphatidylcholine could enter the triacylglycerols. The movement of glycerol from phosphatidylcholine to triacylglycerol was almost as rapid as the flow of fatty acids between these two lipids. On the basis of these observations and also from the similarity in the fatty acid composition of phosphatidylcholine and diacylglycerol in oilseeds subjected to a cold shift (Slack and Roughan, 1978), it was suggested that there was an equilibration between phosphatidylcholine and diacylglycerol. An enzyme (CDP-choline:1,2-diacylglycerol cholinephosphotransferase; choline phosphotransferase, E.C. 2.7.8.2) had previously been described in microsomal membranes from animals (Kanoh and Ohno, 1975; Goracci *et al.*, 1981) which catalyzed the interconversion of diacylglycerol and phosphatidylcholine. This particular enzyme was considered, therefore, as a candidate for the reaction in oilseeds (Roughan and

TABLE V
Positional Distribution of Radioactivity
in the Phosphatidylcholine from Cotyledons
of Safflower Incubated with [¹⁴C]Oleate[a]

Incubation time	Radioactivity (%)	
(min)	*sn*-1	*sn*-2
10	6	94
20	6	94
80	10	90
240	17	83

[a]From Griffiths (1986).

Slack, 1982). It was further speculated that the role of the reaction might be to supply oleoyl-rich diacylglycerols to phosphatidylcholine for desaturation and the channeling of polyunsaturated molecular species from phosphatidylcholine via diacylglycerol to triacylglycerol (Roughan and Slack, 1982). The involvement of a choline phosphotransferase was further suggested from radiolabeling experiments utilizing [^{3}H]glycerol, inorganic ^{32}PO$_4$, and [^{14}C]choline in incubations with developing cotyledons of linseed (Slack *et al.*, 1983). Microsomal preparations from the developing cotyledons of sunflower (Stymne and Stobart, 1984a), safflower (Stobart and Stymne, 1985a), and linseed (Stymne and Stobart, 1985a) also incorporate glycerol from [^{14}C]glycerol phosphate, in the presence of acyl-CoA, into phosphatidylcholine via diacylglycerol. Slack *et al.* (1985) have demonstrated that the choline phosphotransferase in microsomal preparations from developing safflower seeds could operate, to some extent, in a reversible fashion. It has also been shown that safflower microsomes will catalyze the simultaneous movement of diacylglycerol backbone into phosphatidylcholine and the removal of glycerol moieties from phosphatidylcholine to diacylglycerol during the active synthesis of triacylglycerol through the glycerol phosphate pathway (Stobart and Stymne, 1985b). The cofactors for the choline phosphotransferase (CMP, CDP-choline) do not appear to be rate-limiting, in microsomal preparations of safflower, during the operation of the glycerol phosphate pathway (Stobart and Stymne, 1985b). This suggests that these cofactors are present in catalytic amounts and are tightly bound to the microsomal membranes. In this context it is noteworthy that it is possible to incorporate [^{14}C]glycerol into phosphatidylcholine via diacylglycerol to a level which represents some 50% of the microsomal pool of phosphatidylcholine with little change in its mass (A. K. Stobart and S. Stymne, unpublished observations). The reaction, therefore, probably involves no net loss of phosphatidylcholine and hence offers greater energy conservation in a system that has evolved for the efficient production of seed storage oil. The thermodynamics of the interconversion catalyzed by the choline phosphotransferase in developing oilseeds appear to be rather novel and require further study.

At present it is difficult to ascertain the relative contribution of the reversible transfer, between diacylglycerol and phosphatidylcholine, and the acyl exchange in providing oleate for desaturation and the polyunsaturated fatty acid products for the synthesis of triacylglycerol. Although the oleate which enters phosphatidylcholine via the diacylglycerol, which is synthesized in the glycerol phosphate pathway, is readily desaturated to linoleate in microsomes of safflower (Stobart and Stymne, 1985a) the endogenous microsomal phosphatidic acid is as highly unsaturated as the phosphatidylcholine (Griffiths *et al.*, 1985). This observation implies that despite the reversible transfer of diacylglycerol to phosphatidylcholine it is the acyl exchange, coupled to the preferential acylation of glycerol phosphate with linoleate, that is the prime route for linoleate channeling to triacylglycerol. This proposal, however, is difficult to substantiate in *in vitro* systems where both modes of entry of oleate into phosphatidylcholine are coupled. Here, the oleate, which enters position *sn*-2 of *sn*-phosphatidylcholine from diacylglycerol, is desaturated to linoleate, and this can then enter the acyl-CoA pool via acyl exchange for further utilization in the acylation of glycerol phosphate. In this case the acyl-CoA pool is enriched with linoleate partly through the diacylglycerol–phosphatidylcholine interconversion. It is interesting, however, that *in vivo* in cotyledons of safflower and sunflower, the relative entry of oleate into position *sn*-1 of *sn*-phosphatidylcholine via the glycerol phosphate pathway is low compared to that which is accounted for through acyl exchange (Griffiths, 1986; Table V).

The importance of the two modes of entry of oleate into phosphatidylcholine may also differ in different plant species. In certain plants (e.g., high-erucate rape), there must be some selectivity in the diacylglycerol transfer to phosphatidylcholine and/or different metabolic pools of diacylglycerol to account for the high concentration of erucic acid at position *sn*-1 of the *sn*-triacylglycerols (Table III) and its relative absence in the phosphatidylcholine (Sosulski *et al.*, 1981). The biosynthetic pathway and its regulation in oilseeds which accumulate triacylglycerol with high amounts of "unusual" fatty acids requires further attention.

G. Summary

The full sequence of reactions leading to the biosynthesis of linoleate-rich seed oils from acyl-CoA and glycerol phosphate, which is based on our present knowledge, is summarized below and in Fig. 11. Although the evidence is somewhat scarce, the formation of linolenate-rich oils probably follows a similar biosynthetic route.

1. Oleate, which is exported to the cytoplasm from the plastid as oleoyl-CoA, enters position *sn*-2 of the *sn*-phosphatidylcholine in the membranes of the endoplasmic reticulum, by the forward reaction of an acyl-CoA:lysophosphatidyl-choline acyltransferase (reaction I, Fig. 11). The oleate in phosphatidylcholine is desaturated to linoleate *in situ* by an oleoyl-phosphatidylcholine desaturase (reaction II, Fig. 11). The newly formed linoleate is transferred back to the acyl-CoA pool by the reverse reaction of the acyltransferase (reaction I, Fig. 11).

2. The acylation of position *sn*-1 of *sn*-glycerol 3-phosphate occurs with various species of acyl-CoA but with a predominant selectivity for the saturated fatty acids (reaction III, Fig. 11). The subsequent acylation of position *sn*-2 of the *sn*-lysophosphatidic acid has a strong preference for linoleate and excludes almost completely the saturated fatty acids (reaction IV, Fig. 11). The selectivity for linoleate against oleate in the acylation of glycerol phosphate coupled to acyl exchange controls the flow of oleate to phosphatidylcholine and the channeling of linoleate from phosphatidylcholine to phosphatidic acid.

3. The phosphatidic acid is subsequently hydrolyzed by a phosphatidase to yield diacylglycerol (reaction V, Fig. 11).

4. The diacylglycerol moiety can enter, to some extent, the phosphatidylcholine by the forward reaction of a CDPcholine:diacylglycerol cholinephosphotransferase (reaction VI, Fig. 11). The oleate which enters phosphatidylcholine in this way can

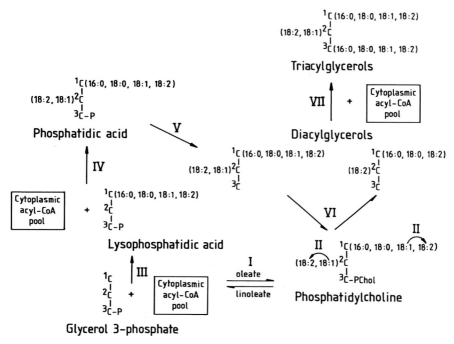

Fig. 11.   Proposed scheme for the biosynthesis of triacylglycerols in linoleate-rich oilseeds.

be further desaturated to linoleate (reaction II, Fig. 11). The entry of species of diacylglycerol into phosphatidylcholine is accompanied by the concomitant transfer of glycerol backbone to diacylglycerol by the reverse reaction of the choline-phosphotransferase (reaction VI, Fig. 11). The two equilibration reactions (acyl exchange and the diacylglycerol–phosphatidylcholine interconversion) will bring about the continuous enrichment of the glycerol backbone with polyunsaturated fatty acids.

5. A diacylglycerol acyltransferase catalyzes the acylation of position *sn*-3 of the *sn*-diacylglycerol and is the ultimate reaction in the synthesis of the triacylglycerols (reaction VII, Fig. 11). The final acylation utilizes the common cytoplasmic pool of acyl-CoA and probably exhibits a broad acyl specificity.

6. The triacylglycerols which are synthesized in the endoplasmic reticulum are released from the membranes into the cytoplasm to form oil bodies.

IV.   TRIACYLGLYCEROLS WITH UNCOMMON FATTY ACIDS

The multiplicity of fatty acid species that are found in plants has led, in some cases, to families which can contain an abundance of certain uncommon fatty acids in their seed oil. Our knowledge of triacylglycerol synthesis in many of these plants is limited, and here we describe just a few examples of oils that are of a particular agricultural and nutritional interest.

A. γ-Linolenic Acid

Certain "medicinal" plants contain γ-linolenic acid (all-*cis*-octadeca-6,9,12-trienoic acid) in their seeds. This is particularly pronounced in seeds of Evening Primrose (*Oenothera biennis*), *Borago officinalis,* and *Ribes* spp. (Hilditch and Williams, 1964; Tetenyi *et al.*, 1974; Traitler *et al.*, 1984). The oils from such species can contain upwards of 10% γ-linolenic acid. A variety of Evening Primrose (Efamol) has been developed for commercial exploitation (Horrobin and Manku, 1983) and is being marketed for therapeutic use. It has been suggested that γ-linolenate oils alleviate many atopic disorders. In many of these conditions it is considered that the Δ^6-desaturase enzyme, which converts dietary (from plant sources) linoleic acid to γ-linolenic acid, is limiting, and hence the synthesis of certain prostaglandin precursors is reduced. In view of the apparent importance of γ-linolenate in nutrition, it is surprising that nothing is yet known about its biosynthesis in plants. The introduction of a further double bond between the carboxyl end of a fatty acid and a preexisting double bond is interesting. In plants which synthesise γ-linolenic acid it has been presumed that a Δ^6-desaturase catalyzes the insertion of a further double bond in linoleic acid. The substrate for linoleic acid synthesis in oilseeds is oleoyl-phosphatidylcholine rather than acyl-CoA (Section III,F,1). The reason for this is unclear and we have surmised, for experimental purposes, that in plants perhaps the insertion of a double bond between the methyl end of a C_{18} fatty acid and an already established double bond requires the intramolecular environment of a complex lipid, whereas the introduction of double bonds between a preexisting double bond and the carboxyl end, as occurs in animals (Holloway, 1983), can utilize the CoA ester directly. The formation of γ-linolenic acid in plants is particularly interesting, since it is the only C_{18}-polyunsaturated fatty acid whose synthesis can be compared directly with an animal counterpart. Investigations *in vivo* with developing cotyledons of borage show that γ-linolenic acid is synthezised from oleate via linoleate in a sequential fashion and rapidly accumulates in phosphatidylcholine, diacylglycerol, and triacylglycerol. Microsomal preparations, however, from the developing cotyledons contained an active Δ^6-desaturase enzyme that catalyzed the conversion of 2-linoleoyl-phosphatidylcholine to 2-γ-linolenoyl-phosphatidylcholine (Stymne and Stobart, 1986a). The evidence, therefore, strongly indicates that the major substrate in plants for the formation of γ-linolenate is phosphatidylcholine and that the acyl-CoA is a most unlikely candidate. An explanation of why higher plants have evolved desaturase enzymes that utilize complex lipids would not appear, therefore, to be based on a direct mechanistic reason, as we originally suggested. One might conjecture that perhaps complex lipid substrates are more efficient in plants because the desaturase enzymes are not then competing directly with other major reactions that utilize acyl-CoA.

It was also observed (Stymne and Stobart, 1986a) that both the endogenous and newly synthesized γ-linolenate resided, almost totally, in position sn-2 of sn-phosphatidylcholine, whereas linoleate was present in high amounts in both positions sn-1 and sn-2. In this respect the synthesis of γ-linolenate appears to differ from the synthesis of linoleate, which is formed from oleate at both positions sn-1 and sn-2 of sn-phosphatidylcholine *in situ* (Section III,F,1). In view of the distinction made by the Δ^6-desaturase for the linoleate in phosphatidylcholine in borage, it is possible that different desaturase enzymes exist that are positionally specific for the fatty acid substrate in the complex lipid. Thus further studies on the synthesis of γ-linolenate and its relationship to complex lipid production should provide valuable information on the desaturase enzymes and their substrates in plants.

B.   Erucic Acid

Older varieties of oilseed rape (*Brassica napus*), turnip rape (*Brassica campestris*), and the related mustards were characterized by the presence of large quantities of erucate (*cis*-docos-13-enoic acid), which is associated with positions sn-1 and 3 of the sn-triacylglycerol (Table III; Appelqvist, 1972). While high-erucate oils are most suitable as specialized lubricants, they resulted in heart lesions in test animals. This led to the demand for and rapid development of rape varieties that were almost completely lacking in erucate (Ackman, 1983). The development of such varieties is one of the most remarkable success stories in oil plant breeding and fosters optimism for the production of new varieties of other oil-producing species with defined triacylglycerols.

Erucic acid is synthesized by the elongation of oleoyl-CoA via eicosenoic acid (Downey and Craig, 1964), and this has been confirmed in other plant species that accumulate these acids (Pollard *et al.*, 1979; Pollard and Stumpf, 1980a,b). High-erucate strains of rape possess the oleoyl-CoA elongase, while it is absent in the low-erucate varieties (Stumpf and Pollard, 1983). It was considered that the synthesis of oleate occurred in the plastid and that its elongation to erucate took place in another modifying compartment (Stumpf and Pollard, 1983). However, evidence was later presented (Agrawal and Stumpf, 1985) which indicated that the elongation of oleate to erucate could occur in a particulate fraction which contained chloroplasts (or plastids) present in the developing seed.

Only one and two gene loci are involved in controlling the absence of erucate in turnip rape and oilseed rape, respectively (Stefansson and Houghen, 1964; Krzymanski and Downey, 1969; Jonsson, 1977). Since an inverse relationship exists between the oleate and erucate content in strains of rape, it is a reasonable assumption that these genes are responsible for the elongation of oleoyl-CoA.

C.   Ricinoleic Acid

The oils of castor bean (*Ricinus communis*) are rich in the oxygenated acid, ricinoleic acid (12-hydroxy-9-octadecanoic acid). Most castor bean oils contain ricinoleate in excess of 90% of the total fatty acids (Hilditch and Williams, 1964). The relative abundance of ricinoleic acid in the oil is particularly interesting, since one of the major aims of plant breeding programs is to produce strains of the more common oil crop species which will produce triacylglycerols containing only one particular fatty acid.

In vitro studies with microsomal preparations from castor bean endosperm (Galliard and Stumpf, 1966) showed that ricinoleate was synthesized from oleoyl-CoA by the action of a membrane-bound mixed-function oxygenase system which required molecular O_2 and NADH. More recently (Moreau and Stumpf, 1981), it was shown that phosphatidylcholine 12-monooxygenase activity was associated with a microsomal membrane fraction. The kinetics of the hydroxylation suggested that an oleoyl-phosphatidylcholine species might be the primary substrate rather than oleoyl-CoA.

V. TRIACYLGLYCEROLS WITH MEDIUM-CHAIN FATTY ACIDS

Oils from coconut and oil palm kernel tissue can contain > 50% medium-chain fatty acids, which consist largely of capric acid (C_{10}) and lauric acid (C_{12}) (Child, 1974). Triacylglycerols with shorter chain fatty acids are important in the commercial production of surfactants and detergents. It is considered desirable to produce coconut and oil palm oils with even greater quantities of lauric acid and to develop oil crops of an annual nature which will accumulate laurate-rich triacylglycerols. Because of the importance of this, the regulation of the termination reactions in shorter chain fatty acid biosynthesis is receiving attention. Preliminary investigations (Oo and Stumpf, 1979) showed that developing endosperm from coconut incorporated labeled acetate and malonate into C_8–C_{18} fatty acids, with the majority of the radioactivity in the C_{12} and C_{14} species. Cell-free preparations, on the other hand, incorporated malonate mostly into palmitate (Oo *et al.*, 1985). Slabas *et al.* (1984) have shown, however, that the concentration of malonyl-CoA and the ratio of acetyl-CoA to malonyl-CoA can markedly affect the *in vitro* products of fatty acid biosynthesis in coconut preparations. This confirms the earlier work by Huang and Stumpf (1971), who employed extracts of potato tissue. The mechanism of medium-chain termination (Slabas *et al.*, 1984) and the relationship of the fatty acid products to triacylglycerol biosynthesis still awaits elucidation. In this respect, it is interesting that a model system for studying medium-chain fatty acid and triacylglycerol biosynthesis has been proposed using seeds of *Cuphea* (Slabas *et al.*, 1982). *Cuphea* oils are characterized by high concentrations of capric and lauric acid (Hilditch and Williams, 1964), and species are available in which either one of these fatty acids can predominate and reach levels as high as 90% of the total fatty acids (Robbelen and Hirsinger, 1982). The triacylglycerols in these oilseeds can consist almost entirely of trilauroyl- or tricaproylglycerols. *Cuphea*, therefore, could be a valuable oil crop of medium-chain triacylglycerols which would help to alleviate the present dependence on oils from coconut and oil palm.

VI. EFFECT OF TEMPERATURE ON FATTY ACID COMPOSITION

It has been known since the 1930s (Hilditch and Williams, 1964) that environmental factors have a strong influence on the fatty acid composition of seed oils. Similar varieties of linseed that were growing, for example, in different areas of the United States, could differ in the quality of their C_{18} unsaturated fatty acids, and this was largely governed by the temperature at the time of oil deposition and seed maturation. It has now been repeatedly observed that membrane and reserve storage lipids in plants grown at lower temperatures are generally enriched in C_{18}-polyunsaturated fatty acids (Canvin, 1965; Slack and Roughan, 1978). The effect of temperature on the increase in such acids in seed oils of a commercial nature has been extensively studied in the laboratory of Mazliak and Trémolières (H. Trémolières et al., 1978; Mazliak, 1979; A. Trémolières et al., 1982). The extent of unsaturation is inversely related to temperature, and this is considered to be an adaptive phenomenon that can regulate the fluidity of particular cell membranes and allow proper physiological function at the lower temperature. An explanation for viscotropic responses in plants has been based on a possible direct effect on the activity and/or the relative amounts of the desaturase enzymes in the membranes. Some years ago it was suggested (Harris and James, 1969a,b) that the higher solubility of oxygen in aqueous solution at lower temperatures may increase desaturase activity by increasing the availability of the oxygen cosubstrate. While the oxygen concentration may be critical in controlling the C_{18} polyunsaturated fatty acids in the lipids of sycamore cell cultures (Rebeille et al., 1980), it appears to be non-rate-limiting in cultures of Tetrahymena (Skriver and Thompson, 1976) and developing safflower cotyledons (Browse and Slack, 1983).

Numerous other explanations have been suggested as possibly responsible for the effect of temperature (Thompson, 1983). An attractive hypothesis, however, has been that the fluidity of the membrane can affect directly the activity of the desaturase enzymes (Brenner, 1984). In this respect we have manipulated the oleate content of microsomal phosphatidylcholine in safflower preparations through acyl exchange (Section III,F,2). A study of the effect of temperature on the activity of the 2-oleoyl-phosphatidylcholine desaturase in these modified membranes indicated that an explanation for an increase in the polyunsaturated fatty acid content that is modulated through membrane fluidity is rather tenuous (Stymne and Stobart, 1986b). Browse and Slack (1983) have suggested that a differential in the temperature response of the fatty acid synthetase and the oleoyl-desaturase, albeit both low at decreased temperature, was more likely to account for the increase in the polyunsaturation of the seed oil. Recently, Trémolières et al. (1982) suggested that the response of oil-producing plants to low temperature is more probably governed at the genetic level than by a direct viscotropic effect on oleate desaturase activity. In this context it is noteworthy that Kasai and Nozawa (1980) provided evidence for an increase in desaturase-enzyme synthesis in *Tetrahymena* subject to low temperature and that in certain poikilothermic fish the control of the Δ^6-desaturase may be through enzyme induction (Torrengo and Brenner, 1976).

VII. CONCLUDING REMARKS

In the introduction to this chapter we referred to M.I. Gurr, who in an earlier contribution in this series (Gurr, 1980) pointed out that, at that time, our understanding of triacylglycerol biosynthesis in plants was rather limited. In the intervening years, as we have endeavored to convey to the present reader, our knowledge in this area has increased considerably. To sustain such progress, however, it is essential that workers establish optimum conditions for triacylglycerol biosynthesis in the *in vitro* membrane systems under study and demonstrate activities which are compatible with the synthesis of triacylglycerols *in vivo*, before embarking on and reporting experimentation designed to elucidate the enzymology and regulation of oil formation. It is necessary to work with precise and well-defined model systems. In this respect we strongly recommend the developing seed cotyledons of safflower, which in our experience are superior to other seeds for the preparation of active membrane fractions. A number of varieties of safflower are available which will synthesize triacylglycerols rich in particular fatty acids (e.g., High-Oleate, High-Linoleate, Intermediate-Linoleate). These varieties are interesting for comparative studies and should prove particularly useful in identifying the microsomal proteins involved in the desaturation of oleate. A further advantage in using safflower is that some understanding of the genetics of many of the fatty acid varieties is also available (Knowles, 1969, 1972, 1982).

The outstanding problems in oil biosynthesis are numerous and offer many exciting challenges which are still of a sufficiently diverse nature to satisfy the interests and demands of most biochemists. A particularly major contribution will be the purification of those proteins which are present in the endoplasmic reticulum of the developing seed and which catalyze the synthesis of triacylglycerol from glycerol phosphate and regulate the acyl quality of the final oil. The solubilization and purification of membrane-bound enzymes is an active area of research and still has many technical problems to overcome (Freedman, 1981). The microsomal membrane from the developing oil seed offers a fundamental system which, because of its specialized nature, is probably most abundant in those enzymes associated with lipid metabolism. In the longer term, success in purifying the enzymes of triacylglycerol synthesis will lead to the preparation of antibodies, mRNA, cDNA, and the ultimate production of cloned genes. These advances will considerably increase our understanding of the regulatory processes involved in triacylglycerol synthesis. Certainly, as the knowledge and techniques become available for interspecies gene transfer (particularly of multiple genes) and its application to practical plant breeding, it should become possible to tailor oil-producing systems to satisfy, precisely, specific industrial and nutritional requirements.

ACKNOWLEDGMENT

The authors are grateful for financial support (past and present) to the Swedish Natural Science Research Council, Swedish Council for Forestry and Agricultural Research, National Swedish Board for Technical Development, and the Nuffield Foundation and The Royal Society, both in the United Kingdom.

REFERENCES

Ackman, A. G. (1983). *In* "High and Low Erucic Acid Rapeseed Oils" (J. K. G. Kramer, F. D. Sauer, and W. J. Pigden, eds.), pp. 85–129. Academic Press, Toronto.
Agrawal, V., and Stumpf, P. K. (1985). *Lipids* **20**, 361–366.
Appleby, R. S., Gurr, M. I., and Nichols, B. W. (1974). *Eur. J. Biochem.* **48**, 209–216.
Appelqvist, L.-Å. (1972). *In* "Rapeseed" (L.-Å. Appelqvist and R. Ohlson, eds.), pp. 1–8. Elsevier, Amsterdam.
Barron, E. J., and Stumpf, P. K. (1962). *Biochim. Biophys. Acta* **60**, 329–337.
Bergfeld, R., Hong, Y.-N., Kühnl, T., and Schopfer, P. (1978). *Planta* **143**, 297–307.
Bertrams, M., and Heinz, E. (1981). *Plant Physiol.* **68**, 653–657.
Block, M. A., Dorne, J.-A., Joyard, J., and Douce, R. (1983). *FEBS Lett.* **164**, 111–115.
Bozzini, A. (1982). *In* "Improvement of Oil-Seed and Industrial Crops by Induced Mutations," pp. 3–16. IAEA, Vienna.
Brenner, R. R. (1984). *Prog. Lipid Res.* **23**, 69–96.
Brindley, D. N. (1978). *Int. J. Obes.* **2**, 7–16.
Brockerhoff, H. (1971). *Lipids* **6**, 942–956.
Browse, J. A., and Slack, C. R. (1981). *FEBS Lett.* **131**, 111–114.
Browse, J. A., and Slack, C. R. (1983). *Biochim. Biophys. Acta* **753**, 145–152.
Browse, J. A. and Slack, C. R. (1985). *Planta* **166**, 74–80.
Call, F. L., and Williams, W. J. (1973). *J. Lab. Clin. Med.* **82**, 663–673.
Canvin, D. T. (1965). *Can. J. Bot.* **43**, 63–69.
Cao, Y.-Z., and Huang, A. H. C. (1986). *Plant Physiol.* **82**, 813–820.
Child, R. (1974). *In* "Coconuts," pp. 45–48. Longmans, London.
Colard, O., Breton, M., and Berziat, G. (1984). *Biochim. Biophys. Acta* **793**, 42–48.
Dennis, D. T., and Miernyk, J. A. (1982). *Annu. Rev. Plant Physiol.* **33**, 27–50.
Downey, R. K., and Craig, B. M. (1964). *J. Am. Oil Chem. Soc.* **41**, 475–478.
Dybing, C. D., and Craig, B. M. (1970). *Lipids* **5**, 422–429.
Fatemi, S. H., and Hammond, E. F. (1977). *Lipids* **12**, 1037–1041.
Finlayson, S. A., and Dennis, D. T. (1980). *Arch. Biochem. Biophys.* **199**, 179–185.
Flesh, I., Ecker, B., and Ferber, E. (1984). *Eur. J. Biochem.* **139**, 431–437.
Freedman, R. B. (1981). *In* "Membrane Structure" (J. B. Finean and R. H. Michell, eds.), pp. 161–214. Elsevier, Amsterdam.
Frentzen, M., Heinz, E., McKeon, T. A., and Stumpf, P. K. (1983). *Eur. J. Biochem.* **129**, 629–636.
Frentzen, M., Hares, W., and Schiburr, A. (1984). *In* "Structure, Function and Metabolism of Plant Lipids" (P. A. Siegenthaler and W. Eichenberger, eds.), pp. 105–110. Elsevier, Amsterdam.
Frey-Wyssling, A., Grieshaber, E., and Mühlethaler, K. (1963). J. *Ultrastruct. Res.* **8**, 506–516.
Galliard, T., and Stumpf, P. K. (1966). *J. Biol. Chem.* **241**, 5806–5812.
Goracci, G., Francescangeli, E., Horrocks, L. A., and Porcellati, G. (1981). *Biochim. Biophys. Acta* **664**, 373–379.

Griffiths, G. (1986). "Aspects of triacylglycerol biosynthesis in developing oilseeds." Ph.D. Thesis, Univ. of Bristol, Bristol, England.

Griffiths, G., Stobart, A. K., and Stymne, S. (1985). *Biochem. J.* **230**, 379–388.

Gunstone, F. D. (1979). *In* "Comprehensive Organic Chemistry" (D. Barton and W. D. Ollis, eds.), p. 645. Pergamon, Oxford.

Gunstone, F. D., and Ilyas-Qureshi, M. (1965). *J. Am. Oil. Chem. Soc.* **42**, 961–965.

Gunstone, F. D., and Norris, F. A. (1983). "Lipids in Foods." Pergamon, Oxford.

Gunstone, F. D., Hamilton, R. J., Padley, F. B. and Ilyas-Qureshi, M. (1965). *J. Am. Oil Chem. Soc.* **42**, 965–970.

Gurr, M. I. (1980). *In* "The Biochemistry of Plants" (P. K. Stumpf and E. E. Conn, eds.), Vol. 4, pp. 205–248. Academic Press, New York.

Gurr, M. I., and James, A. T. (1980). "Lipid Biochemistry: An Introduction." Chapman & Hall, London.

Gurr. M. I., Robinson, M. P., and James, A. T. (1969). *Eur. J. Biochem.* **9**, 70–78.

Gurr, M. I., Blades, J., and Appleby, R. S. (1972). *Eur. J. Biochem.* **29**, 362–368.

Gurr, M. I., Blades, J., Appleby, R. S., Smith, C. G., Robinson, M. P., and Nichols, B. W. (1974). *Eur. J. Biochem.* **43**, 281–290.

Harris, P., and James, A. T. (1969a). *Biochim. Biophys. Acta* **187**, 13–18.

Harris, P., and James, A. T. (1969b). *Biochem. J.* **112**, 325–330.

Harwood, J. L., and Stumpf, P. K. (1972). *Lipids* **7**, 8–19.

Harwood, J. L., Sodja, A., and Stumpf, P. K. (1971). *Lipids* **6**, 851–854.

Hilditch, T. P., and Williams, P. N. (1964). "The Chemical Constitution of Natural Fats." Chapman & Hall, London.

Holloway, P. W. (1983). *In* "The Enzymes" (P. D. Boyer, ed.), 3rd Ed., Vol. 16, pp. 63–83. Academic Press, New York.

Horrobin, D. F., and Manku, M. S. (1983). *In* "Fats for the Future" (S. G. Brooker, A. Renwick, S. F. Hannan, and L. Eyres, eds.), pp. 99–101. Duromark, Auckland, New Zealand.

Huang, A. H. C., and Stumpf, P. K. (1971). *Arch. Biochem. Biophys.* **143**, 412–427.

Ichihara, K. (1984). *Arch. Biochem. Biophys.* **232**, 685–698.

Ichihara, K., and Noda, M. (1980). *Phytochemistry.* **19**, 49–54.

Ichihara, K., and Noda, M. (1982). *Phytochemistry* **21**, 1895–1901.

Irwine, R. F., and Dawson, M. (1979). *Biochem. Biophys. Res. Commun.* **91**, 1399–1405.

Jonsson, R. (1977). *Hereditas* **86**, 159–170.

Joyard, J., and Douce, R. (1977). *Biochim. Biophys. Acta* **486**, 273–285.

Kanoh, H., and Ohno, K. (1975). *Biochim. Biophys. Acta* **380**, 199–207.

Kasai, R., and Nozawa, Y. (1980). *Biochim. Biophys. Acta* **617**, 161–166.

Kennedy, E. P. (1961). *Fed. Proc., Fed. Am. Soc. Exp. Biol.* **20**, 934–940.

Kleinig, H., Steinki, C., Kopp, C., and Zaar, K. (1978). *Planta* **140**, 233–237.

Knowles, P. F. (1969). *J. Am. Oil Chem. Soc.* **46**, 130.

Knowles, P. F. (1972). *J. Am. Oil Chem. Soc.* **49**, 27.

Knowles, P. F. (1982). *In* "Improvements of Oil-Seed and Industrial Crops by Induced Mutations," pp. 89–101. IAEA, Vienna.

Krzymanski, J., and Downey, R. K. (1969). *Can. J. Plant Sci.* **49**, 313–319.

Lin, E. C. C. (1977). *Annu. Rev. Biochem.* **46**, 765–795.

McMahon, V., and Stumpf, P. K. (1964). *Biochem. Biophys. Acta* **84**, 361–364.

Martin, B. A., and Wilson, R. F. (1984). *Lipids* **19**, 117–121.

Mazliak, P. (1979). *In* "Low Temperature Stress in Crop Plants" (J. M. Lyons, D. Graham, and J. K. Raison, eds.), pp. 391–404. Academic Press, London.

Mollenhauer, H. H., and Totten, C. (1971). *J. Cell Biol.* **48**, 395–405.

Moore, T. S., Lord, J. M., Kagawa, T., and Beevers, H. (1973). *Plant Physiol.* **52**, 50–53.

Moreau, R. A., and Stumpf, P. K. (1981). *Plant Physiol.* **67**, 672–676.

Ohlrogge, J. B., Shine, W. E., and Stumpf, P. K. (1978). *Arch. Biochem. Biophys.* **189**, 382–391.

Ohlrogge, J. B., Kuhn, D. N., and Stumpf, P. K. (1979). *Proc. Natl. Acad. Sci. U.S.A.* **76**, 1194–1198.

Oo, K. C., and Stumpf, P. K. (1979). *Lipids* **14**, 132–143.

Oo, K. C., Teh, S. K., Khor, H. T., and Augustine, S. H. (1985). *Lipids* **20**, 205–210.

Pollard, M. R., and Stumpf, P. K. (1980a). *Plant Physiol.* **66**, 641–648.

Pollard, M. R., and Stumpf, P. K. (1980b). *Plant Physiol.* **66**, 649–655.

Pollard, M. R., McKeon, T., Gupta, L. M., and Stumpf, P. K. (1979). *Lipids* **14**, 651–662.

Porra, R. J. (1979). *Phytochemistry.* **18**, 1651–1656.

Rebeille, F., Bligny, R., and Douce, R. (1980). *In* "Biogenesis and Function of Plant Lipids" (P. Mazliak, P. Benveniste, C. Costes, and R. Douce, eds.), pp. 203–206. Elsevier, Amsterdam.

Rest, J. A., and Vaughan, J. G. (1972). *Planta* **105**, 245–262.

Rickers, J., and Spener, F. (1984). *Biochim. Biophys. Acta* **794**, 313–319.

Rochester, C. R., and Bishop, D. G. (1982). *In* "Biochemistry and Metabolism of Plant Lipids" (J. F. G. M. Wintermans and P. J. C. Kuiper, eds.), pp. 57–60. Elsevier, Amsterdam.

Rochester, C. R., and Bishop, D. G. (1984). *Arch. Biochem. Biophys.* **232**, 249–258.

Roughan, P. G., and Slack, C. R. (1977). *Biochem. J.* **162**, 457–459.

Roughan, P. G., and Slack, C. R. (1982). *Annu. Rev. Plant Physiol.* **33**, 97–132.

Robbelen, G., and Hirsinger, F. (1982). *In* "Improvement of Oil-Seed and Industrial Crops by Induced Mutations," pp. 161–170. IAEA, Vienna.

Shewry, P. R., Pinfield, N. J., and Stobart, A. K. (1972). *Phytochemistry* **11**, 2149–2154.

Shine, W. E., Mancha, M., and Stumpf, P. K. (1976). *Arch. Biochem. Biophys.* **173**, 472–479.

Simcox, P. D., Reid, E. E., Canvin, D. T., and Dennis, D. T. (1977). *Plant Physiol.* **59**, 1128–1132.

Skriver, L., and Thompson, G. A. (1976). *Biochim. Biophys. Acta* **431**, 180–188.

Slabas, A. R., Roberts, P. A., Ormesher, J., and Hammond, E. W. (1982). *Biochim. Biophys. Acta* **711**, 411–420.

Slabas, A. R., Harding, J., Hellyer, A., Sidebottom, C., Gwynne, H., Kessel, R., and Tombs, M. P. (1984). *In* "Structure, Function and Metabolism of Plant Lipids" (P. A. Siegenthaler and W. Eichenberger, eds.), pp. 3–10. Elsevier, Amsterdam.

Slack, C. R. (1983). *In* "Biosynthesis and Function of Plant Lipids" (W. W. Thomson, J. B. Mudd, and M. Gibbs, eds.), pp. 40–55. Waverly Press, Baltimore, Maryland.

Slack, C. R., and Browse, J. A. (1984). *In* "Seed Physiology" (D. R. Murray, ed.), Vol. 1, pp. 209–244. Academic Press, Sydney.

Slack, C. R., and Roughan, P. G. (1978). *Biochem. J.* **170**, 437–439.

Slack, C. R., Roughan, P. G., and Balasingham, N. (1978). *Biochem. J.* **170**, 421–433.

Slack, C. R., Roughan, P. G., and Browse, J. (1979). *Biochem. J.* **179**, 649–656.

Slack, C. R., Bertaud, W. S., Shaw, B. P., Holland, R., Browse, J., and Wright, H. (1980). *Biochem. J.* **190**, 551–561.

Slack, C. R., Campbell, L. C., Browse, J. A., and Roughan, P. G. (1983). *Biochim. Biophys. Acta* **754**, 10–20.

Slack, C. R., Roughan, P. G., Browse, J. A., and Gardiner, S. E. (1985). *Biochim. Biophys. Acta* **833**, 438–448.

Sosulski, F., Zadernowski, R., and Babuchowski, K. (1981). *J. Am. Oil Chem. Soc.* **58**, 561–564.

Stefansson, B. R., and Houghen, F. W. (1964). *Can. J. Plant Sci.* **44**, 359–364.

Stobart, A. K., and Stymne, S. (1985a). *Planta* **163**, 119–125.

Stobart, A. K., and Stymne, S. (1985b). *Biochem. J.* **232**, 217–221.

Stobart, A. K., Stymne, S., and Glad, G. (1983). *Biochim. Biophys. Acta* **754**, 292–297.

Stobart, A. K., Stymne, S., and Höglund, S. (1986). *Planta* **169**, 33–37.

Stumpf, P. K., and Pollard, M. R. (1983). *In* "High and Low Erucic Acid Rape Seed Oils" (J. K. G. Kramer, F. D. Sauer, and W. J. Piden, eds.), pp. 131–141. Academic Press, Toronto.

Sturton, R. G., and Brindley, D. N. (1980). *Biochim. Biophys. Acta* **619**, 494–505.

Stymne, S. (1980). "The biosynthesis of linoleic and linolenic acids in plants." Ph.D. Thesis, Swedish Univ. of Agric. Sci., Uppsala.

Stymne, S., and Appelqvist, L.-Å. (1978). *Eur. J. Biochem.* **90**, 223–229.

Stymne, S., and Appelqvist, L.-Å. (1980). *Plant Sci. Lett.* **17**, 287–294.

Stymne, S., and Glad, G. (1981). *Lipids* **16**, 298–305.

Stymne, S., and Stobart, A. K. (1984a). *Biochem. J.* **220**, 481–488.

Stymne, S., and Stobart, A. K. (1984b). *Biochem. J.* **223**, 305–314.

Stymne, S., and Stobart, A. K. (1985a). *Planta* **164**, 101–104.

Stymne, S., and Stobart, A. K. (1985b). *Biochim. Biophys. Acta* **837**, 239–250.

Stymne, S., and Stobart, A. K. (1986a). *Biochem. J.* **240** 385–393.

Stymne, S., and Stobart, A. K. (1986b). *Physiol. Veg.* **24**, 45–51.

Stymne, S., Stobart, A. K., and Glad, G. (1983). *Biochim. Biophys. Acta* **752**, 198–208.

Tamai, Y., and Lands, W. E. M. (1974). *J. Biochem. (Tokyo)* **76**, 847–860.

Tartakoff, A. M. (1983). *Int. Rev. Cytol.* **85**, 221–252.

Tetenyi, P., Hethelyi, I., Okuda, T., and Imre, S. (1974). *Herba Hung.* **13**, 61–71.

Thomas, D. R., Jalil, M. N. H., Ariffin, A., Cooke, R. J., McLaren, I., Yong, B. C. S., and Wood, C. (1983). *Planta* **158**, 259–263.

Thompson, G. A. (1983). *In* "Cellular Acclimatisation to Environmental Change" (A. R. Cossins and P. Sheterline, eds.), pp. 33–53. Cambridge Univ. Press, London and New York.

Torrengo, M. P., and Brenner, R. R. (1976). *Biochim. Biophys. Acta* **424**, 36–44.

Traitler, H., Winter, H., Richli, U., and Ingenbleek, Y. (1984). *Lipids* **19**, 923–928.

Trémolières, A., Dubacq, J. P., and Drapier, D. (1982). *Phytochemistry* **21**, 41–45.

Trémolières, H., Trémolières, A., and Mazliak, P. (1978). *Phytochemistry* **17**, 685–687.

Töregard, B., and Podhala, 0. (1974). *Proc. Int. Rapskongr., 4th, Giessen, F. R. G. pp. 291–300.*

Van den Bosch, H., Van Golden, L. M. G., and Van Deenen, L. L. M. (1972). *Ergeb. Physiol., Biol. Chem. Exp. Pharmakol.* **66**, 13–145.

Vick, B., and Beevers, H. (1977). *Plant Physiol.* **59**, 459–463.

Vick, B., and Beevers, H. (1978). *Plant Physiol.* **62**, 173–178.

Vijay, I. K., and Stumpf, P. K. (1971). *J. Biol. Chem.* **246**, 2910–2917.

Vijay, I. K., and Stumpf, P. K. (1972). *J. Biol. Chem.* **247**, 360–366.

Wanner, G., and Theimer, R. R. (1978). *Planta* **140**, 163–169.

Wanner, G., Formanek, H., and Theimer, R. R. (1981). *Planta* **151**, 109–123.

Weaire, P. J., and Kekwick, R. G. O. (1975). *Biochem. J.* **146**, 425–437.

Wilson, R. F., Weissinger, H. H., Buck, J. A., and Faulkner, J. D. (1980). *Plant Physiol.* **66**, 545–549.

Yamada, M., and Usami, Q. (1975). *Plant Cell Physiol.* **16**, 879–884.

Yamashita, S., Hosaka, K., and Numa, S. (1972). *Proc. Natl. Acad. Sci. U.S.A.* **69**, 3490–3492.

Yatsu, L. Y., and Jacks, T. J. (1972). *Plant Physiol.* **49**, 937–947.

Yatsu, L. Y., Jacks, T. J., and Hensarling, T. P. (1971). *Plant Physiol.* **48**, 673–682.

Yermanos, D. M. (1975). *J. Am. Oil Chem. Soc.* **52**, 115–117.

Zilkey, B. F., and Canvin, D. T. (1972). *Can. J. Bot.* **50**, 323–326.

Galactolipid Synthesis

9

JACQUES JOYARD
ROLAND DOUCE

I. INTRODUCTION

Plastid membranes from eukaryotic algae or higher plants contain galactolipids as major compounds. These polar lipids are also the major components of membranes from cyanobacteria but are absent, in eukaryotes, from extraplastidial membranes. The reader is referred to a previous review (Douce and Joyard, 1980) for a complete discussion of galactolipid distribution and occurrence within plant tissues and cells. Galactolipids have specific physical properties that are responsible for some of the

The Biochemistry of Plants, Vol. 9

characteristic features of plastid membranes. Although these properties might have some important consequences in plastid biogenesis, and especially in galactolipid biosynthesis, we do not intend to discuss this problem. The reader can find valuable informations on this topic in reviews by Quinn and Williams (1983) and Murphy (1986).

Galactolipids are derived from the same three-carbon alcohol as phospholipids (i.e., glycerol). However, a phosphorylated alcohol is attached at the *sn*-3 position of the glycerol backbone in phospholipids, whereas it is a nonphosphorylated galactose moiety in galactolipids. In the IUPAC–IUB nomenclature of lipids, the structures of the two major plastid polar lipids, monogalactosyldiacylglycerol (MGDG) and digalactosyldiacylglycerol (DGDG) are 1,2-diacyl-3-*O*-(β-D-galacto-pyranosyl)-*sn*-glycerol and 1,2-diacyl-3-*O*-(α-D-galactopyranosyl-(1 → 6)-*O*-β-D-galactopyranosyl)-*sn*-glycerol, respectively. In addition, analyses of the proportions, positional distribution, and pairing of fatty acids which esterify the hydroxyl groups available at the *sn*-1 and *sn*-2 positions of the glycerol backbone, have demonstrated a large diversity of the diacylglycerol moiety in galactolipids. This observation reflects in fact the complexity of the biosynthetic pathways involved in galactolipid formation. To understand galactolipid biosynthesis, it is therefore necessary to keep in mind the structure of the molecules present in the membranes. A short description of these structures, presented in Fig. 1, will be made in this introduction. However, for detailed discussions, the reader is referred to previous reviews by Heinz (1977), Douce and Joyard (1980), and Harwood and Russell (1984).

In eukaryotes, but only in few cyanobacteria, galactolipids contain a high amount of polyunsaturated fatty acids. Since the major fatty acid in galactolipids (up to 95% of the total fatty acids in some species) is linolenic acid (all-*cis*-9,12,15-octadecatrienoic acid or $18:3^{\Delta9,12,15}$), the most abundant molecular species of MGDG and DGDG have 18:3 at both *sn*-1 and *sn*-2 positions of the glycerol backbone. Some plants, such as pea, having almost only 18:3 in MGDG, are called "18:3 plants" (Heinz, 1977). In nonphotosynthetic tissues, C_{18} fatty acids are more saturated: linoleic acid (*cis*,*cis*-9,12-octadecenoic acid or $18:2^{\Delta9,12}$) and oleic acid (*cis*-9-octadecenoic acid or $18:1^{\Delta9}$) can be predominant in galactolipids (e.g., in roots, tubers, or etiolated tissues). Other plants, such as spinach, contain appreciable amounts of hexadecatrienoic acid (all-*cis*-7,10,13-hexadecatrienoic acid or $16:3^{\Delta7,10,13}$) in MGDG; they are called "16:3 plants" (Heinz, 1977). Positional distribution of 16:3 in MGDG is highly specific: this fatty acid is only present at the *sn*-2 position of the glycerol backbone and is almost excluded from *sn*-1 position (Fig. 1) (Jamieson and Reid, 1971). Galactolipids contain little saturated fatty acids (at least in eukaryotes), but—when present in galactolipids—palmitic acid (hexadecanoic acid or 16:0) is found only in DGDG. The positional distribution of 16:0 in DGDG does not appear to be highly specific, but could be related to the taxonomic position of the plant, since in green algae (such as *Ulva*) 16:0 is almost exclusively localized at the *sn*-2 position of the glycerol, whereas in maize almost all 16:0 is found at the *sn*-1 position (Heinz, 1977).

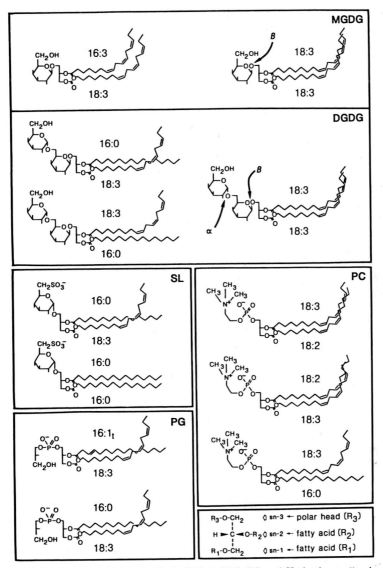

Fig. 1. Structure of plastid glycerolipids. MGDG, DGDG, PG, and SL that have a "prokaryotic" structure are on the left-hand side of the figure, whereas glycerolipids that have a eukaryotic structure are on the right-hand side. [Reprinted with permission from Joyard *et al.* (1987). *In* "Models in Plant Physiology/Biochemistry/Technology" (D.W. Newman and K.G. Wilson, eds.). Copyright CRC Press, Inc., Boca Raton, FL.]

Therefore, on the basis of the structure of their diacylglycerol moiety, galacto-lipids can be separated into two groups: the first one has C_{18} (and to a much lesser extent C_{16}) fatty acids at the sn-1 position of the glycerol backbone, and only C_{16} fatty acids at sn-2 position, whereas the second group has also C_{16} or C_{18} fatty acids at the sn-1 position but only C_{18} fatty acids at sn-2 position. Since the first structure is characteristic for cyanobacteria, it is called a "prokaryotic" structure whereas the second one is called a "eukaryotic" structure, and is present in most glycerolipids, such as phosphatidylcholine (PC), of all eukaryotic cells (see Fig. 1). Galactolipids, and especially MGDG, can have both prokaryotic and eukaryotic structure: in 18:3 plants, such as pea, only the eukaryotic structure is present in MGDG whereas in plants such as *Anthriscus* (16:3 plant) only the prokaryotic structure is present in MGDG. Spinach, another 16:3 plant, is interesting in that it contains both eukaryotic and prokaryotic structure in MGDG (Fig. 1) (Heinz, 1977).

The diversity of the diacylglycerol moiety in galactolipids from algae and higher plants, compared to the more simple situation in cyanobacteria, lead us to focus on one of the main problems in understanding the biosynthetic pathways of these lipids: the origin of diacylglycerol diversity. In fact, the difference observed between cyanobacteria and higher plants lies simply in the existence of several compartments in eukaryotic cells, each compartment having its own capability of synthesizing glycerolipids.

In this review, we will discuss the interacting roles postulated for the different compartments involved, in eukaryotic plant cells, in the synthesis and assembly of the three parts of galactolipid molecules: fatty acids, glycerol, and galactose. The comparison with the situation found in cyanobacteria obviously provides a basis to understand the specific role of plastids in such a biosynthetic process. We will discuss first the formation of the diacylglycerol moiety, which is the substrate for further galactosylation to synthesize MGDG.

II. FORMATION OF DIACYLGLYCEROL THROUGH THE KORNBERG–PRICER PATHWAY

The differences, summarized above, among galactolipid molecules and mostly confined to their diacylglycerol moiety can be due to several reasons. First, the enzymes responsible for the stepwise acylation of sn-glycerol 3-phosphate can be highly specific with respect to the fatty acids used and the positions which are acylated. In addition, compartmentation of these enzymes can also be responsible for this diversity. The enzymes involved (i.e., sn-glycerol-3-phosphate acyltransferase, monoacylglycerol-3-phosphate acyltransferase, and phosphatidate phosphatase) constitute the Kornberg–Pricer pathway, and have been localized in several compartments of the plant cell.

A. Localization within the Plant Cell

Following the observations of Kennedy (1961) and co-workers and Kornberg and Pricer (1953), who established this pathway for the biosynthesis of glycerolipids in bacteria and animal cells, Barron and Stumpf (1962) with avocado mesocarp, Cheniae (1965) and Sastry and Kates (1966) with spinach, demonstrated the formation of diacylglycerol from sn-glycerol 3-phosphate via phosphatidic acid. These experiments established the existence of the enzymes of the Kornberg–Pricer pathway within plant cells. These enzymes were associated with the microsomal fraction and therefore were membrane bound. Some activity was also found in the mitochondrial or chloroplast fraction, but this was attributed to a contamination of the organelles by microsomal membranes. However, it is now well known that microsomal fractions are in fact composite membrane fractions containing all the small vesicles which can pellet at 100,000 g from a 10,000 g supernatant. Therefore, a microsomal localization for an enzyme does not demonstrate a precise site within the cell. By isopycnic centrifugation of cell-free extracts from castor bean endosperm, Vick and Beevers (1977) showed that most of phosphatidic acid synthesis was associated with an endoplasmic reticulum-rich membrane fraction (identified by the presence of CDPcholine:diacylglycerol phosphorylcholine transferase, but see p. 254). Analyses of microsomal membranes from maize roots by Sauer and Robinson (1985) showed a more complex situation. Unlike castor bean endosperm, which is a nongrowing tissue undergoing metabolic and structural modifications associated with gluconeogenesis and which possess only few Golgi apparatus (Moore et al., 1973), maize roots are growing tissues containing numerous Golgi apparatus. Indeed, Sauer and Robinson (1985) achieved a clear separation of three different membrane fractions containing endoplasmic reticulum, Golgi membranes, and plasmalemma, identified by following the distribution of three marker enzymes through the gradient, respectively, antimycine A-insensitive-NADH-cytochrome c oxidoreductase, inosine diphosphatase, and sterylglucoside transferase. Among these membrane fractions, the endoplasmic reticulum and the Golgi membranes were able to acylate sn-glycerol 3-phosphate to almost the same extent. These results demonstrated that the Kornberg–Pricer pathway in plants was indeed not restricted to endoplasmic reticulum.

Analyses on the presence of the Kornberg–Pricer pathway within cell organelles was only made possible by the development of reliable methods to prepare intact plastids or mitochondria, devoid of contamination by other cell membranes. Douce (1970) demonstrated that purified mitochondria from various plant tissues, obtained after isopycnic centrifugation on a sucrose gradient and mostly devoid of plastid or extramitochondrial membranes, were able to incorporate sn-glycerol 3-phosphate into phosphatidic acid. However, in crown-gall tissue, the activity was 10 times higher on a protein basis in the microsomal fraction than in mitochondria. This was confirmed on castor bean endosperm by Vick and Beevers (1977), who found ~10% of the cell-free system acylation activity associated with mitochondria.

Mitochondria are not the only DNA-containing organelles to have the Kornberg–Pricer pathway. Analysis of purified chloroplasts from spinach and of purified etioplasts from maize provided a striking result: Douce and Guillot-Salomon (1970) demonstrated that these intact plastids, devoid of contaminating extraplastidial membranes, were able to incorporate sn-glycerol 3-phosphate into lysophosphatidic acid, phosphatidic acid, and diacylglycerol. These results demonstrated, therefore, the existence of an additional site for phosphatidic acid synthesis in plant cells from green and etiolated tissues. The same was true for *Euglena*: in this phytoflagellate, *sn*-glycerol 3-phosphate acylation activity in chloroplasts represent 60% of the total cell activity, the remaining activity being distributed among microsomal membranes (30%) and mitochondria (10%) as shown by Boehler and Ernst-Fonberg (1976).

All these observations clearly demonstrate the complexity of studies on lipid biosynthesis, and especially *in vivo*. When the kinetics of incorporation of specifically labeled compounds into lipids from either leaves, tissues, seeds, or algae are analyzed, one must keep in mind the multiplicity of sites for phosphatidic acid synthesis. The complexity might be even greater, since numerous membranes (plasmalemma, peroxisomal membranes, etc.) have not yet been clearly investigated for the Kornberg–Pricer pathway.

B. Localization within Plastids

All plastids analyzed so far—proplastids (Journet and Douce, 1985), amyloplasts (Fishwick and Wright, 1980), chromoplasts (Liedvogel and Kleinig, 1979), as well as chloroplasts from algae (Boehler and Ernst-Fonberg, 1976; Jelsema *et al.*, 1982)—are able to acylate *sn*-glycerol 3-phosphate for further metabolism into specific glycerolipids.

It was possible to fractionate chloroplasts to localize precisely the enzymes of the Kornberg–Pricer pathway within the organelle. A slight osmotic shock on purified intact spinach chloroplasts was used for the preparation and isolation of chloroplast subfractions: soluble extract (mostly corresponding to stroma, but containing also all the soluble enzymes of the chloroplast), thylakoids, and envelope membranes (Douce *et al.*, 1973). These fractions have been highly characterized, especially the envelope membranes. The reader is referred to several reviews by Douce and Joyard (1979, 1982, 1984) and Douce *et al.* (1984a) for a complete description of the fractions prepared from spinach chloroplasts. However, when assayed for phosphatidic acid synthesis, thylakoids and envelope membranes—in contrast with intact spinach chloroplasts—were practically unable to incorporate *sn*-glycerol 3-phosphate into membrane lipids (Joyard and Douce, 1977).

This was not due to the lack of fatty acids for acylation, since addition of acyl-CoA had no influence on the incorporation by envelope membranes or thylakoids. However, addition of fatty acid thioesters to the chloroplast extract containing the soluble enzymes induced a marked stimulation of the incorporation of *sn*-glycerol 3-phosphate into lipids. In this case, and in contrast with chloroplasts, neither diacylglycerol nor phosphatidic acid were labeled: most of the *sn*-glycerol 3-phosphate was incorporated into lysophosphatidic acid (Joyard and Douce, 1977). These findings, together with those of Bertrams and Heinz (1976), demonstrated that the first enzyme of the Kornberg–Pricer pathway from chloroplasts, the *sn*-glycerol-3-phosphate acyltransferase, could be recovered as a "soluble" enzyme. In fact, as discussed by Joyard and Douce (1977), this enzyme is probably localized in the stroma, but its presence in the intermembrane space cannot be entirely ruled out (see p. 224). In both cases, however, it is loosely bound to envelope membranes and probably detached during the isolation procedure: the products manufactured by this acyltransferase were indeed recovered in the membranes, but not in the incubation medium (Joyard and Douce, 1977).

This hypothesis is strengthened by the observation that in other cell organelles (e.g., mitochondria) or membranes (microsomal membranes) from animals (for review see Bell and Coleman, 1980) or plants (see above) and prokaryotes (for review see Cronan, 1978), the sn-glycerol-3-phosphate acyltransferase is indeed membrane bound, but can be easily removed in most cases (Yamashita and Murna, 1972; Monroy et al., 1973; Ishinaga et al., 1976; Snider and Kennedy, 1977; Grobovsky et al., 1979; Lightner et al., 1980; Green et al., 1981). However, in some membranes, such as in the outer membrane from rat liver mitochondria, the sn-glycerol-3-phosphate acyltransferase spans the transverse plane of the membrane and is therefore accessible from both the cytosolic side and the intermembrane space of the organelle (Hesler et al., 1985).

The lack of phosphatidic acid and diacylglycerol formation in thylakoids and envelope fractions could be attributed to the lack of sn-glycerol-3-phosphate acyltransferase in pure membrane fractions. Indeed, recombination of chloroplast membranes (envelope or thylakoids) and soluble enzymes from chloroplasts provided striking results: as shown in Table I, addition of the soluble enzymes to envelope membranes induced a strong increase in the incorporation of sn-glycerol 3-phosphate, but was almost without effect on lipid biosynthesis by thylakoids. Analyses of the products revealed that phosphatidic acid and diacylglycerol accounted for > 80% of the sn-glycerol 3-phosphate incorporated into lipids (Fig. 2). These results demonstrated that envelope membranes, but not thylakoids, from mature spinach chloroplasts contained a monoacylglycerol-3-phosphate acyl-transferase responsible for the formation of phosphatidid acid and a phosphatidate

TABLE I

Incorporation of sn-Glycerol 3-Phosphate into Glycerolipids by Combination of Various Fractions from Spinach Chloroplasts[a,b]

Fraction	sn-Glycerol 3-phosphate incorporated (nmol)
Chloroplast extract	10.8
Envelope	0.8
Thylakoids	3.2
Chloroplast extract + envelope	76.0
Chloroplast extract + thylakoids	16.8

[a]Reproduced with permission from Joyard and Douce (1977).

[b]The different fractions were incubated in the presence of sn-[^{14}C]glycerol 3-phosphate and proteins corresponding to 9 mg for chloroplast extract, 4.9 mg for thylakoids, and 0.6 mg for envelope membranes. Assays were for 60 min. For experimental conditions, see Joyard and Douce (1977).

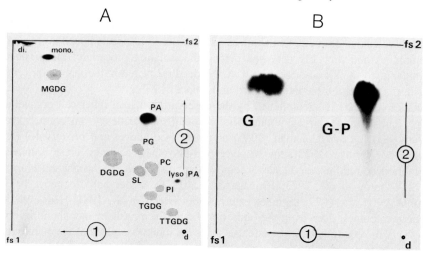

Fig. 2. Incorporation of sn-[^{14}C]glycerol 3-phosphate into envelope membranes. Envelope membranes and a soluble extract from chloroplasts were incubated together as described by Joyard and Douce (1977). After extraction and analyses, the molecules synthesized were identified as lysophosphatidic acid (lyso PA), phosphatidic acid (PA), monoacylglycerol (mono), and diacylglycerol (di) (A) Separation of the envelope glycerolipids by two-dimensional thin-layer chromatography. Solvent systems used: first direction, chloroform–methanol–water (65:25:4, v/v); second direction, chloroform–acetone–methanol–acetic acid–water (100:40:20:20:10, v/v). Envelope lipids are represented as dotted spots, whereas radioactive lipids were revealed by autoradiography. (B) Separation of the polar head groups of newly synthesized glycerolipids. The molecules synthesized by the envelope Kornberg–Pricer pathway were hydrolyzed according to Benson and Maruo (1958), and the polar head groups were separated by two-dimensional paper chromatography in the following solvent system: first direction, phenol–water (100:38, v/v); second direction, methanol–formic acid–water (80:13:7, v/v). Radioactive glycerol (G) and sn-glycerol 3-phosphate (G-P) deriving, respectively, from mono- and diacylglycerol, and from lysophosphatidic and phosphatidic acids. [Reproduced from Joyard (1979).]

phosphatase, yielding diacylglycerol (Joyard and Douce, 1977). These results were confirmed with envelope membranes prepared from pea chloroplasts (Bertrams and Heinz, 1981; Frentzen *et al.*, 1983; Andrews *et al.*, 1985). They were also extended to envelope membranes from proplastids (Journet and Douce, 1985), amyloplasts (Fishwick and Wright, 1980), and chromoplasts (Liedvogel and Kleinig, 1979).

However, the envelope membrane fraction analyzed as described above consists in a mixture of outer and inner envelope membranes (Douce *et al.*, 1973; Douce and Joyard, 1979). The development of methods to subfractionate envelope membranes and then to prepare membrane fractions enriched in outer and inner envelope membranes[1] (Cline *et al.*, 1981; Block *et al.*, 1983a) (Fig. 3) made possible further

---

[1]We want to focus on the problem of cross-contamination of the membrane fractions analyzed. This is most important, especially when topologically close membranes such as the outer and inner envelope membranes are considered. Block *et al.* (1983a) have carefully determined the level of cross-contamination of both membrane fractions by using polyclonal antibodies raised against polypeptides specifically associated with outer and inner envelope membranes (Joyard *et al.*, 1982, 1983). Indeed

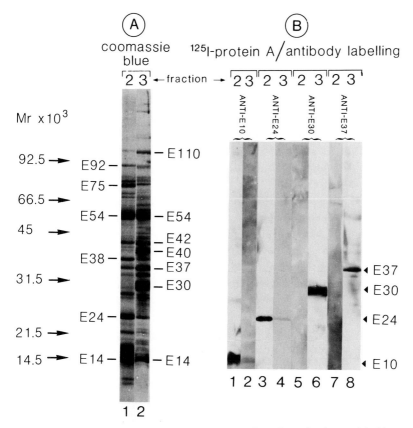

Fig. 3. Electrophoretic and immunochemical analyses of membrane fractions enriched in outer and inner envelope membranes from spinach chloroplasts. (A) Coomassie blue staining of polypeptides from fractions 2 and 3 (Block *et al.*, 1983a) separated by SDS–PAGE at room temperature with a 7.5 to 15% acrylamide gradient. Lane 1; fraction 2; lane 2; fraction 3. (B) Autoradiography of antigen–antibody–[125]I-labeled protein A complexes after electrophoretic transfer of polypeptides from polyacrylamide gels to nitrocellulose sheets. Antibodies to envelope polypeptides E10, E24, E30, and E37 were used. Since E10 and E24 are outer envelope polypeptides whereas E30 is an inner envelope polypeptide (Joyard *et al.*, 1982, 1983), the immunochemical analyses demonstrate that fraction 2 and fraction 3 are, respectively, enriched in outer and inner envelope membrane. [Reproduced with permission from Block *et al.* (1983a).]

immunochemical methods, though money- and time-consuming, are far more reliable than others. The measure of enzymatic activities in different membrane fractions do not always take into account the latency of these activities. A highly confusing method involves the comparison of Coomassie blue-staining intensity of polypeptides having similar electrophoretic mobilities, when analyzed by polyacrylamide gel electrophoresis. In such a system, several different polypeptides might run at the same position in the gel and made quantification of the level of cross-contamination almost impossible.

analyses of the envelope Kornberg–Pricer pathway. Dorne *et al.* (1982a) and Block *et al.* (1983c) reported that both envelope membranes were able to synthesize phosphatidic acid only when supplied with a soluble fraction from chloroplasts. A different situation was found in pea by Andrews *et al.* (1985), who found only little activity in the outer envelope membrane, thus suggesting that in pea only the inner envelope membrane contained the monoacylglycerol 3-phosphate acyltransferase. The existence of such an enzyme, together with the lack of *sn*-glycerol-3-phosphate acyltransferase, on the outer envelope membrane from spinach chloroplasts give some support to the suggestion (see above) that lysophosphatidic acid could be synthesized in the intermembrane space of the envelope, together with a localization in the chloroplast stroma.

The level of *sn*-glycerol 3-phosphate in isolated intact chloroplasts can be strongly reduced by addition of inorganic phosphate to the incubation medium, thus suggesting that the inner envelope phosphate translocator could be involved in *sn*-glycerol 3-phosphate transport into chloroplasts (Sauer and Heise, 1982). If these chloroplasts were really devoid of *sn*-glycerol 3-phosphate after addition of inorganic phosphate, the observation that they are still able to synthesize lyso-phosphatidic acid would suggest an additional site of synthesis in the intermembrane space of the envelope (Sauer and Heise, 1982).

The last enzyme in the Kornberg–Pricer pathway is the phosphatidate phosphatase, which yields diacylglycerol. Block *et al.* (1983c) clearly demonstrated that this enzyme was specifically localized on the inner envelope membrane from spinach chloroplast, an observation confirmed in pea by Andrews *et al.* (1985).

Altogether, these results demonstrate that owing to their envelope membranes, the chloroplasts—and all kinds of plastids—are an important site of phosphatidic acid and diacylglycerol synthesis within the plant cell (see Fig. 4).

A question which remains to be elucidated is whether envelope membranes are the sole site of phosphatidic acid synthesis in chloroplasts from all plant species. For instance, it is not known whether growing thylakoids from developing plastids contain,[2] or are devoid of, the enzymes of the Kornberg–Pricer pathway. In fact, proper characterization of a phosphatidic acid biosynthesis within thylakoids is hampered by several limitations. First, thylakoids are almost never analyzed with

---

[2]The observation that, in higher plant chloroplasts, *sn*-glycerol-3-phosphate acyltransferase was recovered as a soluble enzyme (Bertrams and Heinz, 1976; Joyard and Douce, 1977) has led Jelsema *et al.* (1982) to speculate about a possible inactivation of a thylakoidal Kornberg–Pricer pathway during isolation procedure. Obviously, this hypothesis is unlikely because thylakoids are much more simple membranes to isolate than any other cell membranes and can be obtained, in a rather pure state, in < 30 min by appropriate procedures. In addition, Jelsema *et al.* (1982) take argument of the presence of plastoglobules associated with thylakoids to claim that "localization of lipid-synthetic enzymes to the photosynthetic membranes" occur in higher plants. Such a statement is highly speculative and is not supported by experimental data; for instance, analyses of spinach plastoglobules have clearly demonstrated that these osmiophyllic globules are devoid of any polar lipids (Dorne *et al.*, 1982a). Therefore, the presence of plastoglobules in chloroplasts cannot reflect any polar lipid-biosynthetic activity within thylakoids.

Fig. 4. The envelope Kornberg–Pricer pathway and its tight connection with fatty acid synthesis, in the stroma phase, and galactolipid synthesis, in envelope membranes. The representation of galactolipid transfer does not preclude any system for such a transfer! [Adapted from Joyard (1979).]

respect to contamination by envelope membranes or extraplastidial membranes, which all contain the enzymes of the Kornberg–Pricer pathway. For instance, we have found that almost half of the estimated amount of envelope membranes from spinach chloroplasts was recovered in the thylakoid fraction when intact chloroplasts were fractionated after an osmotic shock (Joyard and Douce, 1976a) (see also p. 241). Furthermore, the inner envelope membrane and thylakoids from pea chloroplasts (Cline *et al.*, 1981) or spinach chloroplasts (Block *et al.*, 1983b) have almost the same polar lipid composition, a situation which highly favors artifactual fusions and therefore may lead to wrong observations for enzyme distribution. In addition, thylakoid and inner envelope membrane have a very tight relationship during the first stages of plastid development (Carde *et al.*, 1982; Carde, 1985; Fig. 5); therefore, these two membrane systems may show the same enzymatic equipment necessary for plastid biogenesis. The stage of development of the plant analyzed is obviously most important: Bertrams and Heinz (1976) found that *sn*-glycerol-3-phosphate acyltransferase was associated with thylakoids from 16- to 20-day-old peas but not with those from 8-day-old shoots, whereas the reverse was observed in tobacco cells analyzed by cytochemical techniques (Brangeon and Forchioni, 1986). Finally, some species specificities cannot be ruled out, and the results obtained in growing *Chlamydomonas* by biochemical (Jelsema *et al.*, 1982) or cytochemical (Michaels *et al.*, 1983) investigations cannot be extended to mature spinach thylakoids without caution.

Therefore, we are convinced that investigations of acylating activities in thylakoids (properly characterized with respect to contaminations by other membranes), at different stages of chloroplast development and prepared from various plant species (higher plants as well as algae) are necessary in order to establish with

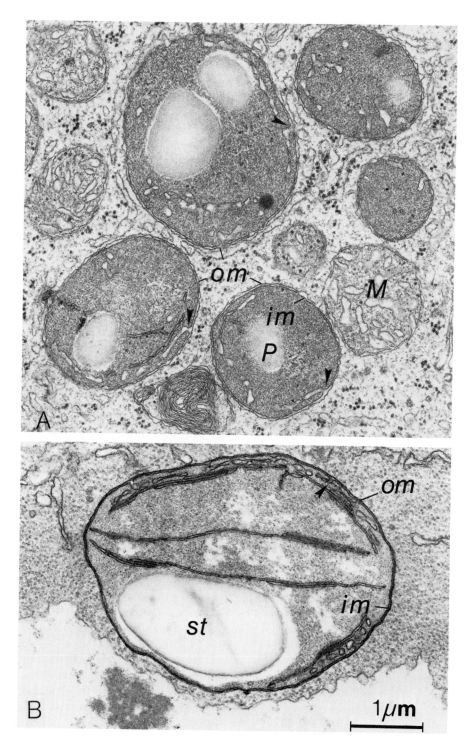

any degree of confidence the role of thylakoid membranes in phosphatidic acid synthesis. Comparison with the situation in prokaryotes may help to elucidate this problem. For instance, in the gram-negative bacterium *Rhodopseudomonas spheroides* grown phototrophically, most of *sn*-glycerol-3-phosphate acyltransferase was associated with cytoplasmic membranes, enzyme activity associated with the intracytoplasmic membranes were within the range attributable to cytoplasmic membrane contamination of this membrane fraction (Cooper and Lueking, 1984).

C.   Connections between the Kornberg–Pricer Pathway and Fatty Acid Synthesis in Plastids

The plastid Kornberg–Pricer pathway is unique within the plant cell in the sense that it is probably the only one to be closely associated with fatty acid synthesis. Ohlrogge *et al.* (1979) have demonstrated that in spinach leaf, all fatty acid synthesis is ACP dependent.[3] After fractionation of spinach leaf protoplasts on a sucrose gradient, they found that all ACP in the cell was associated with chloroplasts, thus demonstrating that in spinach leaf cell fatty acid synthesis was solely associated with chloroplasts (Ohlrogge *et al.*, 1979). Further evidences were obtained by the exclusive localization within chloroplasts of two enzymes necessary for fatty acid synthesis: acetyl-CoA carboxylase (Nikolau *et al.*, 1984) and acetate-CoA ligase (Kuhn *et al.*, 1981). These results, together with the observations of Nothelfer *et al.* (1977) with isolated soybean cells and Vick and Beevers (1978) with castor bean endosperm, demonstrated that in green or etiolated tissues from higher plants, plastids are probably the sole site for *de novo* fatty acid synthesis. In fact, these results were the conclusion of a long series of experiments, starting from Smirnov (1960), showing that plastids possess all the biosynthetic machinery for fatty acid synthesis (for review see Stumpf, this volume, Chapter 5). Some fatty acid synthesis has always been found in microsomal membranes (Weaire and Kekwick, 1975; Cassagne and Lessire, 1978; Sanchez *et al.*, 1982). The observation that in plastids fatty acid synthetase activity was always soluble, in contrast to the microsomal activity which is membrane bound, has led Walker and Harwood (1985) to exclude a plastid contamination for the microsomal fatty acid synthetase activity. Therefore, it is possible that minor additional sites for fatty acid synthesis, ACP-independent, could exist, but this hypothesis (Roughan and Slack,

---

[3]Actually, studies of plant ACP revealed the existence of at least two isoforms expressed in a tissue-specific manner: ACP I and ACP II (Ohlrogge and Kuo, 1985).

Fig. 5.   (A) Aspects of some plastids in a cell of a leaf stem from calamondin (*Citrofortunella mitis*) (unpublished observation of Dr. J. P. Carde). (B) Aspect of a young plastid from a small (2-mm) spinach leaf. Plastid membranes were stained specifically with OsO₄–ferricyanide mixture according to Carde *et al.* (1982). [Adapted with permission from Carde *et al.* (1982).] In both pictures, the inner envelope membrane has very tight relationships with the growing internal membrane system (arrows). P, Plastids; M, mitochondria; om, outer-envelope membrane; im, inner-envelope membrane; st, starch grain.

1982; Walker and Harwood, 1985) needs further analysis to be clearly demonstrated. However, even though the fatty acid synthetase enzymes are soluble in plastids, they are probably closely associated *in vivo* with the inner envelope membrane as suggested by Douce and Joyard (1980), since the enzymes involved in fatty acid metabolism (MGDG, PG, and sulfolipid biosynthesis) or export (acyl-CoA synthesis) are localized on envelope membranes (Douce *et al.*, 1984a,b; Douce and Joyard, 1984). Such a localization would indeed favor the kinetic links between fatty acid synthesis and processing (Douce and Joyard, 1980) (see Fig. 5).

The enzymes involved in fatty acid synthesis have been purified and characterized in the last few years (Caughey and Kekwick, 1982; Hendren and Bloch, 1980; Høj and Mikkelsen, 1982; Schüz *et al.*, 1982; Shimakata and Stumpf, 1982a,b,c, 1983a,b). All the available information of the molecular structures of the fatty acid synthetase complexes of chloroplasts is reviewed by Stumpf (this volume, Chapter 5). The major conclusion of these studies is that the fatty acid synthetase is in fact a nonassociated complex, which is very similar to that of prokaryotes such as *Escherichia coli*. In addition, the enzymes involved in this complex lead to the formation of 16:0-ACP and 18:1-ACP as the major products, when supplied with acetate. Thus, intact chloroplasts, when carefully prepared in order to be physiologically active, can incorporate acetate[4] into fatty acids at very high rates. For

---

[4] In vitro experiments are usually performed with [1-^{14}C]acetate, since this C_2 molecule is the preferred substrate for fatty acid synthesis within chloroplasts (Stitt and ap Rees, 1979; Roughan *et al.*, 1979b). An interesting controversy has developed in the last few years about the origin and synthesis of acetyl units in leaves (see, e.g., Givan, 1983). For instance, Sherratt and Givan (1973) and Stitt and ap Rees (1979) found that almost no $^{14}CO_2$ was incorporated into lipids in isolated pea chloroplasts, in contrast to studies on spinach by Murphy and Leech (1977), who found low, but readily detectable labeled fatty acids from $^{14}CO_2$. In fact, isolated intact plastids have all the enzymes, but one, necessary for CO_2 or starch to be converted into fatty acids: phosphoglycerate mutase has not been detected in all plastids analyzed so far (Givan, 1983; Journet and Douce, 1985), and therefore there is a missing link in the plastid glycolytic pathway that would convert 3-phosphoglycerate into pyruvate. Thus, CO_2 is transformed, during photosynthesis, into triose-phosphate molecules that are exported outside chloroplasts through the phosphate translocator and therefore cannot be used directly for fatty acid synthesis. Another enzyme was thought to be absent from some plastids, namely pyruvate dehydrogenase (lipoamide) complex (Givan, 1983; Murphy and Walker, 1982): this enzymatic complex was present in pea chloroplasts (Elias and Givan, 1979; Williams and Randall, 1979), castor bean endosperm plastids (Reid *et al.*, 1977), or cauliflower bud plastids (Journet and Douce, 1985), but was apparently absent from spinach (Roughan *et al.*, 1979b; Murphy and Stumpf, 1981). Liedvogel (1985) has demonstrated, however, that spinach chloroplasts, in contrast with the previous studies, indeed contained a pyruvate dehydrogenase complex with an activity close to that found in pea chloroplasts, thus demonstrating that pyruvate can be transformed into acetyl-CoA in almost all plant species (Miernyk and Dennis, 1983; Liedvogel, 1986). Therefore, chloroplasts have two enzymatic activities that can provide acetyl-CoA for fatty acid synthesis, namely pyruvate dehydrogenase complex and acetyl-CoA synthetase. Roughan *et al.* (1979b) have shown that pyruvate is incorporated into lipids at lower rates that acetate. In addition, chloroplasts contain much less acetate (< 70 μM) than what was previously supposed (1 mM, Roughan and Slack, 1982) compared to the level of pyruvate (150 μM) (Liedvogel, 1985). In fact, the physiological significance of acetyl-CoA synthetase in plastids could be related to the lack of the glyoxylic cycle in peroxisomes (Gerhardt, 1983): peroxisomes have an acetyl-CoA hydrolase (Brouquisse *et al.*, 1987) and therefore produce acetate, which can be used for fatty acid synthesis in plastids owing to the acetate-CoA lipase. In addition, the plastid stroma is alkaline and can easily accumulate acetate (Jacobson and Stumpf, 1972), which provide support for the mechanism proposed above.

instance, Roughan *et al.* (1979a) measured incorporation rates as high as 1500 nmol of acetate incorporated per hour per milligram chlorophyll, in spinach chloroplasts having high O_2-evolving activity. This rate is almost identical to that measured in whole leaves.

Intact spinach chloroplasts incorporate [1-^{14}C]acetate mostly into 18:1 and 16:0. For instance, Roughan *et al.* (1976) found that between 70 and 80% of the total radioactivity was recovered as unesterified 18:1. Since the plastid fatty acid synthetase form mostly acyl-ACP, these results suggest that an active acyl-ACP thioesterase is present in chloroplasts. The oleoyl-ACP thioesterase is probably specific for acyl-ACP I, since it possessed a tenfold lower K_m for oleoyl-ACP I compared to oleoyl-ACP II (Guerra *et al.*, 1986). This enzyme has been purified and characterized in maturing seeds of safflower (McKeon and Stumpf, 1982) and is present in the stroma phase of chloroplasts (Joyard and Stumpf, 1981). Therefore, the soluble phase of chloroplasts is able to synthesize unesterified fatty acids: 16:0, 18:0, and 18:1. Addition of envelope membranes to the incubation mixture increases the relative proportion of C_{18} fatty acids compared to C_{16} fatty acids (Joyard, 1979), in agreement with the situation observed with intact chloroplasts (Roughan and Slack, 1982). This observation suggests an association of the soluble enzymes involved in fatty acid synthesis, and especially of palmitoyl-ACP elongase (which leads to fatty acid synthesis), with the inner envelope membrane. Such a close relationship would channel fatty acid synthesis toward C_{18} formation, and desaturation into 18:1.

The effect of addition of different metabolites to isolated chloroplasts has been thoroughly reviewed by Roughan and Slack (1982). Among those, *sn*-glycerol 3-phosphate, ATP, and coenzyme A (CoA) are of special interest, since they will channel fatty acids through specific metabolic pathways on envelope membranes.

ATP and CoA, when added to isolated intact chloroplasts, markedly increased the formation of acyl-CoA thioesters (Roughan *et al.*, 1976, 1979a, 1980; Roughan and Slack, 1981) and especially 18:1-CoA. The formation of acyl-CoA by intact chloroplasts is due to an envelope acyl-CoA synthetase (Joyard and Douce, 1977; Roughan and Slack, 1977; Joyard and Stumpf, 1981). Since acyl-CoA synthesis in chloroplasts was stimulated by exogenous ATP and not by endogenous ATP (Roughan *et al.*, 1979a, 1980), it was suggested that acyl-CoA synthetase was localized in a chloroplast compartment freely accessible to external ATP. Indeed, the acyl-CoA synthetase has been localized specifically on the outer envelope membrane (Dorne *et al.*, 1982a; Block *et al.*, 1983c; Andrews and Keegstra, 1983; Heemskerk *et al.*, 1985), in contrast with the envelope acyl-CoA thioesterase (Joyard and Stumpf, 1980), which is localized on the inner envelope membrane (Dorne *et al.*, 1982a; Block *et al.*, 1983c; Andrews and Keegstra, 1983). The presence, on different envelope membranes, of these two enzymes which manipulate acyl-CoA is most interesting. Both enzymes are active at alkaline pH (Joyard and Stumpf, 1980, 1981). The envelope acyl-CoA synthetase converts very efficiently fatty acids into acyl-CoA thioesters; when the incubation mixture contains ATP and CoA, 18:1 is one of the best substrates (but also 18:2 and 18:3) and activities as high as 1.5–3 μmol acyl-CoA formed per hour per milligram protein have been found in envelope membranes (Joyard and Stumpf, 1981). On the other hand, 18:1-CoA is only a poor substrate for acyl-CoA thioesterase (the enzyme is most active with 14:0-CoA), and this enzyme is strongly inhibited by unesterified fatty acids (and especially 18:1), as shown by Joyard and Stumpf (1980). The localization of acyl-CoA synthetase on the outer envelope membrane raises a most interesting question: how are newly synthesized fatty acids (mostly 18:1) transferred across the inner envelope membrane, from their site of synthesis (the plastid stroma) to the outer envelope membrane where they are esterified into acyl-CoA, especially if one considers that free fatty acids behave as strong uncouplers and acyl-CoA as detergents? Up to now, no satisfactory mechanisms have been proposed, although this transport is essential for cell biogenesis.

Addition of sn-glycerol 3-phosphate also channels 16:0-ACP and 18:1-ACP to a specific metabolic pathway: the envelope Kornberg–Pricer enzymes. sn-Glycerol 3-phosphate does not stimulate the rate of acetate incorporation by isolated intact chloroplasts but induces, first, an increase in the amount of 16:0 formed at the expenses of 18:1, and second, a strong stimulation of the amount of diacylglycerol synthesized at the expenses of unesterified fatty acids (Roughan et al., 1979a). This last observation clearly confirms that newly synthesized fatty acids can be incorporated into glycerolipids owing to the envelope Kornberg–Pricer pathway (see also Joyard et al., 1979). In addition to the sn-glycerol 3-phosphate requirement, phosphatidic acid synthesis is also dependent on the availability of acyl-ACP (Frentzen et al., 1983). The pool size of ACP is actually very low (8 μM, Ohlrogge et al., 1979) and is equivalent to only 5 sec of fatty acid synthesis. In addition, Soll and Roughan (1982) have shown that acyl-ACP is regenerated only in the light (at least in leaves, since fatty acid synthesis also occurs in roots). In spinach chloroplasts, their maximum steady-state concentrations are very low (14:0-ACP, 0.33 μM; 16:0-ACP, 1.6 μM; 18:0-ACP, 0.99 μM; 18:1-ACP, 2.08 μM; as calculated by Soll and Roughan, 1982) and clearly represent a balance between synthesis and utilization. Indeed, addition of sn-glycerol 3-phosphate reduced steady-state concentration of acyl-ACP; 18:1-ACP is decreased by 66% upon addition of sn-glycerol 3-phosphate (Soll and Roughan, 1982). Thus, addition of sn-glycerol 3-phosphate channels newly synthesized acyl-ACP toward glycerolipid formation, and acyl-ACP never accumulates in vivo, in agreement with analyses on spinach leaves (Sanchez and Mancha, 1980). Another plastid enzyme, acyl-ACP thioesterase, also requires acyl-ACP as a substrate and therefore competes with the Kornberg–Pricer pathway, which uses only acyl-ACP, but not unesterified fatty acids. Therefore, the respective rates of removal of acyl-ACP by the acyl-ACP thioesterase or by the envelope Kornberg-Pricer pathway would tightly control the formation of glycerolipids or acyl-CoA (from unesterified fatty acids). A very active acyl-ACP thioesterase would channel newly synthesized fatty acids toward acyl-CoA synthesis on the outer envelope membrane, but not toward phosphatidic acid. Conversely, a very active Kornberg–Pricer pathway would favor phosphatidic acid synthesis at the expense of unesterified fatty acids and finally of acyl-CoA formation. In addition, studies of the activity of ACP isoforms in the reactions involved in acyl-ACP metabolism (Guerra *et al.*, 1986) suggest that acyl-ACP I would be preferentially channeled toward acyl-ACP thioesterase and then exported outside the organelle, whereas acyl-ACP II would be preferentially used for glycerolipid biosynthesis on the envelope membranes. Such mechanisms might be partly responsible for the differences observed in chloroplasts from 16:3 and 18:3 plants in the relative importance of unesterified fatty acids or glycerolipids that are synthesized *de novo* from acetate and *sn*-glycerol 3-phosphate (see, e.g., Heinz and Roughan, 1983).

Finally, all these observations clearly demonstrate that the enzymes involved in fatty acid synthesis are closely associated with envelope membranes, where further metabolization of newly synthesized fatty acids occurs: synthesis of acyl-CoA or phosphatidic acid and diacylglycerol synthesis.

D. Properties of the Enzymes of the Envelope Kornberg–Pricer Pathway

1. Acyl-ACP:sn-Glycerol 3-Phosphate Acyltransferase (E.C. 2.3.1.15)

The first enzyme of the envelope Kornberg–Pricer pathway is a so-called "soluble" enzyme which produces lysophosphatidic acid:

sn-glycerol 3-phosphate + acyl-ACP → 1-acyl-*sn*-glycerol 3-phosphate + ACP

Before acyl-ACP was available for such experiments, this enzyme was studied using acyl-CoA as acyl donor (Bertrams and Heinz, 1976, 1979, 1980, 1981; Joyard and Douce, 1977; Joyard, 1979). Bertrams and Heinz (1981) purified the soluble acyl-CoA:*sn*-glycerol-3-phosphate acyltransferase from pea and spinach. In pea chloroplasts, acyltransferase activity was recovered in two isomeric forms having apparent isoelectric points of 6.3 and 6.6. Although the enzyme was not totally pure, as judged by polyacrylamide gel electrophoresis, a molecular weight of ~42,000 was measured for both forms. Spinach acyltransferase was slightly different, since only one peak, with an apparent isoelectric point of 5.2, was obtained; in addition, the molecular weight was slightly higher than that of pea (Bertrams and Heinz, 1981).

Preliminary results obtained with cocoa (*Theobroma cacao*) seed extracts by Fritz *et al.* (1984) suggest that the results obtained with mature spinach chloroplasts could probably be extended to other plastids. These authors have purified from a post microsomal supernatant a soluble *sn*-glycerol-3-phosphate acyltransferase. Further purification was achieved by isoelectric focusing, which revealed a series of five isomeric forms of *sn*-glycerol-3-phosphate activity having isoelectric points close to 5.2. Although the subcellular localization of the activity was not studied, the biochemical properties of *sn*-glycerol-3-phosphate acyltransferase from cocoa (isoelectric point, positional specificity, nature of acyl donors, molecular weight of subunits) suggest that a close relationship exists between these enzymes and those of pea or spinach chloroplasts.

The optimum pH value for lysophosphatidic acid synthesis by the soluble extract from chloroplasts or by purified acyltransferase is around 7.0 to 7.4 (Bertrams and Heinz, 1976, 1981; Joyard, 1979; Joyard et al., 1979). Such a value is somewhat lower than that found in the chloroplast stroma under illumination (Heber and Heldt, 1981). Jelsema et al. (1982) have shown that in *Euglena*, this enzyme is sensitive to *N*-ethylmaleimide and thus probably has SH groups.

Specificities of the *sn*-glycerol-3-phosphate acyltransferase have been studied: for instance, dihydroxyacetone phosphate cannot be used (Bertrams and Heinz, 1981). Acyl-CoA as well as acyl-ACP can be used as acyl donors (Bertrams and Heinz, 1981), but when mixtures of oleoyl-ACP and oleoyl-CoA are offered to the purified acyltransferase, a marked preference for oleoyl-ACP is observed (Frentzen et al., 1983), thus suggesting that ACP thioesters are probably the *in vivo* acyl donors for the enzyme as discussed by Joyard and Douce (1977). However, when either acyl-CoA or acyl-ACP are offered, a striking specificity for *sn*-1 position of the glycerol is observed (Table II) regardless of the fatty acid provided (Joyard, 1979; Bertrams and Heinz, 1981; Frentzen et al., 1983). Furthermore, the *sn*-glycerol 3-phosphate was shown to be more specific for the ACP II isoform, since it possessed a fivefold lower K_m for oleoyl-ACP II as compared to oleoyl-ACP I (Guerra et al., 1986), thus channeling specific fatty acids linked to ACP II toward the envelope Kornberg–Pricer pathway.

The kinetic constants for various substrates of the purified *sn*-glycerol 3-phosphate acyltransferase (Table III) suggest a preferential affinity for 18:1 thioesters (Bertrams and Heinz, 1981; Frentzen et al., 1983; Sauer and Heise,

Table II
Analyses of Fatty Acids Esterified at *sn*-1 and *sn*-2 Positions of Glycerol in Lysophosphatidic Acid[a,b]

Fatty acid used for incorporation	Percentage of radioactivity in	
	1-Acyl-*sn*-glycerol	2-Acyl-*sn*-glycerol
16:0	88	12
18:0	91	9
18:1	90	10

[a]Reproduced from Joyard (1979).

[b]Lysophosphatidic acid was formed by chloroplast soluble proteins from *sn*-[^{14}C]-glycerol 3-phosphate, according to Joyard and Douce (1977). After 1 h incubation, the pH of the incubation mixture was lowered to 6.0 by addition of 0.1 *N* HCl. At this pH value, a soluble phosphatase present in the chloroplast extract (Joyard and Douce, 1977) acts on lysophosphatidic acid to form monoacylglycerol. The lipids were extracted and analyzed by thin-layer chromatography on silica gel plates impregnated with boric acid. By chromatography in chloroform–methanol–water (65:25:4, v/v), 1-acyl-*sn*-glycerol is clearly separated from 2-acyl-*sn*-glycerol. The radioactive spots are detected by autoradiography and their R_f values compared to those of standard molecules.

Table III
Kinetic Constants for Various Substrates of sn-Glycerol 3-Phosphate Acyltransferase[a]

Plant	Acyl donor	K_m (μM)		V (nkat/mg protein)
		Acyl-ACP	sn-Glycerol 3-phosphate	
Spinach	Palmitoyl-ACP	3.2	3150	5.3
	Stearoyl-ACP	3.3	3160	4.3
	Oleoyl-ACP	0.3	31	1.5
Pea	Palmitoyl-ACP	5.6	—[b]	6.5
	Oleoyl-ACP	0.7	—	0.6

[a]Reproduced with permission from Frentzen et al. (1983).
[b]—, Not determined.

1984). Indeed, when mixtures of 16:0- and 18:1-ACP were offered, the purified acyltransferases preferably used oleoyl groups, even with excess 16:0-ACP in the substrate mixture. In addition, the selectivity of acylation of sn-1 position by 18:1 or 16:0 is strongly dependent on sn-glycerol 3-phosphate concentration, since K_m values for sn-glycerol 3-phosphate are very different (Table III). Therefore, the incorporation of 18:1 into lysophosphatidic acid was markedly increased by decreasing the concentration of sn-glycerol 3-phosphate (Frentzen et al., 1983). Since in vivo, the concentration of sn-glycerol 3-phosphate is 0.1–0.3 mM[5] in the chloroplast stroma (Sauer and Heise, 1983, 1984), such a situation would favor oleate incorporation into lysophosphatidic acid, but do not exclude formation of some 1-palmitoyl-sn-glycerol 3-phosphate.

Therefore, analyses of the properties of acyl-ACP:sn-glycerol-3-phosphate acyltransferase suggest that in vivo, this enzyme directs—in pea as well as in spinach—oleic acid (and in some specific conditions such as high sn-glycerol 3-phosphate concentration, palmitic acid) to the sn-1 position of sn-glycerol 3-phosphate. It is interesting to note that the membrane-bound sn-glycerol-3-phosphate acyltransferase from microsomal membranes displays the same positional

[5]Gardiner et al. (1984a) have recalculated these data on the basis of a different chloroplast volume (47 μl/mg chlorophyll) determined by Wirtz et al. (1980) with chloroplasts prepared from protoplasts. The original volume used by Sauer and Heise (1983) was much lower (20 μl/mg chlorophyll). Such a recalculation might be highly confusing, since chloroplast volume (and therefore metabolite concentrations) are susceptible to wide variations according to the osmoticum used (sorbitol, mannitol, sucrose, etc.), to the method used for chloroplast preparation, and to the origin of leaves. Obviously, metabolite levels and chloroplast volumes have to be measured on the same preparations. Indeed, a correct measurement of sn-glycerol 3-phosphate concentration within chloroplasts is important, since variations of this concentration from 20 to 50 μM (Gardiner et al., 1984a) or from 0.1 to 0.3 mM (Sauer and Heise, 1983) might have quite different effects on an enzyme (sn-glycerol-3-phosphate acyltransferase) that have an apparent K_m for this substrate of 30 μM when oleoyl-ACP is offered (see Table II) (Frentzen et al., 1983).

specificity as the plastid enzyme by directing acyl groups from acyl-CoA thioesters to the sn-1 position of sn-glycerol 3-phosphate, but with an opposite selectivity, since palmitoyl groups are preferred (Frentzen $et\ al.$, 1984). However, 80–95% of total fatty acid label formed by isolated intact chloroplasts are 18:1-CoA (Roughan and Slack, 1982), which can be used by extrachloroplastic membranes when added to isolated chloroplast (Roughan $et\ al.$, 1980; Dubacq $et\ al.$, 1984). Therefore, the fatty acid specificity of microsomal acyltransferase allows the formation of both 1-palmitoyl-sn-glycerol 3-phosphate as well as 1-oleoyl-sn-glycerol 3-phosphate (Frentzen $et\ al.$, 1984; Ichihara, 1984).

Finally, it is interesting to note that the plastidial sn-glycerol-3-phosphate acyltransferase directs its product (i.e., lysophosphatidic acid) into the envelope membranes. Therefore, lysophosphatidic acid is not released in the incubation medium but directly into the membrane for further metabolization (Joyard and Douce, 1977). In fact, lysophosphatidic acid is almost a detergent which can easily penetrate within membrane structure and thus become available to further acylation. This is in contrast with the situation observed with phosphatidic acid, which does not behave like a detergent, and therefore, if added exogenously, remains in the soluble phase of the incubation mixture and cannot be metabolized, as shown by Joyard and Douce (1979) (see below).

When isolated envelope membranes are added to preparations containing sn-glycerol-3-phosphate acyltransferase, the level of lysophosphatidic acid is rapidly reduced to a very low level, thus demonstrating a rapid turnover of lysophosphatidic acid (Joyard and Douce, 1977). This observation suggests that, in spinach, sn-glycerol-3-phosphate acyltransferase might be a limiting step in the synthesis of phosphatidic acid. This result was confirmed by Gardiner $et\ al.$ (1984a), who followed acetate incorporation into glycerolipids by isolated chloroplasts from spinach. However, this is not a general feature, since in pea or in $Chenopodium$ $quinoa$, lysophosphatidic acid can be maintained at relatively high concentrations when sn-glycerol 3-phosphate is present in the incubation mixture (Gardiner $et\ al.$, 1984a). Therefore, these observations provide additional evidence for a close association, $in\ vivo$, between envelope membranes and the "soluble" acyltransferase.

2. Acyl-ACP:1-Acyl-sn-Glycerol-3-Phosphate Acyltransferase

This enzyme, which is firmly bound to chloroplast envelope membranes (Joyard and Douce, 1977), catalyzes the formation of phosphatidic acid from lysophosphatidic acid:

1-Acyl-sn-glycerol 3-phosphate + acyl-ACP → 1,2-diacyl-sn-glycerol 3-phosphate + ACP

Unfortunately, purification of this enzyme has not yet been achieved, although it has been solubilized from $Euglena$ chloroplasts (Hershenson $et\ al.$, 1983). In a mixture containing envelope membranes together with the soluble enzymes of chloroplasts, incorporation of sn-glycerol 3-phosphate into phosphatidic acid is almost linear for > 2 h (Joyard and Douce, 1977); in this system, the optimum pH for phosphatidic acid synthesis was alkaline (7.5–8.0) (Joyard, 1979). This enzyme is probably a rate-limiting step in phosphatidic acid synthesis in some plant species, such as pea or *Chenopodium quinoa* (Gardiner *et al.*, 1984a), but not in spinach (Joyard and Douce, 1977; Gardiner *et al.*, 1984a). For instance, Gardiner *et al.* (1984a) have calculated that unlabeled but metabolically active lysophosphatidic acid is present in pea chloroplasts at the beginning of the experiments. They have also shown that in *C. quinoa* chloroplasts, lysophosphatidic acid is maintained to a higher level than that of phosphatidic acid (Gardiner *et al.*, 1984a). This situation is strikingly different from that observed in spinach with either isolated envelope membranes (Joyard and Douce, 1977) or isolated intact chloroplasts (Gardiner *et al.*, 1984a).

Since lysophosphatidic acid formed by the soluble acyltransferase is esterified at the *sn*-1 position of the glycerol backbone, the membrane-bound acyltransferase will direct fatty acids to the available *sn*-2 position. Therefore, this enzyme is unable, *in vivo*, to display any positional specificity for acylation. However, this acyltransferase is apparently specific for palmitic acid, since in our reconstituted system (Joyard, 1979)—which allows the synthesis by envelope membranes of phosphatidic acid from [^{14}C]acetate—phosphatidic acid differed from lysophosphatidic acid by the presence of palmitic acid, which was barely detectable in lysophosphatidic acid (Joyard *et al.*, 1979). The specificity is even more obvious if we consider that C_{18} fatty acids were the major compounds synthesized from [^{14}C]acetate in the system. A similar specificity in directing 16:0 to the *sn*-2 position was also suggested by Bertrams and Heinz (1979) in preliminary experiments using acyl-ACP as acyl donor, and confirmed by an elegant experiment in a further study (Table IV) (Frentzen *et al.*, 1982, 1983). Again, the membrane-bound acyltransferase can use, *in vitro*, acyl-CoA as well as acyl-ACP thioesters, but the physiological donor is more likely acyl-ACP (Frentzen *et al.*, 1983). For instance, Frentzen *et al.* (1983) offered [^{14}C]palmitoyl-ACP together with [^{3}H]palmitoyl-CoA to envelope membranes and 1-oleoyl-*sn*-glycerol 3-phosphate. Regardless of the concentration of 16:0-CoA, at least 85% of the label recovered at the *sn*-2 position of phosphatidic acid was [^{14}C]palmitic acid (Frentzen *et al.*, 1983).

The results obtained with isolated envelope membranes demonstrate that the two acyltransferases have different specificities: the soluble one has a positional specificity and, though being able to use almost any fatty acid, directs mostly C_{18} fatty acids to the *sn*-1 position of the glycerol (see Table II); on the contrary, the membrane-bound acyltransferase cannot exhibit any positional specificity in normal experimental conditions, since only the *sn*-2 position is available for esterification, but directs almost exclusively 16:0 fatty acids to this position (Table IV). Interestingly, in a preliminary work, Frentzen *et al.* (1984) have found that the microsomal 1-acyl-*sn*-glycerol-3-phosphate acyltransferase, in contrast to the envelope enzyme, directs C_{18} fatty acids from acyl-CoA thioesters to the *sn*-2 position of the glycerol backbone.

Table IV
Analyses of Fatty Acids at *sn*-1 and *sn*-2 Positions of Lysophosphatidic Acid and Phosphatidic Acid Formed within Envelope Membranes from Pea and Spinach[a,b]

		Fatty acids incorporated (%)					
		Total		*sn*-1		*sn*-2	
Plant	Product analyzed	18:1	16:0	18:1	16:0	18:1	16:0
Pea	Lysophosphatidic acid	84.9	15.1	84.9	15.1	—	—
	Phosphatidic acid	43.6	54.4	85.8	14.2	3.4	96.6
Spinach	Lysophosphatidic acid	89.0	11.0	89.0	11.0	—	—
	Phosphatidic acid	44.5	55.5	89.9	10.1	3.7	96.3

[a]Reproduced with permission from Frentzen *et al.* (1983).
[b]The reconstituted acyltransferase system (Frentzen *et al.*, 1983) contained *sn*-glycerol 3-phosphate (1 mM), [^{14}C]palmitoyl-ACP, and [^{14}C]oleoyl-ACP in equimolar concentrations (7.5 μM), purified *sn*-glycerol-3-phosphate acyltransferase from pea or spinach (0.6 μg protein), and envelope membranes from pea (80 μg protein) or spinach (20 μg protein). Incubation time: 15 min.

Altogether, these results seem to show that the enzymes responsible for acylation of *sn*-glycerol 3-phosphate in the different compartments of plant cells lead to formation of phosphatidic acid with a specific structure characteristic of the compartment involved.

3. *Phosphatidate Phosphatase (E.C. 3.1.3.4)*

The last enzyme involved in the Kornberg–Pricer pathway converts, in envelope membranes, phosphatidic acid into diacylglycerol (Joyard and Douce, 1977, 1979):

$$1,2\text{-Diacyl-}sn\text{-glycerol 3-phosphate} \rightarrow 1,2\text{-diacylglycerol} + P_i$$

This enzyme differs from the soluble phosphatase, found in the stroma, which dephosphorylates lysophosphatidic acid into monoacylglycerol (at acidic pH), but not phosphatidic acid (Joyard and Douce, 1977; Joyard, 1979). Therefore, the situation in chloroplasts is different from that observed in animal cells, where high amounts of soluble phosphatidate phosphatase are present (for review see Bell and Coleman, 1980). The envelope enzyme is most interesting, since only endogenous phosphatidic acid can be dephosphorylated at high rates (Joyard and Douce, 1979). Indeed, when exogenous phosphatidic acid is provided to envelope membrane vesicles, only a very low percentage is converted into diacylglycerol. Again, the situation is different from animal cells, where the phosphatidate phosphatase can utilize both endogenous or exogenous phosphatidic acid, which might explain partly the variations observed in enzyme activity (Hosaka *et al.*, 1975). Block *et al.* (1983d) have clearly demonstrated that only the inner envelope membrane contains the phosphatidate phosphatase, whereas both outer and inner membranes were able to synthesize phosphatidic acid. The envelope phosphatidate phosphatase has some characteristic features such as an alkaline pH optimum (9.0), a high velocity (at pH

9.0, 0.6 nmol phosphatidic acid can be converted per minute per milligram envelope protein), and a high sensitivity to cations (with 5 mM MgCl$_2$ in the incubation medium, the activity was inhibited by 75%) (Joyard and Douce, 1979). In animal cells, contradictory results were obtained about magnesium effect (Bell and Coleman, 1980), but van Heusden and van den Bosch (1978) reported that magnesium dependence can be only demonstrated with membranes prepared with EDTA. Therefore, the situation differs strikingly from that observed on envelope membranes.

An interesting specificity among different plant species was found by Heinz and Roughan (1983), Frentzen *et al.* (1983), and Gardiner and Roughan (1983): chloroplasts isolated from 18:3 plants have a low phosphatidate phosphatase activity. As shown in Table V, phosphatidic acid accumulates in 18:3 plants rather than diacylglycerol (Gardiner *et al.*, 1984a,b), since the half-life of phosphatidic acid synthesized from acetate in isolated intact chloroplasts was 40 min in 18:3 plants such as pea, but < 2 min in spinach, a 16:3 plant (Gardiner and Roughan, 1983). It is not known whether this reduced level of activity in 18:3 plants is due to a lower amount of enzyme, or to the presence of regulatory molecules which control the activity of the enzyme, unless diacylglycerol formation was due to nonenzymatic breakdown of phosphatidic acid, as suggested by Gardiner *et al.* (1984a) and Kosmac and Feierabend (1985). However, this last hypothesis remains to be demonstrated, and in the absence of any experimental evidences we must consider that even in 18 : 3 plants, diacylglycerol is synthesized by a phosphatidate phosphatase at measurable rates in envelope membranes and therefore should be used for further metabolism, since it does not accumulate *in vivo*. Interestingly, Kosmac and Feierabend (1985) have found that phosphatidate phosphatase was much lower in 70 S ribosome-deficient rye plastids than in normal chloroplasts, resulting in a low synthesis of diacylglycerol, but the activity was not expected to limit glycerolipid biosynthesis in this plant.

Therefore, if diacylglycerol is not used in 18:3 plants for galactolipid biosynthesis, as proposed by Heinz and Roughan (1983), one can suggest that it is at

Table V
Phosphatidic Acid Half-Life in Chloroplasts from Some 16:3 and 18:3 Plant Species[a]

Plant	18:3/16:3-MGDG (% of total MGDG)	Apparent first-order rate constant, k (min^{-1}) $\pm$ SD	Half-life of phosphatidic acid (min)
Pisum sativum	0	0.018 ± 0.002	38.5
Carthamus tinctorius	1	0.018 ± 0.002	38.5
Chenopodium quinoa	8	0.120 ± 0.009	5.8
Solanum nodiflorum	40	0.520 ± 0.060	1.3
Spinacia oleracea	50	0.416 ± 0.019	1.7

[a]Reprinted by permission from Gardiner and Roughan (1983), *Biochemical Journal*, Vol. 210, pp. 949–952. Copyright © 1983 The Biochemical Society, London.

least a substrate for sulfolipid biosynthesis, which occurs in chloroplasts (Haas *et al.*, 1980; Kleppinger-Sparace *et al.*, 1985; Joyard *et al.*, 1986). In support for this suggestion, Joyard *et al.* (1986) have shown that addition of UDP-galactose (UDPgal) to isolated spinach chloroplasts, incubated in presence of $^{35}SO_4^{2-}$ or [1-^{14}C]-acetate, in conditions which promote the functioning of the Kornberg–Pricer pathway, strongly decreased the level of sulfolipid synthesized, thus demonstrating, in 16:3 plants, a competition between galactolipid and sulfolipid synthesis at the level of diacylglycerol.

The studies of the properties of the envelope phosphatidate phosphatase performed either on intact chloroplasts or on envelope membranes clearly demonstrate that this enzyme plays a major regulatory role in glycerolipid metabolism. It would be most interesting to determine whether the differences found among the different plant species are due to a regulation of gene expression in 18:3 plants or to a metabolic regulation at the level of the enzyme itself, both of these regulations being susceptible to variations during growth and development. Finally, in 16:3 plants, diacylglycerol is formed at high rates from phosphatidic acid having a specific structure: 18:1 and 16:0, respectively, at the *sn*-1 and *sn*-2 positions of the glycerol backbone. Therefore, the diacylglycerol thus formed in chloroplasts through the envelope Kornberg–Pricer pathway, (i.e., with C_{16} fatty acids at the *sn*-2 position, a position from which C_{18} fatty acids are apparently excluded), have a characteristic prokaryotic structure. The question which is still unsolved is the origin of the eukaryotic diacylglycerol molecules (having C_{18} fatty acids at both *sn*-1 and *sn*-2 positions). This most important problem will be discussed in the last part of this review.

Finally, the ability of isolated chloroplasts or envelope membranes to form diacylglycerol molecules having only a prokaryotic structure was analyzed only with mature chloroplasts. So far, almost no thorough studies have been done using proplastids or developing plastids. It is well known that such plastids contain the enzymes of the Kornberg–Pricer pathway, and are able to synthesize diacylglycerol and further galactosylate it into MGDG, as shown by Journet and Douce (1985) for plastids from cauliflower buds (Fig. 6), but nothing is known on the positional specificities of the enzymes involved. It is highly likely that results similar to those obtained with mature chloroplasts would probably be obtained, but such studies have to be done to extend the validity of the data obtained with chloroplasts from mature green leaves.

III. LOCALIZATION AND PROPERTIES OF GALACTOSYLTRANSFERASE ACTIVITIES INVOLVED IN GALACTOLIPID BIOSYNTHESIS

The reader is referred to the reviews by Heinz (1977), Harwood (1979), Douce and Joyard (1980), and Roughan and Slack (1982), which describe the early studies on galactolipid formation in whole plants, tissues, organelles, or membranes. In this

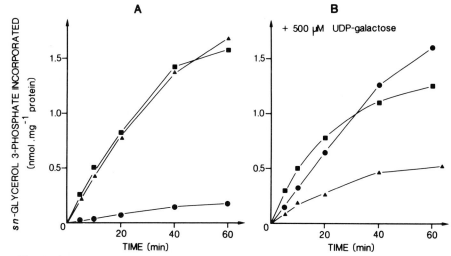

Fig. 6. Synthesis of glycerolipids from *sn*-glycerol 3-phosphate by cauliflower bud plastids. The experimental conditions are described by Journet and Douce (1985). In experiment B, UDP-galactose was added to the incubation medium. The cauliflower bud plastids are nongreen and mostly devoid of internal membranes, however, they behave almost like chloroplasts with respect to glycerolipid synthesis, since phosphatidic acid (■), diacylglycerol (▲) are formed and metabolized into MGDG (●) when UDPgal is present. [Adapted with permission from Journet and Douce (1985).]

part, only the final steps in galactolipid biosynthesis will be considered. In fact, two different galactolipids should be synthesized: MGDG and DGDG. Although the pathway for 18:3/16:3 MGDG formation is rather well known, the mechanism of DGDG synthesis is much less clear if not controversial. It is surprising to observe that almost none of the studies done with intact chloroplasts show any DGDG synthesis. For instance, only 0.1–1% of the total radioactivity is recovered as DGDG after [^{14}C]acetate labeling of intact chloroplasts (Heinz and Roughan, 1983). Therefore, we will discuss the synthesis of these two galactolipids separately.

A. The Formation of MGDG

MGDG synthesis involves the transfer of a galactose from a galactose donor (UDP-galactose, or UDPgal) to an acceptor molecule, (i.e., diacylglycerol). The enzyme involved is a 1,2-diacylglycerol 3-β-galactosyltransferase (E.C. 2.4.1.46):

$$1,2\text{-Diacylglycerol} + \text{UDPgal} \rightarrow 1,2\text{-diacyl-3-galactosyl } sn\text{-glycerol} + \text{UDP}$$
$$\text{(MGDG)}$$

Chloroplasts isolated from 16:3 and 18:3 plants have the same capacity to form MGDG, when supplied with UDP-[^{14}C]galactose (Heinz and Roughan, 1983). Therefore, there is no striking difference between these plant species at the level of galactosyltransferase activity (see below).

1. Localization

Neufeld and Hall (1964) and Ongun and Mudd (1968) first realized that isolated spinach chloroplasts catalyzed the transfer of galactose from UDPgal to an endogenous acceptor, yielding mostly MGDG but also DGDG. Douce (1974) demonstrated that the chloroplast envelope was the major site of UDPgal incorporation into galactolipids, mostly MGDG, and this result was very quickly confirmed by workers using chloroplasts from several plant species (for review see Douce and Joyard, 1980), and extended to nongreen plastids (Liedvogel and Kleinig, 1976, 1977; Fishwick and Wright, 1980; Sandelius and Selstam, 1984; Journet and Douce, 1985). A similar result was obtained in algae such as *Euglena* (Blée and Schantz, 1978a,b; Blée, 1981) or *Chlamydomonas* (Mendolia-Morgenthaler *et al.*, 1985). The incorporation found in other cell organelles or membranes (such as microsomal membranes) was obviously due to contamination of these preparations by small membrane vesicles deriving from the fragile plastids (proplastids, amyloplasts, etioplasts, chloroplasts). The generalization of purification procedures involving Percoll gradients (Morgenthaler *et al.*, 1975; Douce and Joyard, 1982) have definitively convinced most of the workers that all kinds of plastids are probably the sole site of MGDG synthesis, and that their limiting envelope membranes are a key structure in this synthesis. In fact, very high values for MGDG synthesis can be obtained with purified envelope membranes: 45 nmol/mg protein per minute (Joyard and Douce, 1976a),[6] and exceed the corresponding figures for total cell proteins or other membrane fractions (such as thylakoids) by factors of 50–100. Although the specific activity was much lower in thylakoids than in envelope membranes, a large part of the total activity was found in the thylakoid fraction (Douce, 1974). Similar results obtained with pea chloroplasts confirm this distribution (Cline and Keegstra, 1983). Therefore, in the minds of some authors, the existence of additional sites for MGDG remained possible, and (at least for thylakoids) could not be ruled out. Are the activities found in thylakoid fractions due to a genuine galactosyltransferase activity present in thylakoid membranes or only the result of a contamination by envelope membranes? We are convinced that in mature spinach chloroplasts the second hypothesis is more likely. For instance, careful washings of thylakoids prior to incubation lead to a striking decrease of the level of incorporation of galactose from UDPgal into galactolipids (Fig.7). In fact, a large proportion of envelope membranes (>50% in spinach) is recovered in the thylakoid fraction as a contaminant, trapped in the network of thylakoid membranes. Only careful washings can reduce (but probably not eliminate totally) envelope contamination in thylakoid preparations. Therefore, estimation of the level of cross-contamination by appropriate means (such as specific antibodies to proteins known to be solely associated with a given membrane system) should be carefully done prior analyses of galactolipid-synthesizing activities by a membrane system.

[6]Even higher activities were obtained with envelope membranes having a low diacylglycerol content (due to thermolysin treatment of intact chloroplasts) and which contain only diacylglycerol formed through the Kornberg–Pricer pathway (Dorne *et al.*, 1982a,b).

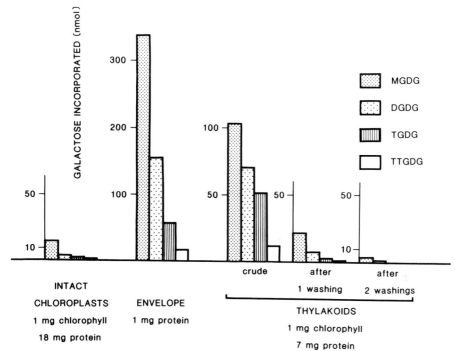

Fig. 7. Synthesis of galactolipids by intact chloroplasts, envelope membranes, and thylakoids from spinach chloroplasts. Intact chloroplasts, envelope membranes and thylakoids were prepared according to Douce *et al.* (1973). Incubation time was 1 h. Both preparations can incorporate galactose from UDPgal into galactolipids. However, careful washing of thylakoids by centrifugation through one or two successive linear sucrose gradients (1 to 1.7 *M*) can reduce strikingly UDPgal incorporation: thylakoids from mature spinach chloroplasts are probably unable to synthesize galactolipids. In addition, this experiment shows that envelope membranes heavily contaminate crude thylakoids (unpublished observation of A. J. Dorne.)

However, again, what is true for mature spinach is not necessarily true for young, developing plastids having structural relationships between their growing thylakoid network and the inner envelope membrane. Although in some young plastids the internal membrane system contains chlorophyll (and is therefore considered as thylakoids), it is not known whether some enzymatic activities associated, in mature chloroplasts, only with envelope membranes could be present in these growing internal membranes at some stages. For instance, in etioplasts, prothylakoids — which derive from and are frequently connected to the inner envelope membrane — are able to synthesize galactolipids to almost the same extent as envelope membranes (Sandelius and Selstam, 1984). Although cross-contamination of prothylakoids by envelope membranes cannot be excluded, the ability of prothylakoids to form galactolipids would not be surprising, since both membrane systems are tightly related within the organelle. The same observation can be made from experiments with chromoplasts. In daffodil, for instance, the numerous convoluted internal membranes are almost impossible to distinguish from the inner envelope membrane, and galactolipid biosynthesis might also occur in these fractions, in addition to those enriched in envelope membranes (Liedvogel and Kleinig, 1976, 1977; Liedvogel et al., 1978). However, clear-cut studies are very difficult to achieve in such plastids, and numerous control experiments are necessary before the existence of additional sites, different from envelope membranes, involved in galactolipid synthesis would be clearly established.

The localization within envelope membranes of the galactosyltransferase which leads to MGDG formation is still controversial, since Cline and Keegstra (1983) found that the activity was located in the outer envelope membrane from pea chloroplasts, whereas in spinach, Dorne et al. (1982a,b) and Block et al. (1983b) (see footnote 1, p. 222) found that MGDG synthesis was concentrated in the inner envelope membrane.

Several hypotheses would explain the discrepancies between these results, but the most interesting one relates to the species differences between pea (an 18:3 plant) and spinach (a 16:3 plant). Indeed, several authors (Heinz and Roughan, 1983; Frentzen and Heinz, 1983; Frentzen, 1986) have suggested that plants such as pea, having only a eukaryotic pathway for MGDG, would have the galactosyltransferase localized in the outer envelope membrane, whereas the enzyme responsible for the synthesis of prokaryotic MGDG would be present only in the inner envelope membrane.[7] In fact, this hypothesis is not supported by the analyses of the data of Cline and Keegstra (1983) and Block et al. (1983b): if the hypothesis was true, spinach (which has eukaryotic and prokaryotic MGDG in equal amounts) should have a similar level of galactosyltransferase activity in both envelope membranes. In fact, Block et al. (1983b) found most of the activity in the inner membrane (Fig. 8); the ratio between activities in the two membrane fractions was 0.14 (OM/IM). On the contrary, in pea (which should be devoid of any activity in a "pure" inner membrane fraction, since it contains only eukaryotic MGDG), Cline and Keegstra (1983) found two maxima of UDPgal incorporation in the linear gradient used to separate the outer from the inner envelope membrane. The controversy is not yet clearly solved. However, Heemskerk et al. (1985), with spinach, and Miquel (1985), with pea, have found in both cases the galactosyltransferase to be associated with inner envelope membrane. For instance, with pea envelope membranes, Miquel (1985) found that the ratio of activity in the outer vs inner membrane fractions was 0.29, a value close to that found in spinach by Block et al. (1983b) (0.14) or by Heemskerk et al. (1985) (0.4). In addition, the possible existence of galactosyltransferase in internal membranes from etioplasts

[7]This hypothesis is closely related to the concept, derived from the endosymbiotic theory, that the inner envelope membrane could derive from the plasma membrane of the prokaryotic ancestor, whereas the outer envelope membrane could derive from the endomembrane system of the protoeukaryote that engulfed the free-living prokaryote. Analyses of the outer envelope membrane (see Douce et al., 1984a, for review) clearly challenge this idea, since its polar lipid composition, except for the presence of phosphatidylcholine in the outer leaflet, is closely related to plastid membranes, and very little to the endomembrane system of the cell (see Douce et al., 1985, for discussion of this problem).

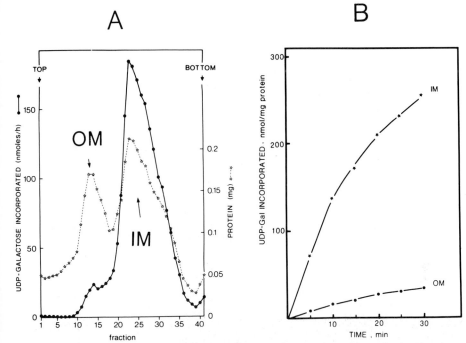

Fig. 8. Galactolipid synthesis is localized in the inner envelope membrane from spinach chloroplasts. Membrane fractions enriched in outer (OM) or inner (IM) envelope membranes from spinach chloroplasts were incubated in the presence of UDPgal and diacylglycerol according to Douce (1974). Only the membrane fraction enriched in inner envelope membrane was able to form galactolipids. (A) Analyses of fractions from a linear sucrose gradient (0.4 to 1.2 M; Block *et al.*, 1983a). (B) Time-course experiments with fraction 2 (OM) and fraction 3 (IM) prepared according to Block *et al.* (1983a). See Fig. 4 for a proper characterization of these fractions. [Reproduced with permission from Douce *et al.* (1984b) (A) and Dorne *et al.* (1982a) (B).]

(Sandelius and Selstam, 1984) or chromoplasts (Liedvogel and Kleinig, 1976) that have originated from the inner envelope membrane give further support for a localization of galactosyltransferase in the inner envelope membrane from mature chloroplasts.

2.   Properties of UDPgal:Diacylglycerol Galactosyltransferase

This enzyme is responsible for the formation of a β-glycosidic bond between galactose and diacylglycerol. This galactosyltransferase has not yet been purified but can be easily solubilized by appropriate detergents. Heinz *et al.* (1978) have shown that the envelope galactosyltransferase is still active in Triton X-100 (0.9% by volume). Several other detergents can be used, but the most efficient detergent for the UDPgal:diacylglycerol galactosyltransferase is probably a nondenaturing zwitterionic detergent: CHAPS (Covès *et al.*, 1986). Covès *et al.* (1986) have fractionated the solubilized envelope membranes by centrifugation on sucrose gradients and chromatography on hydroxyapatite, and achieved a partial purification of the galactosyltransferase. At CHAPS concentrations around 6 mM, about one-third of the galactosyltransferase activity of envelope membranes was solubilized. Dithiothreitol in the solubilization medium was shown to be essential to protect the enzyme (Covès et al., 1986). However, the detection of the activity was possible only in the presence of salts, to maintain a high ionic strength, and lipids, to provide diacylglycerol. However, the purity of the fraction does not allow studies of the kinetic parameters; therefore, these were determined on total envelope membranes.

The galactosyltransferase responsible for MGDG synthesis has a broad pH optimum above 7.5 and up to 9.0 (Joyard and Douce, 1976b; van Besouw and Wintermans, 1978). The same apparent K_m values for UDPgal were obtained for spinach and for pea: $\sim$ 40 μM (van Besouw and Wintermans, 1978; Cline and Keegstra, 1983; Covès, 1987). Cation effect on the enzyme is not yet clear: Joyard (1979) reported only a limited effect of Mg^{2+} and Ca^{2+} (5 mM) on galactose incorporation into MGDG, an observation confirmed by addition of EDTA or NaF (5 mM), which can chelate the high endogenous level of cations (mostly Ca^{2+}) found in isolated envelope membranes (Joyard and Douce, 1976b). Cline and Keegstra (1983) found that addition of 1–4 mM $MgCl_2$ to the reaction mixture resulted in a 60% increase in activity; above 4 mM, no additional increase was obtained. In contrast, Heemskerk et al. (1986a) found a 10-fold stimulation of the UDPgal:diacylglycerol galactosyltransferase in isolated envelope from spinach chloroplasts. In addition, no effect of Ca^{2+} and Ba^{2+}, but a remarkable inhibition by Zn^{2+} and Cd^{2+} were obtained (Heemskerk et al., 1986a). Although addition of salts (NaCl, KCl, KH_2PO_4) was almost without effect on native envelope, a high ionic strength in the assay mixture was necessary for optimal galactosyltransferase activity (Covès et al., 1986).

UDP, but no other nucleotides (ADP, AMP, CDP, CMP, and GTP), has a strong inhibitory effect (K_i = 10–20 μM) on the UDPgal:diacylglycerol galactosyltransferase (van Besouw and Wintermans, 1978; Covès, 1987). UDP, as a competitive inhibitor of the enzyme, prevents accessibility of UDPgal to the active site of the enzyme. This property has been used by Covès (1987): an aryl-azido derivative of UDP (NAP$_4$-UDP) was synthesized according to Guillory (1979) and used on envelope membranes to characterize the UDPgal:diacyl-glycerol galactosyltransferase. Under irradiation with light, NAP$_4$-UDP covalently binds to envelope membranes and inhibits the incorporation of galactose into MGDG (Fig. 9). NAP$_4$-UDP is still a competitive inhibitor (K_i = 300 μM). This specific inhibitor is indeed a potent tool to characterize the active site of UDPgal: diacylglycerol galactosyltransferase.

UDP-glucose cannot be used in higher plants by the envelope galactosyltransferase, but is in fact a competitive inhibitor, with an affinity similar to that of UMP (K_i = 100 μM; van Besouw, 1979). This is in contrast with the prokaryotic situation, since in cyanobacteria, UDP-glucose, and not UDP-galactose, is used to form monoglucosyldiacylglycerol (MgluDG), which is then transformed into MGDG owing to an epimerase activity (Sato and Murata, 1982c) (see also below).

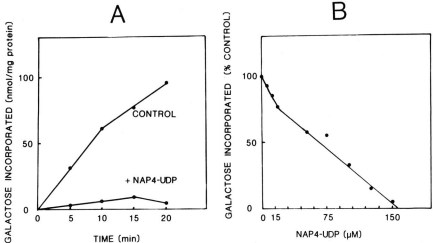

Fig. 9. Effect of NAP₄-UDP on galactose incorporation from UDPgal into galactolipids. After irradiation with white light, NAP₄-UDP binds covalently on envelope membranes and inhibits galactolipid synthesis. (A) Time-course experiment in presence of 150 μM NAP₄-UDP. (B) Effect of NAP₄-UDP concentration on galactolipid synthesis (Covès, 1987).

The sulfhydryl nature of the galactosyltransferase has been established in spinach chloroplast preparations (Chang, 1970; Mudd *et al.*, 1971) and has been confirmed in envelope membranes, since *N*-ethylmaleimide (NEM) and *p*-hydroxymercuribenzoic acid are strong inhibitors of MGDG synthesis (Heemskerk *et al.*, 1986a). After solubilization, the enzyme involved in MGDG synthesis is even more sensitive to sulfhydryl reagents: whereas 5 m*M* NEM were necessary to completely inhibit MGDG synthesis in envelope membranes, only 5–10 μ*M* were necessary after solubilization (Covès *et al.*, 1986). In addition, this inhibition was very rapid: 50% inhibition was obtained after 15 sec in presence of 10 μ*M* NEM (Covès *et al.*, 1986). Therefore, the SH groups of the enzyme are probably close to its active site.

The UDPgal:diacylglycerol galactosyltransferase do not present high specificity for the dialcylglycerol, since molecules having saturated as well as polyunsaturated fatty acids are apparently used with the same velocity (Joyard and Douce, 1976a; Siebertz *et al.*, 1980; Heinz *et al.*, 1979). When UDPgal is provided to isolated envelope membranes, the endogenous diacylglycerol pool (due to the functioning of the galactolipid:galactolipid galactosyltransferase) containing 16:3/18:3 and 18:3/18:3 combinations, is used with a high efficiency (Siebertz *et al.*, 1980). Similar results were obtained with saturated molecular species (Joyard *et al.*, 1979; Joyard, 1979). Heemskerk *et al.* (1986a) have verified this statement by using a system which produces diacylglycerol (available to galactosylation by envelope membranes) from PC by treatment with phospholipase C. PC molecules having characteristic diacylglycerol backbones (10:0/10:0, 16:0/16:0; 18:0/18:0, 16:0/18:1; 18:1/16:0; 18:1/18:1; 18:2/18:3) were used, and no big differences were found in MGDG synthesis. These observations suggest that the UDPgal:

diacylglycerol galactosyltransferase can galactosylate almost any diacylglycerol molecule that is available. As shown above, the envelope Kornberg–Pricer pathway produces almost exclusively "prokaryotic" diacylglycerol molecules having C_{18} fatty acids and C_{16} fatty acids esterified, respectively, at sn-1 and sn-2 positions of the glycerol backbone; therefore, the complete sequence of enzymatic reactions operating on the inner envelope membrane of chloroplasts will lead to the synthesis of MGDG with a prokaryotic structure. However, any eukaryotic diacylglycerol (i.e. having C_{18} fatty acids at both sn-1 and sn-2 positions) would be galactosylated with the same efficiency, thus demonstrating that the diversity of MGDG molecules originates from the formation of diacylglycerol, but not from the galactosyltransferase involved in its galactosylation.

3. MGDG Synthesis in Cyanobacteria

When compared to higher plants, MGDG synthesis in cyanobacteria occurs in a similar but not identical way (Fig. 10). In *Anabaena variabilis* (Sato and Murata, 1982a,b,c), as well as in *Prochloron* (N. Sato and N. Murata, personal communication), a two-step mechanism is involved: (1) formation of monoglucosyldiacylglycerol, owing to a UDP-glucose:diacylglycerol glucosyltransferase, and (2) formation of MGDG by epimerization of the glucosyl group into galactose (Sato and Murata, 1982a,b). The first enzyme is very similar to the envelope UDPgal:

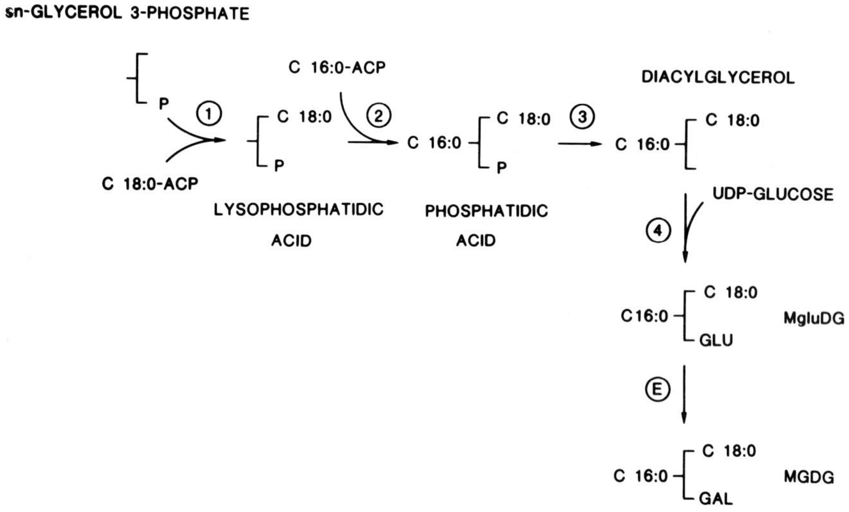

Fig. 10.   Biosynthesis of MGDG in prokaryotes. The only differences between MGDG synthesis in prokaryotes and in chloroplasts are (1) a glycosylation step of diacylglycerol from UDP-glucose, to form monoglucosyldiacylglycerol (MgluDG), which is then epimerized into MGDG (E, epimerase); and (2) the absence of desaturation in most prokaryotes (except in some species such as *Anabaena variabilis*; see Nichols *et al.*, 1965). **1,** Acyl-ACP:*sn*-glycerol-3-phosphate acyltransferase; **2,** acyl-ACP:monoacyl-*sn*-glycerol-3-phosphate acyltransferase; **3,** phosphatidate phosphatase; **4,** UDP-glucose:diacylglycerol glucosyltransferase.

diacylglycerol galactosyltransferase: it is a membrane-bound enzyme with an affinity for UDP-glucose similar to that of the higher plants' enzyme for UDPgal (45 μM). The activity can be strongly stimulated by addition of Mg^{2+}, but Ca^{2+} has an inhibitory effect. Diacylglycerol, too, has a strong effect on the level of MgluDG synthesis, but 1-stearoyl-2-palmitoyl-sn-glycerol (which is supposed to be the natural substrate) is used with a lower efficiency than was 1-oleoyl-2-palmitoyl-sn-glycerol (Sato and Murata, 1982c). It is interesting to note that the glucosyltransferase activity of $A.$ $variabilis$ is similar to that of $Streptococcus$ $faecalis$ (Pieringer, 1968) with respect to its localization, optimum pH, and Mg^{2+} requirement. However, the anomer configurations of the reaction products are different: the β-configuration (analogous to the situation in higher plants MGDG) is formed in the cyanobacteria (Feige et $al.$, 1980) and the α-configuration in $S.$ $faecalis$. MgluDG does not accumulate in the cell (or at least to a very low extent): it cannot be detected in $Anacystis$ $nidulans$ (Allen et $al.$, 1966) or represents between 1 and 3% of the total glycerolipids, respectively, in $Anabaena$ $variabilis$ (Murata and Sato, 1982) and $Prochloron$ (Murata and Sato, 1983). In fact, very quickly, the glucosyl group is transformed into a galactosyl group by epimerization. Sato and Murata (1982a) found that this transformation was complete after a 1 h chase period, following ^{14}C-labeling of intact cells, and was not due to exchange of sugars between labeled MgluDG and nonlabeled MGDG. When isolated membranes were incubated in presence of UDP-glucose and diacylglycerol, no further transformation of MgluDG into MGDG was obtained (Sato and Murata, 1982c), thus demonstrating that no epimerase activity was present in the membrane fraction. In fact, this enzyme is soluble (Sato and Murata, 1982c). Chloroplasts from higher plants apparently lack this epimerase activity, and the transformation probably occurs at the level of UDP-glucose, owing to a soluble cytoplasmic enzyme (Königs and Heinz, 1974; Haas and Heinz, 1980; Bertrams et $al.$, 1981), although in some algae, such as $Euglena$, part of the activity ($\sim$10%) can be associated with membranes (Blée, 1981).

4. Do Alternative Pathways for MGDG Synthesis Exist?

When envelope membranes are incubated in the presence of sn-glycerol 3-phosphate and all the cofactors or enzymes necessary for the whole Kornberg–Pricer pathway to be functional (see above), low but consistent amounts of MGDG are formed in the absence of UDPgal (Joyard and Douce, 1977). The same observation can be made from experiments with isolated intact chloroplasts (see, e.g., Roughan et $al.$, 1979c). Several hypotheses might explain such results: for instance, the existence of low endogenous levels of UDPgal in the incubation mixture, and especially in the stromal phase of chloroplasts. Van Besouw (1979) suggested the possibility of an equilibrium exchange between diacylglycerol and MGDG, but provided little experimental evidence. Heemskerk et $al.$ (1986a) analyzed this problem further. When isolated envelope membranes where incubated in the presence of di-[1-^{14}C]oleoyl-PC (and phospholipase C to form di-[1-^{14}C]oleoyl-glycerol), MGDG can be synthesized even in the absence of UDPgal: up to 200 nmol MGDG per hour per milligram envelope protein can be formed. Furthermore, addition of UDP has only a limited effect on UDPgal-independent MGDG formation, in contrast to UDPgal-dependent synthesis. It is not yet known whether this phenomenon has any physiological significance, since this exchange of diacylglycerol between MGDG and preexisting diacylglycerol does not result in a net increase in MGDG. In addition, the exact nature of the enzyme involved is not known; according to Heemskerk *et al.* (1986a), it could be a specific enzyme, or the result of the functioning of either the galactolipid:galactolipid galactosyltransferase in a reverse way (see below) or a partial reaction due to the UDPgal:diacylglycerol galactosyltransferase.

B. How Is DGDG Synthesized?

DGDG is a major galactolipid in thylakoids as well as in envelope membranes (Douce *et al.*, 1973). When UDPgal is supplied to isolated envelope membranes from higher plant chloroplasts, MGDG formation is readily followed by appearance of DGDG, tri-GDG, tetra-GDG, or even higher homologs (Douce, 1974; Joyard and Douce, 1976b), but in *Euglena*, DGDG (and not MGDG) is the main galactolipid to be synthesized by isolated envelope membranes (Blée, 1981). This last observation was recently extended to *Chlamydomonas* chloroplast envelope (Mendiola-Morgenthaler *et al.*, 1985). These different experiments give apparent support for the stepwise galactosylation of MGDG to form DGDG and higher homologs, as proposed originally by Neufeld and Hall (1964) and Ongun and Mudd (1968). In isolated envelope membranes from spinach chloroplasts, DGDG was formed directly from the endogenous diacylglycerol pool (Joyard and Douce, 1976a) via MGDG, since its diacylglycerol moiety was identical to that present in envelope membranes prior to incubation and since both galactose molecules were labeled with the same intensity (Siebertz *et al.*, 1980).

Van Besouw and Wintermans (1978) observed that DGDG formation could indeed occur even in the absence of UDPgal and proposed a different mechanism to synthesize DGDG, owing to a galactolipid:galactolipid galactosyltransferase:

$$2 \text{ MGDG} \rightarrow \text{DGDG} + \text{diacylglycerol}$$

Such a reaction would explain DGDG (and higher homologs tri- and tetra-GDG) formation in absence of UDPgal, and the presence of high levels of endogenous polyunsaturated diacylglycerol in isolated envelope membranes.

Therefore, two distinct enzymes (Fig. 11) could be responsible for DGDG synthesis: a UDPgal:MGDG galactosyltransferase (as proposed in Neufeld and Hall, 1964; Ongun and Mudd, 1968) or a galactolipid:galactolipid galactosyl-transferase (as proposed by van Besouw and Wintermans, 1978). In fact, the former enzyme has not yet been identified in isolated chloroplasts, in microsomes, or in envelope membranes, whereas the latter can be easily detected in envelope membranes.

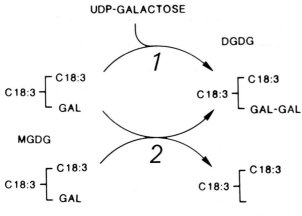

Fig. 11. Possible pathways for DGDG synthesis. Two distinct enzymes could be involved: (1) a UDPgal:MGDG galactosyltransferase, which has not yet been demonstrated, and (2) a galactolipid:galactolipid galactosyltransferase, localized on the outer surface of the outer-envelope membrane.

1. Localization of Galactolipid:Galactolipid Galactosyltransferase

Two of the specific features of isolated envelope membranes (i.e., their high diacylglycerol content and their ability to form galactolipids when incubated with UDPgal) can be abolished when isolated intact chloroplasts are treated with thermolysin prior to envelope membrane purification (Dorne *et al.*, 1982a,b). The polar lipid composition of envelope membranes from nontreated and thermolysin-treated intact spinach chloroplasts is shown in Table VI. Whereas nontreated envelope membranes contained large amounts of diacylglycerol and nonnegligible amounts of tri-GDG and tetra-GDG, envelope membranes from thermolysin-treated chloroplasts were devoid of diacylglycerol, tri-GDG, and tetra-GDG. In addition, the amount of MGDG was much higher in envelope membranes from thermolysin-treated chloroplasts than in those from nontreated chloroplasts. Finally, DGDG, sulfolipid, and phospholipids were present in similar amounts in both preparations (Dorne *et al.*, 1982a,b). Clearly, these results demonstrate that thermolysin treatment of intact spinach chloroplasts prevented the interlipid exchange of galactose during the course of envelope membrane preparation. In addition, Dorne *et al.*, 1982a,b) demonstrated that the lack of UDPgal-dependent galactolipid synthesis was due to the absence of diacylglycerol in these membranes (see Table VI) and not to a destruction of the UDPgal:diacylglycerol galactosyltransferase by thermolysin. Indeed, Dorne *et al.* (1982a,b) clearly demonstrated that this latter enzyme was still active, because thermolysin-treated envelope membranes were able to synthesize MGDG from *sn*-glycerol 3-phosphate, UDPgal, and fatty acids, and even at a higher rate than nontreated envelopes.

Table VI

Lipid Composition of Envelope Membranes from Nontreated and Thermolysin-Treated Intact Spinach Chloroplasts[a,b]

Lipids[c]	Nontreated		Thermolysin treated	
	μg Fatty acids/mg protein	%	μg Fatty acids/mg protein	%
MGDG	154	13.5	454	38.0
DGDG	381	33.5	354	30.0
TGDG	36	3.0	0	0
TTGDG	16	1.5	0	0
SL	86	7.5	85	7.0
PC	172	15.0	160	13.5
PG	102	9.0	108	9
PI	34	3.0	25	2
PE	0	0	0	0
DPG	0	0	0	0
Diacylglycerol	155	13.5	Trace	<0.1
	1136	99.5	1186	99.5

[a]Reproduced with permission from Dorne *et al.* (1982b).

[b]Thermolysin treatment of isolated and purified spinach chloroplasts was done as described by Dorne *et al.* (1982a).

[c]MGDG, Monogalactosyldiacylglycerol; DGDG, digalactosyldiacylglycerol; TGDG, trigalactosyldiacylglycerol; TTGDG, tetragalactosyldiacylglycerol; SL, sulfoquinovosyldiacylglycerol (sulfolipid); PC, phosphatidylcholine; PG, phosphatidylglycerol; PI, phosphatidylinositol; PE, phosphatidylethanolamine; DPG, diphosphatidylglycerol (cardiolipin).

Since Joyard *et al.* (1983) have demonstrated that thermolysin is a nonpenetrating proteolytic enzyme which hydrolyzes, on intact chloroplasts, only polypeptides located on the cytosolic surface of the chloroplast envelope, the results obtained by Dorne *et al.* (1982a,b) clearly establish that the galactolipid:galactolipid galactosyltransferase was located on the cytosolic side of the outer envelope membrane. However, Heemskerk *et al.* (1985), using a specific assay for the galactolipid:galactolipid galactosyltransferase (Heemskerk *et al.*, 1983), found that the activity of the enzyme was predominantly in the inner envelope membrane, in contrast with the data obtained by Dorne *et al.* (1982a,b). In a series of joint experiments, the two laboratories were able to solve this problem and to confirm the localization of the galactolipid:galactolipid galactosyltransferase on the outer envelope membrane (Heemskerk *et al.*, 1986c).

2. Properties of the Galactolipid:Galactolipid Galactosyltransferase

An assay developed by Heemskerk *et al.* (1983) allows the study of the properties of the enzyme, using [^{14}C]MGDG which was made available to the enzyme by addition of deoxycholate and sonication. Very high velocities can be obtained, between 2 and 3 μmol MGDG converted per hour per milligram envelope proteins, provided that enough MGDG was present in the incubation mixture (Heemskerk *et al.*, 1986c). Obviously, when assayed under these conditions, the enzyme presents characteristics identical to those observed for DGDG, tri- and tetra-GDG formation in isolated envelope membranes: it shows a broad optimum pH between 5.9 and 7.0; it is still active at temperatures as low as 4°C and is stimulated by Mg^{2+} and Mn^{2+} (Heemskerk *et al.*, 1983), whereas Cd^{2+} and Zn^{2+} strongly inhibit the enzyme. The galactolipid:galactolipid galactosyltransferase is strongly inhibited by 10 μM *p*-hydroxymercuribenzoic acid, thus suggesting that it has some SH groups. Finally, the enzyme does not follow simple Michaelis–Menten kinetics, since no straight line could be drawn, when the activity (versus MGDG concentration) was presented as a Lineweaver–Burk plot (Heemskerk *et al.*, 1986c).

3.   Is the Galactolipid:Galactolipid Galactosyltransferase Really Involved in DGDG Synthesis in Vivo?

The properties of the enzyme, when measured with the assay developed by Heemskerk *et al.* (1983), obviously correspond to those of DGDG, tri-, and tetra-GDG synthesis by isolated envelope membranes from higher plants. However, the enzyme synthesizes unnatural galactolipids such as tri- and tetra-GDG, which never accumulate *in vivo*. In addition, the enzyme is accessible from the cytosolic side of isolated intact chloroplasts to nonpenetrating proteolytic enzymes such as thermolysin. Therefore, DGDG once synthesized should be transferred from the outer envelope membrane to the other chloroplast membranes, where it does accumulate. On the other hand, Heemskerk *et al.* (1986b) have provided some interesting evidence for a kind of control of UDPgal incorporation into intact chloroplasts by the galactolipid:galactolipid galactosyltransferase. However, we must keep in mind that a high specificity for some molecular species of MGDG should be obtained with this enzyme, since, except for molecules having 18:3/18:3 combinations, DGDG does not contain the same diacylglycerol moiety as MGDG (see Fig. 1). In addition, it is still not understood why, when isolated intact chloroplasts are supplied with [1-^{14}C]acetate, *sn*-glycerol 3-phosphate, and UDPgal, almost no DGDG is formed within the organelle (Heinz and Roughan, 1983). Finally, experiments with envelope membranes from *Euglena* (Blée, 1981), and *Chlamydomonas* (Mendiola-Morgenthaler *et al.*, 1985), showing that DGDG is the major and the first galactolipid to be synthesized from UDPgal do not support the involvement of a galactolipid:galactolipid galactosyltransferase in such a situation. In addition, neither tri-GDG nor tetra-GDG is formed in envelope membranes from *Euglena* (Blée, 1981) or *Chlamydomonas* (Mendiola-Morgenthaler *et al.*, 1985).

All these observations show that further work is needed to understand the way DGDG, having a specific diacylglycerol backbone, is formed in plant cells, and then accumulates in order to be one of the major glycerolipids in green tissues. In addition, it is possible that several mechanisms and/or compartments might be involved, for example to explain the difference between chloroplasts from higher plants and algae.

IV. ORIGIN OF GALACTOLIPID POLYUNSATURATED FATTY ACIDS

The problem of the origin of polyunsaturated fatty acids has been the heart of the debate on galactolipid biosynthesis for years. Roughan (1970) and Williams *et al.* (1975, 1976) proposed that PC was the source of polyunsaturated fatty acids, whereas Siebertz and Heinz (1977), Heinz *et al.* (1979), Siebertz *et al.* (1980) were convinced of a real autonomy for chloroplasts in such a synthesis. In fact, the debate was biased, since different plant species, which proved to be strikingly different at the level of galactolipid structure and biosynthesis, were used in the two sets of experiments.

As discussed in the introduction (see Fig. 1), there are two major species of MGDG in plant tissues: those having 18:3/18:3 fatty acids and those having 18:3/16:3 ones. The studies done with isolated envelope membranes, and extended to whole chloroplasts, have clearly demonstrated that MGDG molecules having a "prokaryotic" structure (18:3/16:3) were derived from a diacylglycerol having 18:1/16:0 fatty acids which was formed through the envelope Kornberg–Pricer pathway (Joyard and Douce, 1977; Frentzen *et al.*, 1983). However, these findings do not provide a complete answer to the question of the origin of polyunsaturated fatty acids: MGDG molecules synthesized by envelope membranes have diacylglycerol backbones with 18:1/16:0, but with neither 18:3/16:3 or 18:3/18:3.

To this problem an additional puzzling observation is added: isolated chloroplasts from 16:3 and 18:3 plants are able to galactosylate diacylglycerol with UDPgal to almost the same rate (Heinz and Roughan, 1983); but, when supplied with a mixture of [^{14}C]acetate, *sn*-glycerol 3-phosphate, and UDPgal, chloroplasts from 16:3 plants have high rates of MGDG formation whereas those from 18:3 plants only synthesize MGDG at rates 20 times lower (Heinz and Roughan, 1983) (Fig. 12). In contrast, chloroplasts isolated from acetate-labeled leaves contain high amounts of labeled MGDG (for review see Roughan and Slack, 1982).

In fact, the two problems, origin of polyunsaturated fatty acids and formation of prokaryotic or eukaryotic MGDG, are closely related.

A. Possible Origin of Eukaryotic MGDG

Careful *in vivo* studies of the kinetics of incorporation of radioactive precursors, such as acetate, are at the origin of the PC hypothesis (Roughan, 1970). The reader is referred to several reviews by Heinz (1977), Douce and Joyard (1980), and Roughan and Slack (1982) for a detailed presentation of the arguments in favor of such a hypothesis. This hypothesis involves (1) the synthesis of PC, (2) its transfer (or that of diacylglycerol) to chloroplasts, and (3) the integration of the diacylglycerol part into MGDG and DGDG, on envelope membranes.

Endoplasmic reticulum is often considered to be the sole site for PC synthesis (for review see Moore, 1982), but additional sites—such as Golgi membranes

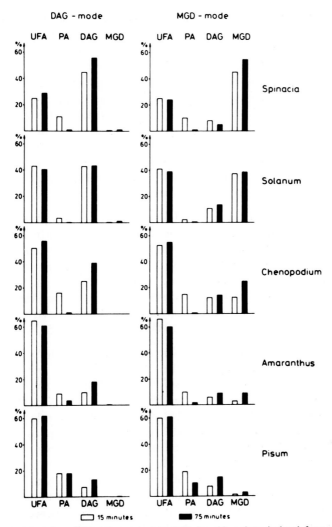

Fig. 12. Incorporation of labeled acetate into lipids by chloroplasts isolated from several plant species. In the left-hand part of the figure, only the metabolites necessary for fatty acid synthesis in the stroma and for the functioning of the envelope Kornberg–Pricer pathway were present in the incubation medium. In the right-hand part, UDPgal was added to the total incubation mixture. It is obvious that in 16:3 plants, diacylglycerol is formed at high rates and readily galactosylated into MGDG when UDPgal is supplied. On the contrary, much lower amounts of diacylglycerol are formed in 18:3 plants, and galactosylation occurs at much lower rates. This is in contrast to the situation where chloroplasts are supplied with UDP-[14C]galactose, which are able to form MGDG at similar rates, regardless of the origin of the chloroplasts (16:3 or 18:3 plants), as shown in Table VII. DAG, Diacylglycerol; MGD, monogalactosyldiacylglycerol; UFA, unesterified fatty acids; PA, phosphatidic acid. [Reproduced with permission from Heinz and Roughan (1983).]

(Montague and Ray, 1977; Sauer and Robinson, 1985) or the aleurone grain–oleo-some complex in seeds (Wilkinson et al., 1984)—have been demonstrated. On the contrary, chloroplasts, and especially their envelope membranes (Joyard and Douce, 1976c), are apparently devoid of cholinephosphotransferase. Therefore, PC is formed in extraplastidial membranes. Fatty acyl groups necessary for such a synthesis are provided by plastids, since these organelles are considered to be the sole site for fatty acid synthesis. PC synthesis is due to the enzymes of the Kornberg–Pricer pathway and to a CDPcholine:diacylglycerol cholinephospho-transferase (Slack et al., 1977). It is not the purpose of this review to detail PC synthesis, and the reader is referred to Moore (1982) for such details. After their synthesis, PC molecules are manipulated by the following reactions, resulting in a mixing of fatty acids or diacylglycerol parts of PC:

1. The acyl-CoA:1-acyl-sn-glycerol-3-phosphorylcholine acyltransferase (lyso-PC acyltransferase, E.C. 2.3.1.23), which acylates the sn-2 position of lyso-PC to form PC, is also able to catalyze the exchange of fatty acids between 18:1-CoA and unsaturated fatty acids esterified at the sn-2 position of PC (Stymne and Glad 1981; Stymne and Stobart, 1984; Murphy et al., 1984, 1985; Moreau and Stumpf, 1982).

2. The CDPcholine:diacylglycerol cholinephosphotransferase catalyzes revers-ible exchange of the diacylglycerol moiety between PC and diacylglycerol (Slack et al., 1983, 1985; Stymne and Stobart, 1984; Justin et al., 1985).

These reactions, especially the last one, are most important for triacylglycerol synthesis in seeds (Roughan and Slack, 1982; Slack, 1983), but they probably operate in a similar fashion in leaves (see, however, Martin and Wilson, 1984, for triacylglycerol synthesis in envelope membranes). The lyso-PC acyltransferase is not yet characterized (Moore, 1982) but plays an important role in glycerolipid metabolism, since it synthesizes 1-acyl-2-oleoyl-PC, which proves to be the true substrate for oleoyl desaturase, and not oleoyl-CoA as previously proposed (for reviews see Roughan and Slack, 1982; Murphy et al., 1985). In addition, there is a tight functional coupling between the acyltransferase and desaturase, which are apparently specific for the sn-2 position of PC, thus leading to the formation of 1-acyl-2-linoleoyl-PC (Murphy et al., 1984, 1985). The presence of a linoleoyl group at the sn-1 position is possibly due to a series of deacylation–reacylation–acyl exchange mechanisms (Murphy et al., 1985). Finally, despite extensive studies, almost nothing is known of the mechanism of desaturation (Roughan and Slack, 1982).

Therefore, the coordinated functioning of the enzymes of the Kornberg–Pricer pathways, of CDPcholine:diacylglycerol cholinephosphotransferase, oleoyl-CoA: lyso-PC acyltransferase, and oleate desaturase is responsible for the synthesis of PC having 18:2 at the sn-2 position of the glycerol backbone. At the sn-1 position, C_{16} or C_{18} fatty acids are esterified. Since these molecules have indeed a eukaryotic structure (see Fig. 1) which is also present in some galactolipid molecules, it was tempting to speculate that in MGDG, the C_{18}/C_{18} diacylglycerol was provided by PC.

Indeed, there is obviously a major difference in [^{14}C]acetate incorporation into MGDG between 18:3 chloroplasts isolated before or after labeling: chloroplasts isolated from acetate-labeled leaves contained MGDG with labeled C_{18} fatty acids at both sn-1 and sn-2 positions, whereas [^{14}C]acetate-labeled chloroplasts did not contain such a molecular species (Heinz and Roughan, 1983). These results are consistent with a long series of labeling experiments with leaves or isolated chloroplasts (a nonexhaustive list includes: Roughan, 1970; Trémolières and Mazliak, 1970; Slack and Roughan, 1975; Williams et al., 1975, 1976; Hawke et al., 1974a,b; Williams, 1980; Trémolières et al., 1980; Jones and Harwood, 1980; Hawke and Stumpf, 1980; Ohnishi and Yamada, 1982; Simpson and Williams, 1979). In addition, they have been extended by several authors (Gardiner et al., 1984a,b; Dubacq et al., 1983, 1984; Kosmac and Feierabend, 1985). All these observations suggest a direct participation of PC from an extraplastidial compartment to the formation of MGDG molecules having a eukaryotic structure.

However, as far as MGDG synthesis is concerned, the main problem—which is not yet solved—is that of diacylglycerol transfer from PC to MGDG. Up to now, only one alternative apparently occurs: (1) either PC is transported to chloroplasts and diacylglycerol should be formed within envelope membranes to be available for galactosylation (unless diacylglycerol exchange occurs between PC and MGDG) or (2) diacylglycerol (or phosphatidic acid) formed outside plastids is transported to chloroplasts for further metabolization into MGDG.

Several hypotheses have been proposed for glycerolipid transfer within the cell: the membrane flow (Morré, 1975) between the endomembrane system, or the transfer of molecules owing to specific proteins (Mazliak and Kader, 1980; Kader et al., 1982). This last mechanism involves phospholipid transfer (exchange) protein(s) which have been purified (Julienne et al., 1983). Protein-mediated PC transfer from liposomes (Julienne et al., 1981; Ohnishi and Yamada, 1982) or from microsomes (Dubacq et al., 1984) to isolated chloroplasts has been demonstrated, and envelope membranes behave like efficient acceptors for PC molecules (Miquel et al., 1984). However, only limited amounts of C_{18} fatty acids were apparently transferred to MGDG molecules (Ohnishi and Yamada, 1982), and therefore it is not yet known whether this process indeed reflects a physiological situation. This is particularly true in that PC only accumulates in the outer envelope from pea (Cline et al., 1981) or spinach (Dorne et al., 1982a; Block et al., 1983b) chloroplasts. In addition, a mild treatment of isolated intact chloroplasts by phospholipase C resulted in the almost total degradation of chloroplast PC (Dorne et al., 1985) (Fig. 13). Since under these conditions the treated chloroplasts were still intact and have an apparently normal outer envelope (as judged by electron-microscopic studies) (Fig. 14), Dorne et al. (1985) concluded that PC molecules were almost concentrated in the outer leaflet of the outer envelope membrane. This accumulation does not fit with a precursor–product relationship between PC and MGDG if we consider the postulated role for PC as a diacylglycerol precursor for MGDG. In fact, a very interesting question is whether PC distribution within the outer envelope membrane represents a stable configuration or only reflects a transient state (Dorne et al.,

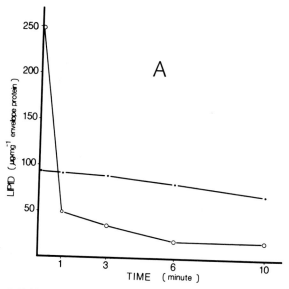

Fig. 13. Phospholipid content of envelope membranes isolated from phospholipase C-treated intact spinach chloroplasts. (A) Time-course experiment. Isolated intact chloroplasts (1 mg chlorophyll per milliliter) were treated for different times with 0.3 phospholipase C units per milliliter, at 12°C. (B) Effect of phospholipase C concentration. Isolated intact chloroplasts (1 mg chlorophyll per milliliter) were treated for 5 min with different phospholipase C concentrations, at 12°C. After incubation, the chloroplasts were purified on Percoll; the intact chloroplasts were then recovered for preparation of their envelope membranes and analysis of the phospholipid content. [Reproduced with permission from Dorne *et al.* (1985), *The Journal of Cell Biology,* 1985, **100,** 1690–1697, by copyright permission of The Rockefeller University Press.]

1985). Transmembrane movements are usually a very slow process (half-time value of 7–8 h at physiological temperatures, see Op den Kamp, 1979, 1981), although in some specific conditions they can be very rapid (see, e.g., Rothman and Kennedy, 1977; Bishop and Bell, 1985). Although almost nothing is known on transmembrane movements in envelope membranes, the asymmetry in PC distribution is maintained by an apparent lack of transmembrane diffusion. It is, however, possible that PC is metabolized as soon as it reaches the inner layer of the outer envelope membrane. Up to now this has not been demonstrated, since envelope membranes apparently lack a phospholipase C-type activity which would produce diacylglycerol (Joyard and Douce, 1976a; Roughan and Slack, 1977). However, this problem should be considered very cautiously, since the outer envelope membrane contains several acyltransferase activities (Block *et al.*, 1983b; Heemskerk *et al.*, 1986c) which could exchange either fatty acids or diacylglycerol moieties between PC and MGDG.

    If envelope membranes are indeed unable to transfer the diacylglycerol part of PC to MGDG, then, one might consider that diacylglycerol or phosphatidic acid (and

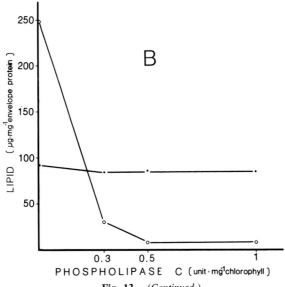

Fig. 13. (*Continued.*)

not PC) has to be transferred from microsomal membranes to chloroplasts. Up to now, there is no evidence for such a transfer.

Finally, a last possibility for establishing an "eukaryotic" structure in MGDG is that the chloroplast itself, and more likely the envelope membranes, could be involved in this formation. Again, there is no evidence for this, but we must keep in mind that envelope membranes contain several acyltransferases or galactosyltransferases which could be involved in an interlipid mixing of fatty acids or diacylglycerol, when functioning in a reverse way, in a similar fashion to that previously discussed for PC synthesis and desaturation in "microsomal" membranes.

In conclusion, the PC hypothesis is very attractive and might provide a simple explanation for the lack of synthesis, in isolated chloroplast but not in whole leaves, of MGDG molecules having C_{18}/C_{18} fatty acids. However, there are still some missing links to be discovered (Fig. 15). On the other hand, one must consider that the original purpose of the PC hypothesis was to provide a general explanation for the formation of all diacylglycerol moieties of plant cell glycerolipids. Obviously this is not the case, and we must acknowledge a striking role for plastids in the formation, metabolization, and even possibly export of diacylglycerol molecules having C_{18}/C_{16} fatty acid combinations.

B. MGDG as a Substrate for Desaturation

The MGDG molecules synthesized by chloroplasts, owing to their envelope membranes, contain 18:1 and 16:0 at *sn*-1 and *sn*-2 positions of the glycerol

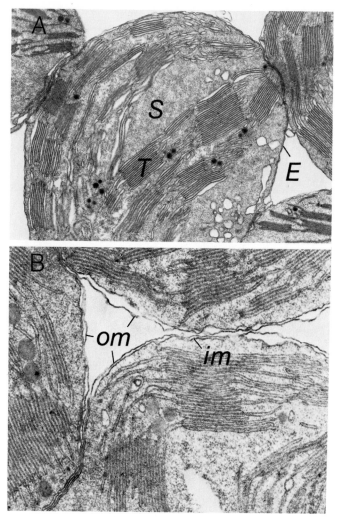

Fig. 14. Electron micrographs of phospholipase C-treated intact chloroplasts. Isolated intact chloroplasts (1 mg chlorophyll per milliliter) were treated with 0.5 units phospholipase C per milliliter at 12°C for 5 min. For further experimental details see Dorne *et al.* (1985). E, Envelope; S, stroma; T, thylakoids; OM, outer envelope membrane; IM, inner envelope membrane. (A) × 8,250; (B) × 19,500. [Reproduced with permission from Dorne *et al.* (1985), *The Journal of Cell Biology*, 1985, **100**, 1690–1697, by copyright permission of The Rockefeller University Press.]

backbone. Evidences from both *in vivo* and *in vitro* studies have led Heinz *et al.* (1979) and Siebertz *et al.* (1980) to conclude "that all lipids carrying at least part of their C_{16} fatty acids at position 2 of glycerol, are made in the chloroplast." These results provided support and extended the data obtained by Siebertz and Heinz (1977) showing desaturation of C_{16} and C_{18} fatty acids on MGDG molecules in

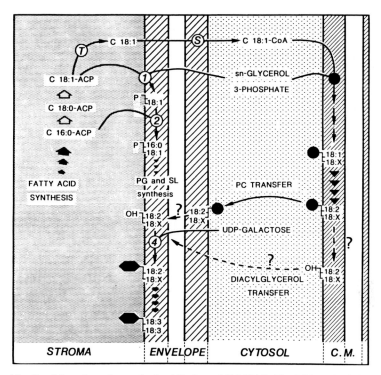

Fig. 15. Possible pathway for synthesis of "eukaryotic" MGDG. The PC hypothesis involves the following events: synthesis of fatty acids in the plastid stroma, export of acyl-CoA outside the organelle, synthesis of PC on extraplastidial membranes together with a series of deacylation– reacylation reactions which are operating in close relationship with oleoyl desaturase. PC is then either transported to envelope membranes where it is transformed into diacylglycerol or transformed into diacylglycerol which is then transported to envelope membranes, but none of these mechanisms has been yet demonstrated convincingly. Then, in envelope membranes, diacylglycerol is galactosylated into MGDG, which is further desaturated into 18:3/18:3 MGDG. T, Acyl-ACP thioesterase; S, acyl-CoA synthetase; C. M., cytoplasmic membranes; PG, phosphatidylglycerol; SL, sulfolipid. The numbers are identical to those of Fig. 10 (except for 4, UDPgal:diacylglycerol galactosyltransferase.)

Anthriscus (for reviews see Heinz, 1977; Douce and Joyard, 1980). In addition, after a short labeling time with $^{14}CO_2$ *in vivo* of spinach leaves, Heinz *et al.* (1979) and Siebertz *et al.* (1980) demonstrated that MGDG as well as DGDG molecules contained *de novo* synthesized saturated or monounsaturated fatty acids. Then, the time-dependent changes in the pattern of galactolipid molecular species indicated that desaturation operates at the level of MGDG and DGDG molecules (Fig. 16). The extent of *de novo* synthesis of polyunsaturated fatty acids was monitored by α-oxidation and gas chromatography analyses of degradation products (Siebertz *et al.*, 1980). In addition, the same results were obtained with the other plastid glycerolipids (phosphatidylglycerol or PG, and sulfolipid). These results provided further evidence for MGDG to be a substrate for C_{16} and C_{18} fatty acid desaturation.

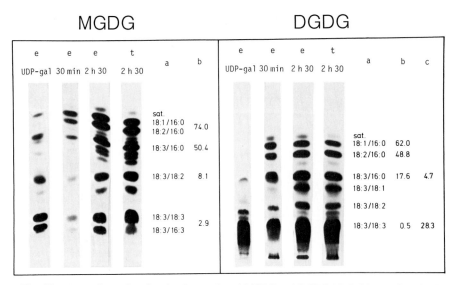

Fig. 16. Autoradiography of molecular species of MGDG and DGDG labeled *in vivo* in spinach leaves. After 30 min or 2 h 30 min of labeling with $^{14}CO_2$, chloroplasts were purified and envelope membranes (e) and thylakoids (t) separated. Galactolipids were then purified and analyzed by silver nitrate thin-layer chromatography. The UDP-[^{14}C]galactose-labeled species from an *in vitro* incubation of envelope membranes are included in each panel (left lane) for comparison. (a) Characterization of the species according to the fatty acid pairing; (b) percentage of label in fatty acids after 2 h 30 min labeling; (c) ratio of label in outer/inner galactose of DGDG after 2 h 30 min (Siebertz *et al.*, 1980) These results indicate that desaturation of fatty acids occurs on MGDG and DGDG, after assembly of the three parts of the molecule. [Reproduced wih permission from Heinz *et al.* (1979).]

Unfortunately, direct desaturation of semisynthetic MGDG molecules (Heinz *et al.*, 1979) or saturated MGDG molecules synthesized *in vitro* (Joyard *et al.*, 1979) could not be observed with isolated envelope membranes. In fact, such a desaturation could be observed when intactness of chloroplasts was preserved during the incubation: using highly active isolated spinach chloroplasts, Roughan *et al.* (1979c) were able to synthesize MGDG molecules (owing to the envelope enzymes) in which polyunsaturated fatty acids were detected. After 60 min incubation, 70% of [^{14}C]oleate originally at *sn*-1 position of MGDG was desaturated to [^{14}C]linoleate, and 45% of [^{14}C]linoleate was further desaturated to [α-^{14}C]linolenate (Roughan *et al.*, 1979c). Further evidence was immediately provided by several groups (Jones and Harwood, 1980; Williams, 1980; Williams *et al.*, 1982, 1983; Williams and Khan, 1982; Gardiner *et al.*, 1982, 1984a,b; Drapier *et al.*, 1982; Dubacq *et al.*, 1983; Gardiner and Roughan, 1983). For instance, with chloroplasts isolated from a 16:3 plant (*Solanum nodiflorum*), Heinz and Roughan (1983) analyzed, by reversed thin-layer chromatography, the fatty acids esterified at *sn*-1 and *sn*-2 positions obtained after lipase treatment of MGDG. The complete series of radioactive 16:0, 16:1, 16:2, and 16:3 was found at the *sn*-2 position, whereas at the sn-1 position 18:1, 18:2, and 18:3 were present (Fig. 17). These results indeed summarized the coordinated functioning of intact chloroplasts from 16:3 plants in the *de novo* MGDG synthesis: the synthesis of fatty acids from acetate, the acylation of sn-glycerol 3-phosphate (owing to the envelope Kornberg–Pricer pathway), the transfer of galactose from UDPgal into MGDG (owing to envelope membranes), and the desaturation of C_{18} fatty acids (at sn-1 position) and of C_{16} fatty acids (at sn-2 position) on the newly synthesized MGDG molecule (Fig. 18). The same thing probably holds true with DGDG molecules (Heinz *et al.*, 1979; Siebertz *et al.*, 1980).

C.   Comparison between Chloroplasts from 16:3 and 18:3 Plants

Although chloroplasts from 16:3 and 18:3 plants have apparently a very different physiology and contain galactolipids with distinct diacylglycerol structure, they appear to be rather similar:

1. Chloroplasts from 16:3 and 18:3 plants are able to synthesize fatty acids from [^{14}C]acetate at almost the same rate (Table VII); in both cases, 18:1 is the major fatty acid synthesized, but 16:0 is also formed in appreciable amounts. Although

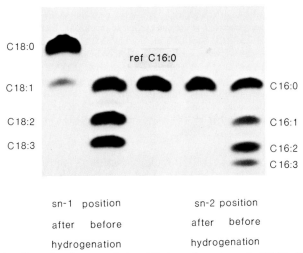

Fig. 17.   Autoradiography of a reversed-phase thin-layer chromatography (TLC) of fatty acids from MGDG synthesized and desaturated by isolated intact chloroplasts from *Solanum nodiflorum* leaves. The fatty acids esterified at the sn-1 and sn-2 positions were obtained after lipase hydrolysis of MGDG. Hydrogenation confirmed the identification of individual fatty acids. The front of the solvent after the TLC run is at the bottom of the figure. Note that on each position of the glycerol moiety, fatty acids with one, two, and three double bonds have been synthesized, thus demonstrating that desaturation occurs after MGDG synthesis. [Reproduced from Heinz and Roughan (1983), with permission.]

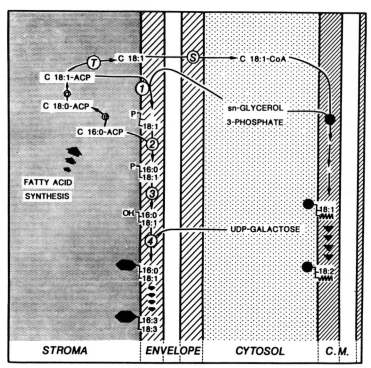

Fig. 18. Synthesis of 18:3/16:3 MGDG. The three parts of MGDG molecules (fatty acids, glycerol, and galactose) are assembled on the inner envelope membrane. Fatty acids are synthesized in the plastid stroma to form acyl-ACP. Acyl-ACP are used by the envelope Kornberg–Pricer pathway to form phosphatidic acid (from sn-glycerol 3-phosphate provided by the cytosol) and then diacylglycerol. Diacylglycerol is then galactosylated with UDPgal to form MGDG. UDPgal is probably formed in the cytosol and should be transported inside the chloroplast unless the galactosyltransferase, localized on the inner envelope membrane, is accessible from the intermembrane space, which would facilitate accessibility of the substrate to the enzyme. Galactolipids having 16:0 and 18:1 fatty acids are formed and have to be desaturated into 16:3/18:3. For abbreviations, see legend to Fig. 15.

fatty acid synthesis yields acyl-ACP molecules, a large part of newly synthesized fatty acids are recovered as unesterified fatty acids owing to a soluble acyl-ACP thioesterase, and can be exported outside chloroplasts probably owing to an outer envelope acyl-CoA synthetase; however, the level of fatty acids channeled towards the acyl-ACP thioesterase is higher in 18:3 plants than in 16:3 plants.

2. Chloroplasts from all plant species, owing to their envelope membranes, are able to acylate sn-glycerol 3-phosphate, leading to the formation of 1-oleoyl-2-palmitoyl-sn-glycerol 3-phosphate, but probably not of 1,2-dioleoyl-sn-glycerol 3-phosphate, and to dephosphorylate phosphatidic acid into the corresponding diacylglycerol. However, the activity of the envelope phosphatidate phosphatase is apparently much lower in 18:3 chloroplasts than in 16:3 chloroplasts (Table VII).

Table VII

Comparison of Chloroplasts from 16:3 and 18:3 Plants[a]

	Spinacia oleracea	Solanum nodiflorum	Chenopodium album	Chenopodium quinoa	Amaranthus lividus	Carthamus tinctorius	Pisum sativum
% 18:3/16:3 in MGDG	50	40	nd[b]	8	nd	1	0
Acetate incorporated (nmol/mg chlorophyll/h)	1470	719	1647	1210	855	905	1066
In presence of sn-glycerol 3-phosphate							
Unesterified fatty acids (% total acetate incorporated)	25	43.5	51	53.6	64.5	71.2	60
Phosphatidic acid + diacylglycerol (% total acetate incorporated)	56.5	46.5	41	31	18	19.5	25
Half-life of phosphatidic acid (min)	1.7	1.3	nd	5.8	nd	38.5	38.5
In presence of sn-glycerol 3-phosphate + UDPgal MGDG (% total acetate incorporated)	45	37.5	10	11.3	2.5	2.0	1.3
In presence of UDPgal MGDG (nmol UDPgal incorporated/ mg chlorophyll/h)	nd	26	34	nd	32	nd	21
Desaturation of fatty acids in MGDG							
sn-1 Position	+++	++++	+++		++		++
sn-2 Position	+++	++++	?		−[b]		−

[a] This table was obtained from data published by Heinz and Roughan (1982, 1983), Gardiner et al. (1984a,b), and Gardiner and Roughan (1983).

[b] −, Not detected; nd, not determined.

3. Chloroplasts from 16:3 and 18:3 plants are able to galactosylate diacylglycerol with UDPgal to form MGDG to almost the same extent; however, diacylglycerol molecules formed through the envelope Kornberg–Pricer pathway are apparently used to a lesser extent in 18:3 chloroplasts than in 16:3 chloroplasts (Table VII).

4. Chloroplasts from all plant species are able to synthesize glycerolipids other than MGDG owing to their envelope membranes. Isolated intact chloroplasts are able to synthesize CDP-diacylglycerol, phosphatidylglycerol phosphate, and PG from the phosphatidic acid formed through the envelope Kornberg–Pricer pathway in *Euglena* (Chammaï, 1980) as well as in higher plants (Sparace and Mudd, 1982; Mudd and de Zacks, 1981; Andrews and Mudd, 1985; Roughan, 1985); in addition, they are able to synthesize sulfolipid (Haas *et al.*, 1980; Kleppinger-Sparace *et al.*, 1985; Joyard *et al.*, 1986) (Fig. 19).

5. Chloroplasts from 16:3 and 18:3 plants contain desaturases; however, it is not yet known whether they can desaturate all glycerolipid molecules that are present in the chloroplasts (eukaryotic MGDG, but also PG and sulfolipid). However, chloroplasts from 16:3 and 18:3 plants are able to desaturate 18:1 into 18:3 at the *sn*-1 position of MGDG (although 16:3 chloroplasts are apparently more efficient), but at the *sn*-2 position, desaturation is very specific: in 16:3 plants, 16:0 MGDG is readily desaturated into 16:3 MGDG, but not in 18:3 plants (Table VII).

Thus, chloroplasts from 16:3 or 18:3 plants show almost the same enzymatic equipment for glycerolipid biosynthesis, although in 18:3 plants, some enzymes have a lower specific activity. They differ apparently in the ability of desaturation–at the *sn*-2 position of the glycerol backbone of MGDG—of monounsaturated into disaturated fatty acids. They may also differ at the level of regulation of the enzymes and channeling of the different precursors for glycerolipid biosynthesis: the rate of removal of these precursors may channel them to their transformation into strikingly different molecules. For instance, in chloroplasts from 18:3 plants, the rate of removal of newly synthesized acyl-ACP by the *sn*-glycerol-3-phosphate acyltransferase is rather low, and a large part of the molecules are transformed into unesterified fatty acids by acyl-ACP thioesterase, for further metabolism, probably outside the plastids. In addition, since the phosphatidate phosphatase activity is low, little diacylglycerol is available for galactolipid and sulfolipid synthesis, and phosphatidic acid is available for PG synthesis. In contrast, in 16:3 chloroplasts, a large proportion of newly synthesized fatty acids is channeled toward the enzymes of the Kornberg–Pricer pathway and galactosyltransferases, and therefore MGDG is synthesized in much larger amounts than are PG or sulfolipid (Fig. 20).

D. Conclusion

The experiments with whole leaves and isolated chloroplasts have led to a rather complex model for the formation of polyunsaturated fatty acids for MGDG. Oleic acid esterified at the *sn*-2 position of PC can be readily desaturated into 18:2, in

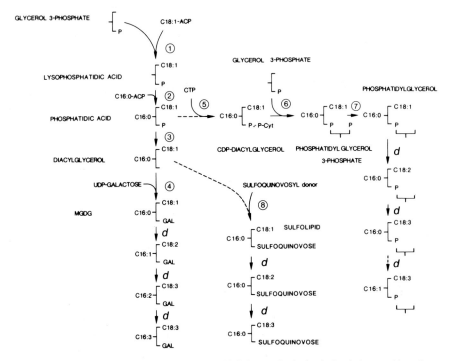

Fig. 19. Pathway for MGDG, PG, and sulfolipid synthesis in isolated intact chloroplasts. Phosphatidic acid formed by acyl-ACP:*sn*-glycerol-3-phosphate acyltransferase (**1**), and acyl-ACP: monoacylglycerol-3-phosphate acyltransferase (**2**) is used either for diacylglycerol synthesis (**3**) or PG synthesis (**7**). Diacylglycerol formed by the phosphatidate phosphatase (**3**) is then galactosylated into MGDG or used for sulfolipid synthesis (**8**). PG synthesis from phosphatidic acid involves the formation of CDP-diacylglycerol (**5**) and phosphatidylglycerol phosphate (**6**). Such a pathway leads to the synthesis of glycerolipids having 16:0/18:1 fatty acids, and the molecules are then desaturated. All these steps (except for sulfolipid synthesis and desaturation) have been demonstrated in isolated envelope membranes.

almost all plants or tissues analyzed so far (Roughan and Slack, 1982). This observation can even be extended to algae (such as *Chlamydomonas*) which are devoid of PC, but contain as a substitute diacylglyceryl trimethyl homoserine, or DGTS (Eichenberger, 1982; Sato and Furuya, 1984), since in *Chlamydomonas* oleoyl desaturation occurs on DGTS (Schlapfer and Eichenberger, 1983). It is not yet known how the diacylglycerol moiety of PC (or DGTS) is transferred to MGDG, if the PC hypothesis is actually the right answer to the most interesting problem of the origin of C_{18}/C_{18}-containing MGDG. There are several good pieces of evidence for such a hypothesis, but a definite answer is still missing.

On the contrary, the formation of C_{18}/C_{16} MGDG is now clearly understood: the assembly of fatty acids, glycerol, and galactose occurs in the chloroplast envelope membrane, and further desaturation of 16:0 and 18:1 also occurs in isolated

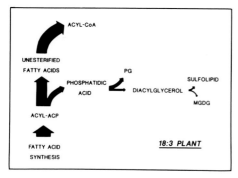
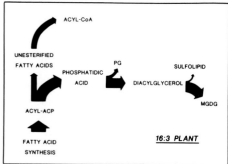

Fig. 20. Comparison between channeling of fatty acids synthesized in chloroplasts from 16:3 and 18:3 plants. In 16:3 plants, the envelope Kornberg–Pricer pathway is most active and leads to the synthesis of prokaryotic MGDG, PG, and sulfolipid. The acyl-ACP thioesterase channels fatty acids to the acyl-CoA synthetase for further metabolism outside plastids. In 18:3 plants, the envelope Kornberg–Pricer pathway is much less active than in 16:3 plants; therefore, most of newly synthesized acyl-ACPs are channeled toward acyl-ACP thioesterase and acyl-CoA synthetase.

chloroplasts, on the *de novo*-synthesized glycerolipid molecule. The same observation can be made from cyanobacteria, which—although devoid of PC—are able to synthesize MGDG having (in some cases) polyunsaturated fatty acids.

From these observations, a quite interesting picture emerges: the ability of different membrane systems or organelles to synthesize *de novo* at least part of their glycerolipids and to desaturate (on glycerolipid molecules themselves) their fatty acids, suggests the existence of multiple sites for glycerolipid biosynthesis within the cell, in contrast to the central role in membrane biogenesis postulated for endoplasmic reticulum a few years ago. Additional evidence for the existence of multiple sites for desaturation in a plant cell is provided by experiments on isolated Acer cells, in suspension cultures. Rebeillé *et al.* (1980) have shown that at low O_2 concentration in the medium (10 μM) all the membrane glycerolipids (phospholipids and galactolipids) are strongly enriched in 18:1. When air is provided to the culture in order to reach saturation levels of O_2 (250 μM), a stoichiometric disappearance of 18:1 and appearance of 18:2 was observed in all glycerolipids and at the same rate, and no predominance of any site for desaturation was observed.

IV.   FUTURE PROSPECTS

Galactolipid synthesis is a general property of the plant kingdom: it can be followed in green or etiolated tissues, in algae or in cyanobacteria. It is an essential event of plastid biogenesis and therefore has to be regulated: requirements for galactolipid biosynthesis are not the same in greening tissues, in mature spinach chloroplasts, in amyloplasts, or other specialized plastids (e.g., from 18:3 or 16:3 plants). Numerous studies have been done on light, temperature, greening, and growth effects on galactolipid levels in plant tissues (e.g., the proceedings of several meetings held on plant lipids: Tevini and Lichtenthaler, 1977; Appelqvist and Liljenberg, 1979; Mazliak et al., 1980; Wintermans and Kuiper, 1982; Siegenthaler and Eichenberger, 1984). But up to now, very little is known about the mechanisms that regulate galactolipid biosynthesis under different environmental or physiological conditions. For example, in storage tissues or in roots, expansion of the amyloplast envelope keeps pace with the growth of the starch granule inside it, and vice versa (Brianty et al., 1979). We believe that the development of such studies is most important.

Plants grow under rapidly changing environmental conditions, and the amounts of metabolites necessary for galactolipid biosynthesis might vary accordingly. For instance, the levels of acyl-ACP, sn-glycerol 3-phosphate, and UDPgal are susceptible to variations during dark–light transitions in leaves, but in roots requirements for such metabolites might be different. Therefore, the events which are responsible for galactolipid biosynthesis are probably strongly dependent on the biosynthetic pathways involved in carbohydrate metabolism, since these events (photosynthesis, glycolysis, pentose pathway, respiration) are the major factors which control the plant cell life (Douce, 1985).

However, although the impressive flexibility of plant cell metabolism can meet its changing energy requirements during dark–light transitions or when rapid, transitory, or short-term changes in the environmental conditions occur, such metabolic processes cannot explain the differences in the capacities of galactolipid synthesis by 16:3 or 18:3 chloroplasts or the differentiated galactolipid synthesis that is necessary in etiolated or greening tissues, in roots or in petals, when compared to green leaves. In fact, regulation at the level of gene expression is probably the major control process in the whole plant: light (and probably other environmental factors) affects the transcription of genes, and genes can also be selectively expressed during development, in different tissues or in different plants (Broglie et al., 1984; Coruzzi et al., 1984). In greening tissues, in roots, in 16:3 or 18:3 plants, the differences in the channeling of the different precursors necessary for galactolipid biosynthesis could be due to variation of transcriptional controls of individual genes encoding for all the enzymes involved in glycerolipid metabolism. All these enzymes are probably encoded by nuclear DNA. For instance, Blée (1981) has shown that *Euglena* tissues cultivated in the presence of chloramphenicol (which prevents translation by chloroplastic ribosomes) and a mutant devoid of plastid DNA (*Euglena gracilis* W$_3$BUL; Schiff, 1973) were both able to synthesize galactolipids. In higher plants, the same result was obtained: Dorne et al. (1982c) have shown that a mutant line of *Hordeum vulgare* "albostrians" contains galactolipids but lacks plastid ribosomes and therefore is unable to synthesize plastid proteins (Börner et al., 1976). The reduced amount of galactolipids found in the "albostrians" mutant, when compared with normal green leaves, only reflects the expansion of plastid membranes (in the mutant, the plastid internal membranes were almost entirely absent and only envelope membranes were present) but not a reduced capacity of lipid biosynthesis (Dorne et al., 1982c). Additional evidence was provided by Kosmac and Feierabend (1984, 1985), who compared the glycerolipid content of green and 70 S ribosome-deficient leaves from rye (*Secale cereale* L.) and the biosynthetic capabilities of their plastids. However, rye and barley are both 18:3 plants, and it would be most interesting to determine whether in true 16:3 plants the same results would be obtained.

Finally, differentiation of all plastid types (proplastids, amyloplasts, chromoplasts, etioplasts, chloroplasts, etc.) in 18:3 as well as in 16:3 plants, and their developmental transitions are associated with, or dependent on, marked changes in the specific enzymatic equipment of plastid envelope membranes. This is true also for the enzymes involved in glycerolipid biosynthesis. We believe that tissue-specific, plant-specific, or light-regulated expression of genes encoding for envelope proteins are major events in plastid biogenesis and constitute a challenging goal for plant research in the next decade.

REFERENCES

Allen, C. F., Hirayama, O., and Good, P. (1966). *In* "Biochemistry of Chloroplasts" (T. W. Goodwin, ed.), pp. 195–200. Academic Press, London.

Andrews, J., and Keegstra, K. (1983). *Plant Physiol.* **72**, 735–740.

Andrews, J., and Mudd, J. B. (1985). *Plant Physiol.* **79**, 259–265.

Andrews, J., Ohlrogge, J. B., and Keegstra, K. (1985). *Plant Physiol.* **78**, 459–465.

Appelqvist, L.-Å., and Liljenberg, C., eds. (1979). "Advances in the Biochemistry and Physiology of Plant Lipids." Elsevier, Amsterdam.

Barron, E. J., and Stumpf, P. K. (1962). *Biochim. Biophys. Acta* **60**, 329–337.

Bell, R. M., and Coleman, R. A. (1980). *Annu. Rev. Biochem.* **49**, 459–487.

Benson, A.A., and Maruo, B. (1958). *Biochim. Biophys. Acta* **41**, 328–333.

Bertrams, M., and Heinz, E. (1976). *Planta* **132**, 161–168.

Bertrams, M., and Heinz, E. (1979). *In* "Advances in the Biochemistry and Physiology of Plant Lipids" (L.-Å. Appelqvist and C. Liljenberg, eds.), pp. 139–144. Elsevier, Amsterdam.

Bertrams, M., and Heinz, E. (1980). *In* "Biogenesis and Function of Plant Lipids" (P. Mazliak, P. Benveniste, C. Costes, and R. Douce, eds.), pp. 67–72. Elsevier, Amsterdam.

Bertrams, M., and Heinz, E. (1981). *Plant Physiol.* **68**, 653–657.

Bertrams, M., Wrage, K., and Heinz, E. (1981). *Z. Naturforsch., 36C,* 62–70.

Bishop, W. R., and Bell, R. M. (1985). *Cell (Cambridge, Mass.)* **42**, 51–60.

Blée, E. (1981). Thèse Doct. Univ. de Strasbourg.

Blée, E., and Schantz, R. (1978a). *Plant Sci. Lett.* **13**, 247–256.

Blée, E., and Schantz, R. (1978b). *Plant Sci. Lett.* **13**, 257–267.

Block, M. A., Dorne, A.-J., Joyard, J., and Douce, R. (1983a). *J. Biol. Chem.* **258**, 13273–13280.

Block, M. A., Dorne, A.-J., Joyard, J., and Douce, R. (1983b). *J. Biol. Chem.* **258**, 13281–13286.

Block, M. A., Dorne, A.-J., Joyard, J., and Douce, R. (1983c). *FEBS Lett.* **153**, 377–381.

Block, M. A., Dorne, A.-J., Joyard, J., and Douce, R. (1983d). *FEBS Lett.* **169**, 111–115.

Boehler, B. A., and Ernst-Fonberg, M. L. (1976). *Arch. Biochem. Biophys.* **175**, 229–235.

Börner, T., Schumann, B., and Hagemann, R. (1976). *In* "Genetics and Biogenesis of Chloroplasts and Mitochondria" (W. Bücher, W. Neupert, W. Sebald, and S. Werner, eds.), pp. 41–48. North-Holland Publ., Amsterdam.

Brangeon, J., and Forchioni, A. (1986). *In* "Regulation of Chloroplast Differentiation" (G. Akoyounoglou and H. Senger, eds.), pp. 141–146. Alan R. Liss, New York.

Brianty, L. G., Hughes, C. E., and Evers, A. D. (1979). *Ann. Bot. (London)* **44**, 641–658.

Broglie, R., Coruzzi, G., Fraley, R. T., Rogers, S. G., Horsch, R. B., Niedermeyer, J. G., Fin, C. L., Flick, J. S., and Chua, N.-H. (1984). *Science* **224**, 838–843.

Brouquisse, R., Nishimura, N., Gaillard, J., and Douce, R. (1987). *Plant Physiol.* (in press).
Carde, J.-P. (1985). *Eur. J. Cell Biol.* **34**, 18–26.
Carde, J.-P., Joyard, J., and Douce, R. (1982). *Biol. Cell.* **44**, 315–324.
Cassagne, C., and Lessire, R. (1978). *Arch. Biochem. Biophys.* **191**, 146–151.
Caughey, I., and Kekwick, R. G. O. (1982). *Eur. J. Biochem.* **123**, 553–561.
Chammaï, A. (1980). Thèse Doct. Univ. de Strasbourg.
Chang, S. B. (1970). *Phytochemistry* **9**, 1947–1948.
Cheniae, G. M. (1965). *Plant Physiol.* **40**, 235–243.
Cline, K., and Keegstra, K. (1983). *Plant Physiol.* **71**, 366–372.
Cline, K., Andrews, J., Mersey, B., Newcomb, E. H., and Keegstra, K. (1981). *Proc. Natl. Acad. Sci. U.S.A.* **78**, 3595–3599.
Cooper, C. L., and Lueking, D. R. (1984). *J. Lipid Res.* **25**, 1222–1232.
Coruzzi, G., Broglie, R., Edwards, C., and Chua, N.-H. (1984). *EMBO J.* **3**, 1671–1679.
Covès, J. (1987). Thèse Doct. Univ. de Grenoble.
Covès, J., Block, M. A., Joyard, J., and Douce, R. (1986). *FEBS Lett.* **208**, 401–406.
Cronan, J. E., Jr. (1978). *Annu. Rev. Biochem.* **47**, 163–189.
Dorne, A.-J., Block, M. A., Joyard, J., and Douce, R. (1982a). *In* "Biochemistry and Metabolism of Plant Lipids" (J. F. G. M. Wintermans and P. J. C. Kuiper, eds.), pp. 153–164. Elsevier, Amsterdam.
Dorne, A.-J., Block, M. A., Joyard, J., and Douce, R. (1982b). *FEBS Lett.* **145**, 30–34.
Dorne, A.-J., Carde, J.-P., Joyard, J., Börner, T., and Douce, R. (1982c). *Plant Physiol.* **69**, 1467–1470.
Dorne, A.-J., Joyard, J., Block, M. A., and Douce, R. (1985). *J. Cell Biol.* **100**, 1690–1697.
Douce, R. (1970). Thèse Doct. Univ. de Paris.
Douce, R. (1974). *Science* **183**, 852–853.
Douce, R. (1985). "Mitochondria in Higher Plants." Academic Press, New York.
Douce, R., and Guillot-Salomon, T. (1970). *FEBS Lett.* **11**, 121–126.
Douce, R., and Joyard, J. (1979). *Adv. Bot. Res.* **7**, 1–116.
Douce, R., and Joyard, J. (1980). *In* "The Biochemistry of Plants, Vol. 4, Lipids: Structure and Function" (P. K. Stumpf, ed.), pp. 321–362. Academic Press, New York.
Douce, R., and Joyard, J. (1982). *In* "Methods in Chloroplast Molecular Biology" (M. Edelman, R. Hallick, and N.-H. Chua, eds.), pp. 239–256. Elsevier, Amsterdam.
Douce, R., and Joyard, J. (1984). *In* "Topics in Photosynthesis, Vol. 5, Chloroplast Biogenesis" (N. Baker and J. Barber, eds.) pp. 71–132. Elsevier, Amsterdam.
Douce, R., Holtz, R. B., and Benson, A. A. (1973). *J. Biol. Chem.* **248**, 7215–7222.
Douce, R., Block, M. A., Dorne, A.-J., and Joyard, J. (1984a). *Subcell. Biochem.* **10**, 1–84.
Douce, R., Dorne, A.-J., Block, M. A., and Joyard, J. (1984b). *In* "Chloroplast Biogenesis" (R. J. Ellis, ed.), pp. 193–224. Cambridge Univ. Press, London and New York.
Douce, R., Joyard, J., Dorne, A.-J., and Block, M. A. (1985). *In* "New Developments and Methods in Membrane Research and Biological Energy Transduction" (L. Packer, ed.), pp. 179–205. Plenum, London.
Drapier, D., Dubacq, J. P., Trémolières, A., and Mazliak, P. (1982). *Plant Cell Physiol.* **23**, 125–135.
Dubacq, J.-P., Drapier, D., and Trémolières, A. (1983). *Plant Cell Physiol.* **24**, 1–9.
Dubacq, J.-P., Drapier, D., Trémolières, A., and Kader, J. C. (1984). *Plant Cell Physiol.* **25**, 1197–1204.
Eichenberger, W. (1982). *Plant Sci. Lett.* **24**, 91–95.
Elias, B. A., and Givan, C. V. (1979). *Plant Sci. Lett.* **17**, 115–122.
Feige, G. B., Heinz, E., Wrage, K., Cochems, N., and Poncelar, E. (1980). *In* "Biogenesis and Function of Plant Lipids" (P. Mazliak, P. Benveniste, C. Costes, and R. Douce, eds.), pp. 135–140. Elsevier, Amsterdam.
Fishwick, M. J., and Wright, A. J. (1980). *Phytochemistry* **19**, 55–59.
Frentzen, M (1986). *J. Plant Physiol.* **124**, 193–209.
Frentzen, M., and Heinz, E. (1983). *Biol. I.U.Z.* **6**, 178–187.

Frentzen, M., Heinz, E., Mckeon, T. A., and Stumpf, P. K. (1982). *In* "Biochemistry and Metabolism of Plant Lipids" (J. F. G. M. Wintermans and P. J. C. Kuiper, eds.), pp. 141–152. Elsevier, Amsterdam.

Frentzen, M., Heinz, E., McKeon, T. A., and Stumpf, P. K. (1983). *Eur. J. Biochem.* **129,** 629–636.

Frentzen, M., Hares, W., and Schiburr, A. (1984). *In* "Structure, Function and Metabolism of Plant Lipids" (P.-A. Siegenthaler and W. Eichenberger, eds.), pp. 105–110. Elsevier, Amsterdam.

Fritz, P. J., Kauffman, J. M., Robertson, C. A., and Wilson, M. R. (1984). *Biochemistry* **23,** 3374 (abstr.).

Gardiner, S. E., and Roughan, P. G. (1983). *Biochem. J.* **210,** 949–952.

Gardiner, S. E., Roughan, P. G., and Slack, C. R. (1982). *Plant Physiol.* **70,** 1316–1320.

Gardiner, S. E., Roughan, P. G., and Browse, J. (1984a). *Biochem. J.* **224,** 637–643.

Gardiner, S. E., Heinz, E., and Roughan, P. G. (1984b). *Plant Physiol.* **74,** 890–896.

Gerhardt, B. (1983). *Planta* **159,** 238–246.

Givan, C. V. (1983). *Physiol. Plant.* **57,** 311–316.

Green, P. R., Merrill, A. M., Jr., and Bell, R. M. (1981). *J. Biol. Chem.* **256,** 11151–11159.

Grobovsky, L. V., Hershenson, S., and Ernst-Fonberg, M. L. (1979). *FEBS Lett.* **102,** 261–264.

Guillory, R. J. (1979). *Curr. Top. Bioenerg.* **9,** 267–414.

Guerra, D.J., Ohlrogge, J.B., and Frentzen, M. (1986). *Plant Physiol.* **82,** 448–453.

Haas, R., and Heinz, E. (1980). *In* "Biogenesis and Function of Plant Lipids" (P. Mazliak, P. Benveniste, C. Costes, and R. Douce, eds.), pp. 19–28. Elsevier, Amsterdam.

Haas, R., Siebertz, H. P., Wrage, K., and Heinz, E. (1980). *Planta* **148,** 238–244.

Harwood, J. L. (1979). *Prog. Lipid Res.* **18,** 55–86.

Harwood, J. L., and Russell, N. J. (1984). "Lipids in Plants and Microbes." Allen & Unwin, London.

Hawke, J. C., and Stumpf, P. K. (1980). *Arch. Biochem. Biophys.* **203,** 296–306.

Hawke, J. C., Rumsby, M. G., and Leech, R. M. (1974a). *Phytochemistry* **13,** 403–413.

Hawke, J. C., Rumsby, M. G., and Leech, R. M. (1974b). *Plant Physiol.* **53,** 555–561.

Heber, H., and Heldt, H. W. (1981). *Annu. Rev. Plant Physiol.* **32,** 139–168.

Heemskerk, J. W. M., Bögemann, G., and Wintermans, J. F. G. M. (1983). *Biochim. Biophys. Acta* **754,** 181–189.

Heemskerk, J., Bögemann, G., and Wintermans, J. F. G. M. (1985). *Biochim. Biophys. Acta* **835,** 212–220.

Heemskerk, J. W. M., Jacobs, F. H. H., Bögemann, G., Scheijen, M. A. M., and Wintermans, J. F. G. M. (1986a). *In* Heemskerk, J. W. M. (1986). Ph.D. Thesis, Univ. of Nijmegen, Netherlands.

Heemskerk, J. W. M., Jacobs, F. H. H., Wolfs, W. J., and Wintermans, J. F. G. M. (1986b). *In* Heemskerk, J. W. M. (1986). Ph.D. Thesis, Univ. of Nijmegen, Netherlands.

Heemskerk, J. W. M., Wintermans, J. F. G. M., Joyard, J., Block, M. A., Dorne, A.-J., and Douce, R. (1986c). *Biochim. Biophys. Acta* **877,** 281–289.

Heinz, E. (1977). *In* "Lipids and Lipid Polymers In Higher Plants," pp. 102–120. Springer-Verlag, Berlin and New York.

Heinz, E., and Roughan, P. G. (1982). *In* "Biochemistry and Metabolism of Plant Lipids" (J. F. G. M. Wintermans and P. J. C. Kuiper, eds.), pp. 169–182. Elsevier, Amsterdam.

Heinz, E., and Roughan, P. G. (1983). *Plant Physiol.* **72,** 273-279.

Heinz, E., Bertrams, M., Joyard, J., and Douce, R. (1978). *Z. Pflanzenphysiol.* **87,** 325–331.

Heinz, E., Siebertz, H. P., Linscheid, M., Joyard, J., and Douce, R. (1979). *In* "Advances in the Biochemistry and Physiology of Plant Lipids" (L.-Å. Appelqvist and C. Liljenberg, eds.), pp. 99–120. Elsevier, Amsterdam.

Hendren, R. W., and Bloch, K. (1980). *J. Biol. Chem.* **255,** 1504–1508.

Hershenson, S., Boehler-Kohler, B. A., and Ernst-Fonberg, M. L. (1983). *Arch. Biochem. Biophys.* **223,** 76–84.

Hesler, C.B., Carroll, M.A., and Haldar, D. (1985). *J. Biol. Chem.* **260,** 7452–7456.

Høj, P. B., and Mikkelsen, J. D. (1982). *Carslberg Res. Commun.* **47,** 119–141.

Hosaka, K., Yamashita, S., and Numa, S. (1975). *J. Biochem. (Tokyo)* **77,** 501–509.

Ichihara, K. (1984). *Arch. Biochem. Biophys.* **232,** 685–698.

Ishinaga, M., Nishihara, M., and Kito, M. (1976). *Biochim. Biophys. Acta* **450**, 269–272.

Jacobson, B. S., and Stumpf, P. K. (1972). *Arch. Biochem. Biophys.* **153**, 656–663.

Jamieson, G. R., and Reid, E. H. (1971) *Phytochemistry* **10**, 1837–1844.

Jelsema, C. L., Michaels, A. S., Janero, D. R., and Barrnett, R. J. (1982). *J. Cell Sci. Ser. D* **58**, 469–488.

Jones, A. V. M., and Harwood, J. L. (1980). *Biochem. J.* **190**, 851–854.

Journet, E.-P., and Douce, R. (1985). *Plant Physiol.* **79**, 458–467.

Joyard, J. (1979). Thèse Doct. Univ. de Grenoble.

Joyard, J., and Douce, R. (1976a). *Biochim. Biophys. Acta.* **424**, 126–131.

Joyard, J., and Douce, R. (1976b). *Physiol. Veg.* **14**, 31–48.

Joyard, J., and Douce, R. (1976c). *C. R. Hebd. Seances Acad. Sci. Ser. D* **282**, 1515–1518.

Joyard, J., and Douce, R. (1977). *Biochim. Biophys. Acta* **486**, 273–285.

Joyard, J., and Douce, R. (1979). *FEBS Lett.* **102**, 147–150.

Joyard, J., and Stumpf, P. K. (1980). *Plant Physiol.* **65**, 1039–1043.

Joyard, J., and Stumpf, P. K. (1981). *Plant Physiol.* **67**, 250–256.

Joyard, J., Chuzel, M., and Douce, R. (1979). *In* "Advances in the Biochemistry and Physiology of Plant Lipids" (L.-Å. Appelqvist and C. Liljenberg, eds.), pp. 181-186. Elsevier, Amsterdam.

Joyard, J., Grossman, A. R., Bartlett, S. G., Douce, R., and Chua, N.-H. (1982). *J. Biol. Chem.* **257**, 1095–1101.

Joyard, J., Billecocq, A., Bartlett, S. G., Block, M. A., Chua, N. -H., and Douce, R. (1983). *J. Biol. Chem.* **258**, 10000–10006.

Joyard, J., Blée, E., and Douce, R. (1986). *Biochim. Biophys. Acta* **879**, 78–87.

Joyard, J., Dorne, A.-J., and Douce, R. (1987). *In* "Models in Plant Physiology/Biochemistry/Technology" (D. W. Newman and K. G. Wilson, eds.). CRC Press, Boca Raton, Florida. In press.

Julienne, M., Vergnolle, C., and Kader, J. C. (1981). *Biochem. J.* **197**, 763–766.

Julienne, M., Vergnolle, C., and Kader, J. C. (1983). *C. R. Hebd. Seances Acad. Sci. Ser. D* **296**, 617–620.

Justin, A. M., Demandre, C., Trémolières, A., and Mazliak, P. (1985). *Biochim. Biophys. Acta* **836**, 1–7.

Kader, J. C., Douady, D., and Mazliak, P. (1982). *In* "Phospholipids" (J. N. Hawthorne and G. B. Ansell, eds.), pp. 279–311. Elsevier, Amsterdam.

Kennedy, E. P. (1961). *Fed. Proc. Fed. Am. Soc. Exp. Biol.* **20**, 934–940.

Kleppinger-Sparace, K. F., Mudd, J. B., and Bishop, D. G. (1985). *Arch. Biochem. Biophys.* **240**, 859–865.

Königs, B., and Heinz, E. (1974). *Planta* **118**, 159–169.

Kornberg, A., and Pricer, W. (1953). *J. Biol. Chem.* **204**, 329–343.

Kosmac, U., and Feierabend, J. (1984). *Z. Pflanzenphysiol.* **114**, 377–392.

Kosmac, U., and Feierabend, J. (1985). *Plant Physiol.* **79**, 646–652.

Kuhn, D. N., Knauf, M., and Stumpf, P. K. (1981). *Arch. Biochem. Biophys.* **209**, 441–450.

Liedvogel, B. (1985). *Z. Naturforsch.* **40c**, 182–188.

Liedvogel, B. (1986). *J. Plant Physiol.* **124**, 211–222.

Liedvogel, B., and Kleinig, H. (1976). *Planta* **129**, 19–21.

Liedvogel, B., and Kleinig, H. (1977). *Planta* **133**, 249–253.

Liedvogel, B., and Kleinig, H. (1979). *Planta* **144**, 467–471.

Liedvogel, B., Kleinig, H., Thompson, J. A., and Falk, H. (1978). *Planta* **141**, 303–309.

Lightner, V. A., Larson, T. J., Tailleur, P., Kantor, G. D., Raetz, C. R. H., Bell, R. M., and Modrich, P. (1980). *J. Biol. Chem.* **255**, 9413–9420.

McKeon, T., and Stumpf, P. K. (1982). *J. Biol. Chem.* **257**, 12141–12147.

Martin, B. A., and Wilson, R. F. (1984). *Lipids* **19**, 117–121.

Mazliak, P., and Kader, J. C. (1980). *In* "The Biochemistry of Plants. Vol. 4. Lipids: Structure and Function" (P. K. Stumpf, ed.), pp. 283–300. Academic Press, New York.

Mazliak, P., Benveniste, P., Costes, C., and Douce, R. (1980). "Biogenesis and Function of Plant Lipids." Elsevier, Amsterdam.

Mendiola-Morgenthaler, L., Eichenberger, W., and Boschetti, A. (1985). *Plant Sci. Lett.* **41,** 97–104.
Michaels, A. S., Jelsema, C. L., and Barrnett, R. J. (1983). *J. Ultrastruct. Res.* **82,** 35–51.
Miernyk, J. A., and Dennis, D. T. (1983). *J. Exp. Bot.* **34,** 712–718.
Miquel, M. (1985). Thèse Doct. Univ. de Paris.
Miquel, M., Block, M. A., Joyard, J., Dorne, A.-J., Dubacq, J.-P., Kader, J. C., and Douce, R. (1984). *In* "Structure, Function and Metabolism of Plant Lipids" (P. A. Siegenthaler and W. Eichenberger, eds.), pp. 295–298. Elsevier, Amsterdam.
Monroy, G., Chroboczek-Keller, H., and Pullman, M. E. (1973). *J. Biol. Chem.* **248,** 2845–2852.
Montague, M. J., and Ray, P. M. (1977). *Plant Physiol.* **59,** 225–230.
Moore, T. S. (1982). *Annu. Rev. Plant Physiol.* **33,** 235–259.
Moore, T. S., Lord, J. M., Kagawa, T., and Beevers, H. (1973). *Plant Physiol.* **52,** 50–53.
Moreau, R. A., and Stumpf, P. K. (1982). *Plant Physiol.* **69,** 1293–1297.
Morgenthaler, J.-J., Marsden, M. P. F., and Price, C. A. (1975). *Arch. Biochem. Biophys.* **168,** 289–301.
Morré, D. J. (1975). *Annu. Rev. Plant Physiol.* **26,** 441–481.
Mudd, J. B., and de Zacks, R. (1981). *Arch. Biochem. Biophys.* **209,** 584–591.
Mudd, J. B., McManus, T. T., Ongun, A., and Mc Cullogh, T. E. (1971). *Plant Physiol.* **48,** 335–339.
Murata, N., and Sato, N. (1982). *In* "Biochemistry and Metabolism of Plant Lipids" (J. F. G. M. Wintermans and P. J. C. Kuiper, eds.), pp. 165–168. Elsevier, Amsterdam.
Murata, N., and Sato, N. (1983). *Plant Cell Physiol.* **24,** 133–138.
Murphy, D. J. (1986). *Biochim. Biophys. Acta* **864,** 33–94.
Murphy, D. J., and Leech, R. M. (1977). *FEBS Lett.* **77,** 164–168.
Murphy, D. J., and Stumpf, P. K. (1981). *Arch. Biochem. Biophys.* **212,** 730–739.
Murphy, D. J., and Walker, D. A. (1982). *Planta* **156,** 84–88.
Murphy, D. J., Mukherjee, K. D., and Woodrow, I. E. (1984). *Eur. J. Biochem.* **139,** 373–379.
Murphy, D. J., Woodrow, I. E., and Mukherjee, K. D. (1985). *Biochem. J.* **225,** 267–270.
Neufeld, E. F., and Hall, C. W. (1964). *Biochem. Biophys. Res. Commun.* **14,** 503–508.
Nichols, B. W., Harris, R. V., and James, A. T. (1965). *Biochem. Biophys. Res. Commun.* **20,** 256–262.
Nikolau, B. J., Wurtele, E. S., and Stumpf, P. K. (1984). *Arch. Biochem. Biophys.* **235,** 555–561.
Nothelfer, H. G., Barckhaus, R. H., and Spener, F. (1977). *Biochim. Biophys. Acta* **489,** 370–380.
Ohlrogge, J. B., and Kuo, T. M. (1985). *J. Biol. Chem.* **260,** 8032–8037.
Ohlrogge, J. B., Kuhn, D. A., and Stumpf, P. K. (1979). *Proc. Natl. Acad. Sci. U.S.A.* **76,** 1194–1198.
Ohnishi, J.-I., and Yamada, M. (1982). *Plant Cell Physiol.* **23,** 767–773.
Ongun, A., and Mudd, J. B. (1968). *J. Biol. Chem.* **243,** 263–275.
Op den Kamp, J. A. F. (1979). *Annu. Rev. Biochem.* **48,** 47–71.
Op den Kamp, J. A. F. (1981). *In* "New Comprehensive Biochemistry" (J. B. Finean and R. H. Michell, eds.), Vol. 1, pp. 83–126. Elsevier, Amsterdam.
Pieringer, R. A. (1968). *J. Biol. Chem.* **243,** 4894–4903.
Quinn, P. J., and Williams, W. P. (1983). *Biochim. Biophys. Acta* **737,** 223–266.
Rebeillé, F., Bligny, R., and Douce, R. (1980). *Biochim. Biophys. Acta* **620,** 1–9.
Reid, E. E., Thompson, P., Lyttle, C. R., and Dennis, D. T. (1977). *Plant Physiol.* **59,** 842–848.
Rothman, J. E., and Kennedy, E. P. (1977). *Proc. Natl. Acad. Sci. U.S.A.* **74,** 1821–1825.
Roughan, P. G. (1970). *Biochem. J.* **117,** 1–8.
Roughan, P. G. (1985). *Biochim. Biophys. Acta* **835,** 527–532.
Roughan, P. G., and Slack, C. R. (1977). *Biochem. J.* **162,** 457–459.
Roughan, P. G., and Slack, C. R. (1981). *FEBS Lett.* **135,** 182–186.
Roughan, P. G., and Slack, C. R. (1982). *Annu. Rev. Plant Physiol.* **33,** 97–132.
Roughan, P. G., Slack, C. R., and Holland, R. (1976). *Biochem. J.* **158,** 593–601.
Roughan, P. G., Holland, R., and Slack, C. R. (1979a). *Biochem. J.* **184,** 193–202.
Roughan, P. G., Rolland, R., Slack, C. R., and Mudd, J. B. (1979b). *Biochem. J.* **184,** 565–569.
Roughan, P. G., Mudd, J. B., McManus, T. T., and Slack, C. R. (1979c). *Biochem. J.* **184,** 571–574.
Roughan, P. G., Holland, R., and Slack, C. R. (1980). *Biochem. J.* **188,** 17–24.
Sanchez, J., and Mancha, M. (1980). *Phytochemistry* **19,** 817–820.

Sanchez, J., Jordan, B. R., Kay, J., and Harwood, J. L. (1982). *Biochem. J.* **204,** 463–470.

Sandelius, A. S., and Selstam, E. (1984). *Plant Physiol.* **76,** 1041–1046.

Sastry, S. S., and Kates, M. (1966). *Can. J. Biochem.* **44,** 459–467.

Sato, N., and Murata, N. (1982a). *Biochim. Biophys. Acta* **710,** 271–278.

Sato, N., and Murata, N. (1982b). *Biochim. Biophys. Acta* **710,** 279–289.

Sato, N., and Murata, N. (1982c). *Plant Cell Physiol.* **23,** 1115–1120.

Sauer, A., and Heise, K.-P. (1982). *In* "Biochemistry and Metabolism of Plant Lipids" (J. F. G. M. Wintermans and P. J. C. Kuiper, eds.), pp. 187–190. Elsevier, Amsterdam.

Sauer, A., and Heise, K.-P. (1983). *Z. Naturforsch.* **38C,** 399–404.

Sauer, A., and Heise, K.-P. (1984). *Z. Naturforsch.* **39C,** 593–599.

Sauer, A., and Robinson, D. G. (1985). *J. Exp. Bot.* **36,** 1257–1266.

Schiff, J. A. (1973). *Adv. Morphog.* **10,** 265–311.

Schlapfer, P., and Eichenberger, W. (1983). *Plant Sci. Lett.* **32,** 243–252.

Schüz, R., Ebel, J., and Hahlbrock, K. (1982). *FEBS Lett.* **140,** 207–209.

Sherratt, D., and Givan, C. V. (1973). *Planta* **113,** 147–152.

Shimakata, T., and Stumpf, P. K. (1982a). *Plant Physiol.* **69,** 1257–1262.

Shimakata, T., and Stumpf, P. K. (1982b). *Proc. Natl. Acad. Sci. U.S.A.* **79,** 5808–5812.

Shimakata, T., and Stumpf, P. K. (1982c). *Arch. Biochem. Biophys.* **218,** 77–91.

Shimakata, T., and Stumpf, P. K. (1983a). *Arch. Biochem. Biophys.* **220,** 39–45.

Shimakata, T., and Stumpf, P. K. (1983b). *J. Biol. Chem.* **258,** 3592–3598.

Siebertz, H. P., and Heinz, E. (1977). *Z. Naturforsch.* **31C,** 193–205.

Siebertz, H. P., Heinz, E., Joyard, J., and Douce, R. (1980). *Eur. J. Biochem.* **l08,** 177–185.

Siegenthaler, P. A., and Eichenberger, W. (1984). "Structure, Function and Metabolism of Plant Lipids." Elsevier, Amsterdam.

Simpson, E. E., and Williams, J. P. (1979). *Plant Physiol.* **63,** 674–676.

Slack, C. R. (1983). *In* "Biosynthesis and Function of Plant Lipids" (W. W. Thomson, J. B. Mudd, and M. Gibbs, eds.), pp. 40–55. Am. Soc. Plant Physiol., Rockville, Maryland.

Slack, C. R., and Roughan, P. G. (1975). *Biochem. J.* **152,** 217–228.

Slack, C. R., Roughan, P. G., and Balasingham, N. (1977). *Biochem. J.* **162,** 289–296.

Slack, C. R., Campbell, L. C., Browse, J. A., and Roughan, P. G. (1983). *Biochim. Biophys. Acta* **754,** 10–20.

Slack, C. R., Roughan, P. G., Browse, J. A., and Gardiner, S. E. (1985). *Biochim. Biophys. Acta* **833,** 438–448.

Smirnov, B. P. (1960). *Biokhimiya (Moscow)* **25,** 545–555.

Snider, M., and Kennedy, E. P. (1977). *J. Bacteriol.* **130,** 1072–1083.

Soll, J., and Roughan, P. G. (1982). *FEBS Lett.* **146,** 189–192.

Sparace, S. A., and Mudd, J. B. (1982). *Plant Physiol.* **70,** 1260–1264.

Stitt, M., and ap Rees, T. (1979). *Phytochemistry* **18,** 1905–1911.

Stymne, S., and Glad, G. (1981). *Lipids* **16,** 298–305.

Stymne, S., and Stobart, A. K. (1984). *Biochem. J.* **220,** 481–488.

Tevini, M., and Lichtenthaler, H. K. (1977). "Lipids and Lipid Polymers in Higher Plants." Springer Verlag, Berlin and New York.

Trémolières, A., and Mazliak, P. (1970). *Physiol. Vég.* **8,** 135–150.

Trémolières, A., Dubacq, J.-P., Müller, M., Drapier, D., and Mazliak, P. (1980). *FEBS Lett.* **114,** 135–138.

van Besouw, A. (1979). Ph.D. Thesis, Univ. of Nijmegen, Netherlands.

van Besouw, A., and Wintermans, J. F. G. M. (1978). *Biochim. Biophys. Acta* **529,** 44–53.

van Heusden, G. P. H., and van den Bosch, H. (1978). *Eur. J. Biochem.* **84,** 405–412.

Vick, B., and Beevers, H. (1977). *Plant Physiol.* **59,** 459–463.

Vick, B., and Beevers, H. (1978). *Plant Physiol.* **62,** 173–178.

Walker, K. A., and Harwood, J. L. (1985). *Biochem. J.* **226,** 551–556.

Weaire, P. J., and Kekwick, R. G. O. (1975). *Biochem. J.* **146,** 439–445.

Wilkinson, M. C., Laidman, D. L., and Galliard, T. (1984). *Plant Sci. Lett.* **35,** 195–199.

Williams, J. P. (1980). *Biochim. Biophys. Acta* **618,** 461–472.

Williams, J. P., and Khan, M. U. (1982). *Biochim. Biophys. Acta* **713,** 177–184.

Williams, J. P., Watson, G. R., Khan, M. U., and Leung, S. P. K. (1975). *Plant Physiol.* **55,** 1038–1042.

Williams, J. P., Watson, G. R., and Leung, S. P. K. (1976). *Plant Physiol.* **57,** 179–184.

Williams, J. P., Khan, M. U., and Mitchell, K. (1982). *In* "Biochemistry and Metabolism of Plant Lipids" (J. F. G. M. Wintermans and P. J. C. Kuiper, eds.), pp. 183–186. Elsevier, Amsterdam.

Williams, J. P., Khan, M. U., and Mitchell, K. (1983). *In* "Biosynthesis and Function of Plant Lipids" (W. W. Thomson, J. B. Mudd, and M. Gibbs, eds.), pp. 28–39. Am. Soc. Plant Physiol., Rockville, Maryland.

Williams, M., and Randall, D. D. (1979). *Plant Physiol.* **64,** 1099–1103.

Wintermans, J. F. G. M., and Kuiper, P. J. C. (1982). "Biochemistry and Metabolism of Plant Lipids." Elsevier, Amsterdam.

Wirtz, W., Stitt, M., and Heldt, H. W. (1980). *Plant Physiol.* **66,** 187–193.

Yamashita, S., and Murna, S. (1972). *Eur. J. Biochem.* **31,** 565–573.

Sulfolipids | *10*

J. BRIAN MUDD
KATHRYN F. KLEPPINGER-SPARACE

I. INTRODUCTION

In Volume 4, Chapter 11 of "The Biochemistry of Plants" the subject of "Sulfolipids" was comprehensively reviewed by John Harwood (1980). Many aspects of this review are still relevant today. The purpose of the present article is to examine the literature that has appeared since Harwood's summary and to reexamine some of the older findings in the light of more recent results.

Any discussion of sulfolipids in plants should pay tribute to the discovery and determination of the structure of sulfoquinovosyldiacylglycerol (SQDG) by Andy Benson and co-workers (1959). In the years following the discovery of SQDG, considerable research was conducted in Benson's laboratories. The significant progress that was made has been reviewed by Benson (1963). The structure and fatty acid composition had been determined and several studies on the synthesis and turnover in intact organisms had been completed. The pathway of synthesis was not clarified, but two significant suggestions were made. First, there had been identification of UDP-sulfoquinovose in $^{35}SO_4$- labeled *Chlorella*, Shibuya *et al.*

(1963), and this gave rise to the suggestion that SQDG was formed by the reaction of diacylglycerol (DG) and the nucleotide sugar in a process analogous to that responsible for the formation of the galactolipids. Second, an analogy had been drawn between compounds such as sulfolactate and sulfolactaldehyde and corresponding phosphorylated intermediates in the glycolytic sequence, and a "sulfoglycolytic" pathway for the synthesis of sulfoquinovose (SQ) had been suggested.

Although several laboratories have worked on SQDG since the original investigations of Benson and co-workers, and much new and interesting information has been presented, the details of biosynthesis, particularly of SQ itself, still resist clarification. The discussion of the biosynthesis of the molecule can be conveniently divided into consideration of the DG moiety and the SQ moiety.

II. STRUCTURE OF SULFOLIPIDS

In this review the term "sulfolipid" is used to include any lipid component that contains a sulfur moiety. In earlier literature on plants the term sulfolipid has been used synonymously with SQDG. Although several interesting sulfolipids have been found in algae and diatoms, this review will concentrate mainly on the sulfolipids of higher plants.

In addition to SQDG (Fig. 1), there have been indications of other sulfolipids in higher plants mainly by detection in experiments where radioactive precursors have been used. Mudd et al. (1980) have reported the existence of a sulfolipid in roots which becomes labeled when seedlings are supplied radioactive sulfate. This lipid behaves on thin-layer chromatography as less polar than SQDG. The same lipid has been observed in root tissue by L. J. de Kok (personal communication). In isolated chloroplasts the incorporation of $^{35}SO_4$ gives rise to a group of sulfolipids which are less polar than SQDG. These lipids have not been identified (Fig. 2). There is a strong possibility that they are artifacts of the chloroplast system, but similar lipids have been found in some in vivo systems.

Shibuya and Hase (1965) found that when Chlorella protothecoides was incubated with $^{35}SO_4$ an unidentified lipid in addition to SQDG was formed in quantities equivalent to SQDG. This lipid was less polar than SQDG, chromatographing in the region of pigments. Shibuya et al. (1965) found that Lemna perpusilla also produced large quantities of a radiolabeled sulfolipid which chromatographed with the pigment fraction. Opute (1974a,b) found that the diatoms of the genus Navicula produce two sulfolipids in addition to SQDG, both being less polar than SQDG. Under certain conditions one of these lipids behaves as a precursor of SQDG. It is difficult to identify these lipids when the only detection method is radioactivity. It is possible that the unknown sulfolipids of the algae will turn out to be the same as those encountered in higher plants.

There is considerable variation in the fatty acid composition of SQDG from various sources. By comparison with the other lipids in the chloroplast, SQDG has a high proportion of saturated fatty acids, much more than the galactolipids and

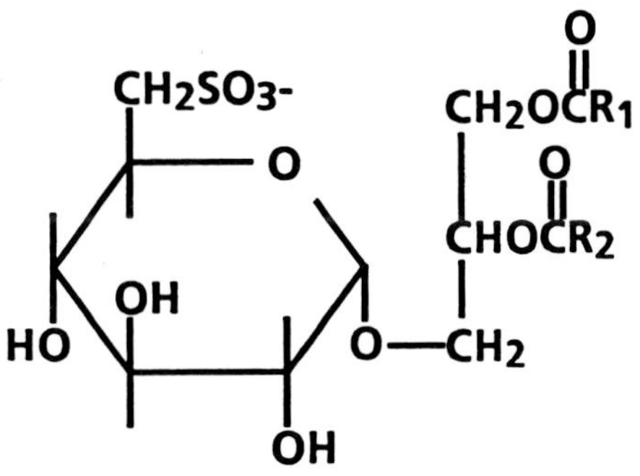

Fig. 1. The structure of SQDG.

about the same as phosphatidylglycerol (PG). The similarity in fatty acid composition of SQDG and PG has given rise to the speculation that these two lipids may have common intermediates, but as the pathway for PG synthesis has been worked out (Andrews and Mudd, 1985) and further work on SQDG biosynthesis has been completed, this idea has become unacceptable. The fatty acid composition of SQDG can be as much as 76% saturated fatty acid and is usually not less than 40%. Some examples are presented in Table I.

The ratio of SQDG to other lipids of the chloroplast is somewhat variable in higher plants. Ratios of galactolipids (MGDG + DGDG) to SQDG in the range 5–20 have been reported for maize, spinach, barley, and antirrhinum. In lower plants the ratio can be much more in favor of SQDG. For example, the galactolipid–SQDG ratio in *Dunaliella* is 2.8 (Evans and Kates, 1984) and in *Fucus vesiculosus* it is 0.8 (Radunz, 1969). It is not known whether these differences in composition reflect differences in physiology.

III. ORIGIN OF THE DIACYLGLYCEROL MOIETY

O'Brien and Benson (1963) analyzed the fatty acid composition of SQDG in alfalfa and found that palmitic acid was 43% and linolenic acid was 47% of the fatty acids. The relatively high content of saturated fatty acid has been found characteristic of SQDG in higher plants and algae (Radunz, 1969). When the positional distribution of fatty acids was first determined it was discovered that the palmitic acid was primarily localized at the *sn*-2 position of the glycerol backbone. In fact the predominant molecular species of SQDG contained the 16-carbon fatty acid at the *sn*-2 position and the 18-carbon unsaturated fatty acid at the *sn*-1 position in the cases of *Spinacia oleracea* and *Pteridium aquilinum* (Tulloch *et al.*, 1973). Extension of these results, however, revealed cases where the palmitic acid was primarily located at the *sn*-1 position (Heinz, 1977). These results were puzzling when considering the origin of the DG moiety, since DG synthesized in the chloroplast has palmitate as the only fatty acid residue at the *sn*-2 position.

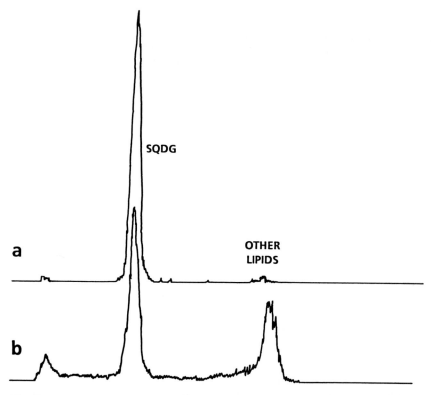

Fig. 2. Thin-layer chromatograhy of ^{35}S-labeled lipids made by incorporation of $^{35}SO_4^{2-}$ by isolated chloroplasts. The lipids were chromatographed on silica gel thin-layer plates using the solvent system chloroform:acetone:methanol:acetic acid:water, 50:20:20:10:15:5. Reaction mixtures contained the basic components 0.3 M sorbitol, 2 mM MgCl$_2$, 0.5 mM DTT, 33 mM Tricine-KOH, pH 7.9, and 100 μM $^{35}SO_4^{2-}$ in 1.0 ml. (a) Basic reaction mixture plus 2 mM ATP and 0.2 mM P$_i$. (b) Basic reaction mixture with no further additions. The origin of the chromatogram is at the left. Scanning of radioactivity was done with a Berthold LB 2832 linear analyzer.

By using radioactive precursors, one hopes to determine the order of labeling of molecular species of SQDG. These experiments have shown in *Vicia faba* that 30–50% of the labeled molecular species contain two double bonds, whereas this molecular species accounts for only 8% of the mass (Heinz and Harwood, 1979). The distribution of molecular species was unchanged over the experimental period of 2 h. In a similar study using *Hordeum vulgare,* it was found that 80% of the label (from either acetate or sulfate) was found in the combined anenoic, monoenoic, and dienoic molecular species, which accounted for only 5% of the mass (Nichols and Harwood, 1979). These results led the authors to conclude that SQDG was not a substrate for desaturation reactions of the fatty acid moieties, and that the final fatty acid composition of SQDG would be determined by acyl exchange rather than *de*

TABLE I
Fatty Acid Composition of SQDG from Various Plants

Plant names[a]	Fatty acids (%)							
	14:0	16:0	18:1	18:2	18:3	20:5	22:0	24:0
Oscillatoria chalybea (1)	—	61.0	5.4	6.5	15.3	—	—	—
Fucus vesiculosus (1)	12.7	36.7	19.7	3.2	13.4	—	4.3	9.3
Bactrachospermum moniliforme (1)	13.3	45.3	4.9	1.7	2.0	12.5	—	4.8
Dryopteris filix-mas (1)	—	46.8	23.3	3.2	19.8	—	3.0	—
Antirrhinum majus (1)	2.2	46.9	4.4	5.7	36.8	—	—	—
Dunaliella C9 (2)	0.6	76.7	2.6	8.0	8.5	—	—	—
Dunaliella D11 (2)	0.5	75.8	3.2	7.5	9.6	—	—	—
Spinacia oleracea (3)	—	38.9	1.9	6.1	45.0	—	—	—
Sinapis (3)	—	42.7	1.6	5.0	49.9	—	—	—

[a]References: 1. Radunz (1969); 2. Evans and Kates (1984); 3. Tulloch et al. (1973).

novo synthesis. However, examination of the data showing fatty acid labeling in the individual lipids indicated some accumulation of label in linoleic acid at the same time that unsaturated fatty acids accumulate in MGDG (Heinz and Harwood, 1977). The lack of rapid formation of the unsaturated molecular species of SQDG may be attributed to the age of the plant material or the time of exposure of the experimental material. Siebertz et al. (1980) have reported that spinach labeled for 30 min with radioactive CO_2 shows only saturated molecular species of SQDG labeled. After 2.5 h a full array of molecular species of increasing desaturation had become labeled. In this context it should be noted that in the experiments of Leech et al. (1973) using sections of maize leaves of different ages, the youngest sections (closest to the basal meristem) had a fatty acid composition with low amounts of 18:3 and high amounts of 18:0, whereas the older sections had low amounts of 18:0 and higher amounts of 18:3. It is possible that this change in composition was achieved by desaturation of the 18:0 while in ester linkage to the SQDG.

Whereas these experiments with intact plant material probably demonstrate de novo synthesis, experiments with intact chloroplasts may show SQDG synthesis from preexisting DG precursors. The report of Haas et al. (1980) showed that when isolated chloroplasts were incubated with radioactive sulfate, the molecular species of SQDG formed were the same after 20, 40, and 150 min incubation. It is likely that DG present in the isolated chloroplasts (generated by the MGDG dismutation reaction; Heemskerk et al., 1983) is the acceptor for the SQ which is synthesized from the administered radioactive sulfate.

Questions concerning the pool of DG from which the lipid moiety of SQDG is drawn have been raised. Of the four glycerolipids in chloroplasts, only SQDG synthesis is not known in detail, but estimates of the origin of the DG moiety can be made from analysis of selected plant species. The analyses done by Bishop et al. (1985) chose both "18:3" and "16:3" plants. It is accepted widely that the difference between these two groups of plants is in the capability to use DG synthesized in the chloroplast directly for the synthesis of glycerolipids. In the case of the "16:3" plants the DG with 16:0 at the sn-2 position is a substrate for galactosylation. Subsequent desaturation generates MGDG with 16:3 at the sn-2 position. In the case of the "18:3" plants, newly synthesized fatty acids are exported to the cytoplasmic compartment where they are used in the synthesis of PC. The DG moiety of PC is returned to the chloroplast where it is used as a precursor of MGDG. After desaturation reactions the MGDG which is generated has 18:3 at both sn-1 and sn-2 positions. The fatty acid composition of DGDG reveals significant percentages of 16:0 at the sn-1 position, and this is thought to be contributed from PC with DG moieties containing 16:0 at the sn-1 position. While the galactolipids can therefore have two different sources of the DG moiety, PG synthesized in the chloroplast has only one source of the DG moiety and that is from the chloroplast itself. The DG moiety synthesized in the chloroplast always has 16:0 at the sn-2 position. By analyzing the fatty acid composition and distribution in SQDG from "18:3" and "16:3" plants one can ascertain the origin of the DG moiety. It can be generalized that when the MGDG has 16-carbon fatty acid at the sn-2 position, then SQDG will also have 16-carbon fatty acid at the sn-2 position. SQDG also has a significant percentage of 16:0 at the sn-1 position (Table II). It is postulated that this arises from the same DG pool used in the synthesis of DGDG. These analyses therefore suggest that the DG pool used in the synthesis of SQDG is the same pool (or pools) available for the synthesis of MGDG and DGDG. The overall scheme for the utilization of DG moieties in the synthesis of chloroplast glycerolipids is shown in Fig. 3.

IV. ORIGIN OF THE HEAD GROUP

The origin of the SQ moiety of SQDG has eluded the investigations of many researchers. Davies et al. (1966) demonstrated that cysteic acid did not affect the uptake of radioactive sulfate by *Euglena* cells but significantly decreased the proportion of sulfate found in SQDG. This result was followed by a demonstration that cysteic acid labeled in both the carbon and sulfur moieties was incorporated into SQDG without change in the ratio of label in carbon and sulfur. On the basis of these results it was proposed that the cysteic acid was a direct precursor to SQ by a series of reactions analogous to the glycolytic sequence. Harwood (1975) demonstrated with alfalfa plants that radioactive cysteic acid was incorporated with higher efficiency than sulfate into SQDG. It was also found that cysteic acid decreased the incorporation of sulfate into SQDG. These results were therefore very similar to the results with *Euglena* and it was concluded that in the case of higher plants cysteic acid was incorporated intact into the SQ moiety.

On the other hand, Mudd et al. (1980) analyzed the capability of spinach seedlings to synthesize SQDG from a number of precursors and found that cysteic acid was not as good a precursor as sulfate. In these experiments cysteine very effectively lowered the uptake of sulfate by the root system but did not alter the

TABLE II
Positional Distribution of Fatty Acids in SQDG[a,b]

	Fatty acid (%)					
	16:0	16:1	18:0	18:1	18:2	18:3
Spinach						
Total	50	+	2	2	4	42
sn-1	34	+	3	6	7	50
sn-2	67	+	1	+	3	29
Tobacco						
Total	44	1	5	3	10	37
sn-1	36	4	11	6	11	32
sn-2	53	2	1	2	7	36
Wheat						
Total	28	2	3	1	5	61
sn-1	46	+	6	3	6	39
sn-2	8	1	1	1	4	84
Cucumber						
Total	35	1	6	3	2	53
sn-1	60	2	11	3	31	21
sn-2	2	+	+	8	3	86

[a]From Bishop et al. (1985), with permission.
[b]SQDG samples were degraded with *Rhizopus arrhizus* lipase.
The free fatty acid from sn-1 and the lyso-SQDG were resolved
by thin-layer chromatography and analyzed separately.

proportion of radiolabel found in SQDG. Cysteate, however, lowered neither the
uptake of sulfate nor the proportion found in SQDG. In contrast to these results,
when radiolabeled cysteate was used as the precursor, sulfate and cysteine lowered
the uptake and eliminated the incorporation into SQDG. Analysis of the sulfur-
containing compounds found in the plant tissue indicated that in the cases of both
cysteate and cysteine, incorporation into SQDG could be attributed to the sulfate
formed from these precursors. Thus in the case of spinach there is no support for the
idea that cysteate is incorporated intact into the SQ moiety of SQDG. It was further
noted by Mudd et al. (1980) that the analogy between the sulfonic acid group and
the phosphate ester called for in the "sulfoglycolytic" pathway is inaccurate. For
example, the necessary first step is the transamination of cysteate to generate
sulfopyruvate. This step is very well catalyzed by the transaminase that normally
uses aspartate, indicating the recognition of the sulfonic acid moiety as analogous
to a carboxyl group. The reduction of sulfopyruvate to sulfolactate by malate
dehydrogenase attests to the same recognition. It was therefore not surprising that
SQ was neither a substrate for nor an inhibitor of glucose-6-phosphate dehydrogenase (Mudd et al., 1980), which would have been expected if the sulfonic acid
group were analogous to phosphate.

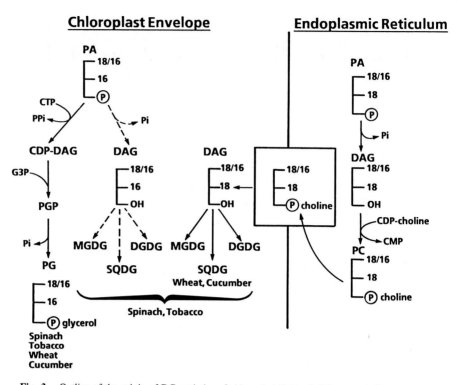

Fig. 3. Outline of the origin of DG moieties of chloroplast lipids. Solid arrows indicate pathways present in both 16:3 and 18:3 plants. Dashed arrows indicate pathways present in 16:3 plants only. [From Bishop *et al.* (1985), with permission.]

If the sulfonic acid group is not introduced at the level of a three-carbon precursor, what other possibilities can be entertained? One piece of information is that SQ can be synthesized by the reaction of sulfite with glucoseenide (dehydration at the 5,6 positions of glucose) (Lehmann and Benson, 1964). An analogous reaction in animal tissue generates the carbon–sulfur bond in cysteic acid (Sass and Martin, 1972). In this case PAPS reacts with α-amino acrylic acid. One might conceive of activated sulfate (APS or PAPS) reacting with an intermediate like UDP- or ADP-glucoseenide.

A summary of the *in vitro* experiments suggests that the incorporation into SQDG may proceed by the pathway outlined in Fig. 4.

V. BIOSYNTHESIS OF SULFOLIPIDS *IN VIVO*

The initial study of the metabolism of sulfolipids was done by Benson and his co-workers, usually in algal systems. Several of these studies were concerned with

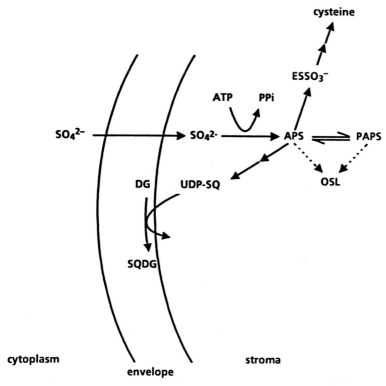

Fig. 4. Hypothesis for the incorporation of $^{35}SO_4^{2-}$ into sulfolipids. Sulfate is taken up by the chloroplasts and activated to APS. APS can be used for several purposes: (1) the incorporation into cysteine; (2) conversion to PAPS; (3) conversion to one of the other sulfolipids; (4) incorporation into UDP-SQ. The UDP-SQ is used as donor to DG which is generated in the chloroplast envelope.

the turnover of SQDG. It has been suggested that the metabolism of SQDG can provide the sulfur which is subsequently incorporated into protein. When *Chlorella ellipsoidea* was allowed to incorporate $^{35}SO_4$ into the SQDG and then transferred to a sulfur-deficient medium, the loss of radioactivity in the SQDG had a half-time of about 10 h. In sulfur-sufficient medium with adequate bicarbonate, the labeled SQDG was stable over 24 h, but in CO_2-free air there was turnover of the SQDG (Miyachi and Miyachi, 1964). Similar results were reported by Opute for the diatom *Navicula muralis* (Opute, 1974a). In this case SQDG degraded when cells were incubated in sulfur-free medium, but in sulfur-sufficient medium SQDG increased slightly at the expense of an unidentified sulfolipid.

Experiments using sulfur-deficient media have not been done with higher plants. In the results reported by Mudd *et al.* (1980), the incorporation of $^{35}SO_4$ into sulfolipids was allowed for a period of 6 h and then the fate of the radiolabeled lipids followed for an additional 138 h. The [^{35}S]SQDG of the cotyledons showed no indication of turnover. Similarly the ^{35}S-labeled sulfolipids of the roots and hypocotyls, which contained at least one other sulfolipid besides SQDG, showed no decline in radioactivity.

When lipid-synthesizing studies are done using intact tissue with either acetate or bicarbonate as the radioactive precursor, small amounts of SQDG are synthesized. For example, when $^{14}CO_2$ was supplied to *Vicia faba* leaves for 70 min, only 0.8% of the labeled lipid was SQDG. Exposure to [^{14}C]acetate for 6 h gave 8% of the label in lipid as SQDG (Heinz and Harwood, 1977). In both of these cases the majority of labeling is in phosphatidylcholine (PC). Gardiner *et al.* (1982) have also presented data on the labeling of SQDG from [^{14}C]acetate in intact leaves and leaf disks. In this case also the majority of the label was found in PC. However, the label in SQDG was from 0.5 to 1.3% in pumpkin, sunflower, and broad bean, while it was 2.8–9.3% in tomato, parsley, and spinach. It was noticeable that an additional difference between these groups of plants was in the much heavier labeling of DGDG in the plants where SQDG was heavily labeled. This may indicate some shared features of the DG pool for the synthesis of SQDG and DGDG.

VI. BIOSYNTHESIS OF SULFOLIPIDS *IN VITRO*

The synthesis of SQDG by isolated chloroplasts was first reported by Haas *et al.* (1980). Using carrier-free sulfate, rates of 1.7 pmol/h/mg chlorophyll were observed. The reaction mixture contained Mg^{2+}, Mn^{2+}, and 10 mM bicarbonate and was illuminated. In darkness no SQDG was produced. It is not clear that the SQDG was synthesized *de novo* in these experiments; indeed the authors deduced from the labeling of molecular species of SQDG that probably endogenous DG was being used as acceptor for the SQ. The experiments demonstrated that the isolated chloroplasts contained all of the enzymes necessary to generate SQ from sulfate, but one can only guess how many enzymes this may involve.

Hoppe and Schwenn (1981) prepared a cell-free system capable of synthesizing SQDG from *Chlamydomonas reinhardii*. The homogenate synthesized SQDG from radioactive sulfate in the presence of 5 mM ATP and 10 mM Mg^{2+}. The synthesis was greatly stimulated by the addition of glutathioine (GSH) to the reaction mixture. Stimulation by an NADPH-generating system was slight (17% above control), suggesting that reductive steps of sulfur assimilation were not required for the incorporation of sulfate into SQDG. The presence of cysteic acid in the reaction mixture did not significantly lower the incorporation of radioactive sulfate into SQDG, suggesting that cysteic acid was not on the pathway to SQ. Both PAPS and sulfite were also tested as precursors of SQDG. Incorporation from sulfite appeared to be superior to PAPS, but the incorporation of sulfite was linear through 0.2 mM and it did not have the expected effect of diminishing the incorporation from PAPS. PAPS appeared to behave as a legitimate substrate, having an apparent K_m of 25 μM. Incorporation was stimulated by thiol compounds such as DTT and GSH. It is not clear whether they were active as reductants or as carriers of the activated sulfate. When the homogenate was separated into a membrane fraction and a soluble fraction, the reconstitution of these fractions gave synthetic activity, but when the fractions were tested alone most of the activity was in the soluble protein fraction. The latter result is somewhat confusing since one would expect DG acceptors to be endogenous to the membrane fraction.

Further experiments on the synthesis of sulfolipids by chloroplasts from higher plants have been reported by Kleppinger-Sparace *et al.* (1985). In a basic reaction mixture the synthesis of SQDG required illumination and intact chloroplasts. Rates of incorporation were up to 1 nmol/h/mg chlorophyll. In these experiments there were other lipidic materials labeled from radioactive sulfate. These lipids behave as less polar than SQDG. In all probability they are artifacts of the chloroplast system, but since they are affected by reaction conditions differently, the study of the other sulfolipids (OSL) may lead to an understanding of the precursors for SQDG. For example, it was reported that the addition of ATP and inorganic phosphate has essentially no effect on the synthesis of SQDG but greatly inhibits the synthesis of OSL (Fig. 2). Isolation of potential intermediates such as APS and PAPS under these two experimental conditions may distinguish between the pathway to SQDG and to OSL. Although nucleoside triphosphates have little effect on SQDG synthesis, it was found that UDP-galactose inhibited the incorporation of sulfate into SQDG. This was interpreted as utilization of the DG acceptor for MGDG synthesis rather than the synthesis of SQDG (Fig. 5).

Subsequent experiments have shown that the requirement for illumination in order to assay SQDG synthesis can be replaced either by ATP or by systems that generate ATP in the chloroplast (the dihydroxyacetone phosphate shuttle). It has also been demonstrated that APS and PAPS can both be used by the chloroplast system for the synthesis of SQDG (K. F. Kleppinger-Sparace and J. B. Mudd, unpublished observations).

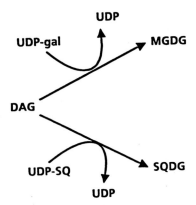

Fig. 5. Inhibition of incorporation of $^{35}SO_4^{2-}$ into SQDG by UDPgal is interpreted as depletion of the DG pool by formation of MGDG.

VII. PROPERTIES AND FUNCTIONS OF SULFOLIPIDS

SQDG is the most acidic lipid in the chloroplast structure. Oldani *et al.* (1975) have examined the relationship between molecular area and surface pressure, of monolayers of SQDG. At a given surface pressure, SQDG occupied only half the area occupied by MGDG. This result illustrates the importance of the fatty acid substituents on the surface area occupied by lipid molecules, since the fatty acid substituents of the MGDG are much more unsaturated than those of SQDG. When considering the behavior of SQDG in the lipid bilayer under physiological conditions, the physicochemical characteristics of packing and phase transition consequent on the fatty acid composition should be taken into consideration.

Radunz and Berzborn (1970) prepared antibodies to SQDG and tested the effect of these antibodies on lamellar systems, isolated from chloroplasts. The results showed that the antibodies were adsorbed to the lamellar systems but there was no direct agglutination until the addition of rabbit γ-globulin antiserum. The conclusion was reached that the SQDG molecules were located at the bottom of depressions on the thylakoid surface. After disruption of the lamellar system by sonication, the antibodies caused direct agglutination. It was assumed that the mechanical disruption had exposed SQDG which was on the inner surface of the thylakoids. Radunz *et al.* (1984) have reported that antiserum to SQDG inhibits electron transport in chloroplasts. They suggest that binding of the antibody causes a conformational change in proteins cooperating in electron transport.

Very little work has been reported on the reconstitution of chloroplast membrane proteins with SQDG. The report of Livne and Racker (1969) demonstrated that CF_1 from spinach chloroplasts was particularly well protected from heat inactivation by SQDG. Galactolipid and phospholipid were greatly inferior to SQDG in this respect. Chain (1985) has reconstituted the cytochrome b_6–f complex isolated from spinach with various lipids. DGDG, PG, and PC were active in reconstitution, while MGDG and SQDG were not.

Pick *et al.* (1984, 1985) have been studying the activation of the chloroplast CF_0–CF_1 ATPase. The activation is greatest with MGDG. However, it has been found that the purified CF_0–CF_1 ATPase retains SQDG to the highest levels of protein purification. For the enzymes from *Dunaliella salina* and from spinach there are 5 and 20 mol SQDG per mole of enzyme, respectively. The SQDG is not covalently linked but is very tightly bound.

The lipid analysis of components of the thylakoid system have revealed a lateral heterogeneity of lipids. Comparison of appressed thylakoid regions with whole thylakoids shows some differences in the percentage of SQDG (Murphy and Woodrow, 1983; Gounaris *et al.*, 1983). It is not clear whether the differences in lipid composition are functionally related to the different photosynthetic functions in different regions of the thylakoids. The fatty acid composition of SQDG was much more saturated in the appressed regions than in the intact thylakoids (Murphy and Woodrow, 1983). It is not known whether this difference is related to function or biosynthetic origin.

The results of Murata have recently called attention to the possible involvement of saturated molecular species of PG in the relationship to chilling injury (Murata, 1983). There seem to be exceptions to the generalization that chilling-sensitive species will have relatively high content of saturated molecular species of PG. These exceptions include many solanaceous species (Roughan, 1985). Several authors have considered the possibility that in the cases where PG cannot account for the chilling sensitivity, the role could be taken over by SQDG, since it also has a rather high content of saturated fatty acids. Murata and Hoshi (1984) have determined the fatty acid composition of SQDG from a number of species. They concluded that saturated molecular species of SQDG were not closely correlated with chilling sensitivity. However, in the case of two chilling-sensitive species, maize and rice, 19 and 14%, respectively, of the molecular species were dipalmitoyl. In considering the phase transition which lipids may undergo at chilling temperatures, it seems premature to discount a contribution from SQDG in addition to that from PG.

An interesting attempt has been made to obtain cells of *Chlorella pyrenoidosa* deficient in SQDG. Sinensky (1977) attempted to do this by growing the algae with cysteine as the sole sulfur source. The rationale was that the alga would have sufficient sulfur-containing amino acids from the cysteine supplement but would not have sulfur in the oxidation state necessary for the synthesis of SQDG. The cysteine-grown cells were found to have lowered capacity to assimilate bicarbonate photosynthetically, and this was traced to a deficiency in phosphoribulokinase. Results were also presented to show that the cysteine-grown cells did not contain significant quantities of SQDG. These results are surprising in at least two respects: first, one might have expected the algal cells as those of higher plants to be capable of generating sulfate from cysteine, and second, the effect on a soluble enzyme of a membrane lipid component is unexpected. It would be reassuring if the report could be confirmed in other laboratories.

VIII. CONCLUSIONS

Progress in studies of the biosynthesis of SQDG has been slow. Several laboratories have ventured into this area and come away without achieving the goal: a detailed description of intermediates between sulfate (or cysteic acid) and SQDG.

Now that SQDG biosynthesis can be studied routinely using isolated chloroplasts, more progress should be made. The use of $[^{35}S]APS$ and $[^{35}S]PAPS$ should facilitate experimentation. However, the development of a broken-chloroplast system would be very advantageous in showing the relative importance of enzyme activities in the stroma, thylakoid, and envelope. Even with the intact chloroplast, it is now appropriate to search for water-soluble intermediates between $^{35}SO_4^{2-}$, $[^{35}S]APS$, and $[^{35}S]PAPS$, and SQDG. Such intermediates may be present in very low steady state concentrations, but it should be possible to find them.

Sulfolipids other than SQDG are now worthy of further study. These sulfolipids include those which are found *in vivo* and therefore probably are not artifacts, such as those found in root tissue. The other sulfolipids labeled when chloroplasts are incubated with ^{35}S precursors may be artifacts, but we should recall that the discovery of SQDG itself followed from radiolabeling with $^{35}SO_4^{2-}$. Perhaps there are new sulfolipids and structures yet to be found in plant tissue.

REFERENCES

Andrews, J., and Mudd, J. B. (1985). *Plant Physiol.* **79**, 259–265.

Benson, A. A. (1963). *Adv. Lipid Res.* **1**, 387–394.

Benson, A. A., Daniel, H., and Wiser, R. (1959). *Proc. Natl. Acad. Sci. U.S.A.* **45**, 1582–1587.

Bishop, D. G., Sparace, S. A., and Mudd, J. B. (1985). *Arch. Biochem. Biophys.* **240**, 851–858.

Chain, R. K. (1985). *FEBS Lett.* **180**, 321–325.

Davies, W. H., Mercer, E. I., and Goodwin, T. W. (1966). *Biochem. J.* **98**, 369–373.

Evans, R. W., and Kates, M. (1984). *Arch. Microbiol.* **140**, 50–56.

Gardiner, S. E., Roughan, P. G., and Slack, C. R. (1982). *Plant Physiol.* **70**, 1316–1320.

Gounaris, K., Sundby, C., Andersson, B., and Barber, J. (1983). *FEBS Lett.* **156**, 170–174.

Haas, R., Siebertz, H. P., Wrage, K., and Heinz, E. (1980). *Planta* **148**, 238–244.

Harwood, J. L. (1975). *Biochim. Biophys. Acta* **398**, 224–230.

Harwood, J. L. (1980). *In* "The Biochemistry of Plants, Vol. 4, Lipids: Structure and Function" (P. K. Stumpf, ed.), pp. 301–320. Academic Press, New York.

Heemskerk, J. W., Bogemann, G., and Wintermans, J. F. G. M. (1983). *Biochim. Biophys. Acta* **754**, 181–189.

Heinz, E. (1977). *In* "Lipids and Lipid Polymers in Higher Plants" (M. Tevini and H. K. Lichtenthaler, eds.), pp. 102–120. Springer-Verlag, Berlin and New York.

Heinz, E., and Harwood, J. L. (1977). *Hoppe-Seyler's Z. Physiol. Chem.* **358**, 897–908.

Hoppe, W., and Schwenn, J. D. (1981). *Z. Naturforsch. C* **36C**, 820–826.

Kleppinger-Sparace, K. F., Mudd, J. B., and Bishop, D. G. (1985). *Arch. Biochem. Biophys.* **240**, 859–865.

Leech, R. M., Rumsby, M. G., and Thomson, W. W. (1973). *Plant Physiol.* **52**, 240–245.

Lehmann, J., and Benson, A. A. (1964). *J. Am. Chem. Soc.* **86**, 4469–4472.

Livne, A., and Racker, E. (1969). *J. Biol. Chem.* **244**, 1332–1338.

Miyachi, S., and Miyachi, S. (1964). *Plant Physiol.* **41**, 479–486.

Mudd, J. B., Dezacks, R., and Smith, J. (1980). *In* "Biogenesis and Function of Plant Lipids" (P. Mazliak, P. Benveniste, C. Costes, and R. Douce, eds.), pp. 57–66. Elsevier/North-Holland, Amsterdam.

Nicholls, R. G., and Harwood, J. L. (1979). *Phytochemistry* **18**, 1151–1154.

Murata, N. (1983). *Plant Cell Physiol.* **24**, 81–86.

Murata, N., and Hoshi, H. (1984). *Plant Cell Physiol.* **25**, 1241–1245.

Murphy, D. J., and Woodrow, I. A. (1983). *Biochim. Biophys. Acta,* **725**, 104–112.

O'Brien, J. S., and Benson, A. (1963). *J. Lipid Res.* **5**, 432–436.

Oldani, D., Hauser, H., Nichols, B. W., and Phillips, M. C. (1975). *Biochim. Biophys. Acta* **382**, 1–9.

Opute, F. L. (1974a). *J. Exp. Bot.* **25**, 798–809.

Opute, F. L. (1974b). *J. Exp. Bot.* **25**, 823–835.

Pick, U., Gounaris, K., Admon, A., and Barber, J. (1984). *Biochim. Biophys. Acta* **764**, 12–20.

Pick, U., Gounaris, K., Weiss, M., and Barber, J. (1985).*Biochim. Biophys. Acta* **808**, 415–420.

Radunz, A. (1969). *Hoppe-Seyler's Z. Physiol. Chem.* **350**, 411–417.

Radunz, A., Bader, K. P., and Schmid, G. H. (1984). *In* "Structure, Function and Metabolism of Plant Lipids" (P.-A. Siegenthaler and W. Eichenberger, eds.), pp. 479–484. Elsevier, Amsterdam.

Radunz, A., and Berzborn, R. (1970). *Z. Naturforsch., B* **25B**, 412–419.

Roughan, P. G. (1985). *Plant Physiol.* **77**, 740–746.

Sass, N. L., and Martin, W. G. (1972). *Proc. Soc. Exp. Biol. Med.* **139,** 755–761.
Shibuya, I., and Hase, E. (1965). *Plant Cell Physiol.* **6,** 267–283.
Shibuya, I., Yagi, T., and Benson, A. A. (1963). "Studies on Microalgae and Photosynthetic Bacteria,"
    pp. 627–636. Univ. of Tokyo Press, Tokyo.
Shibuya, I., Maruo, B., and Benson, A. A. (1965). *Plant Physiol.* **40,** 1251–1256.
Siebertz, H. P., Heinz, E., Joyard, J., and Douce, R. (1980). *Eur. J. Biochem.* **108,** 177–185.
Sinensky, M. (1977). *J. Bacteriol.* **129,** 516–524.
Tulloch, A. P., Heinz, E., and Fischer, W. (1973). *Hoppe-Seyler's Z. Physiol. Chem.* **354,** 879–889.

Lipid-Derived Defensive Polymers and Waxes and Their Role in Plant–Microbe Interaction

11

P. E. KOLATTUKUDY

I. INTRODUCTION

This chapter summarizes only the progress achieved since the preparation of Chapter 18 in Volume 4 of this series (Kolattukudy, 1980a). No attempt has been made to be comprehensive, and many scattered reports on various aspects and confirmatory results have not been included. Thus I confine this chapter to fatty acid chain elongation, the discovery of a novel biochemical decarbonylation reaction which yields an alkane and CO from a fatty aldehyde, the chemistry and biochemistry of suberin, and the biochemistry and molecular biology of cutinases which might play a significant role in plant–microbe interaction.

The Biochemistry of Plants, Vol. 9

II. HYDROCARBON BIOSYNTHESIS

Experimental evidence previously discussed (Kolattukudy, 1980a) showed that hydrocarbons are synthesized by elongation of a preformed acid to the appropriate length such as C_{30} or C_{32} followed by the loss of the carboxyl carbon.

A. Fatty Acid Elongation

Elongation of fatty acids by cell-free preparations from epidermal cells, where alkanes are known to be generated (Kolattukudy, 1980a), has been demonstrated (Buckner and Kolattukudy, 1972; Agrawal *et al.* 1984; Agrawal and Stumpf, 1985). Thus microsomal preparations generate $> C_{20}$ acids from acyl-CoA using malonyl-CoA and NADPH as substrates. Although very long acids with chain lengths approaching those of the alkanes can be generated by some of the cell-free preparations, chain length distribution of the products generated *in vitro* does not often correspond to that of the alkanes. Since the epidermis generates many classes of lipids, each with its own characteristic chain length distribution, it is likely that different chain-elongating enzyme systems are involved in their synthesis. The different cell-free preparations so far studied probably contain more than one elongating system. The results obtained with the crude solubilized preparations (Agrawal and Stumpf, 1985; Lessire *et al.*, 1985a) supported the previous conclusions that multiple elongating systems are present in the membrane preparations (Kolattukudy, 1980a). Resolution of two elongating activities by sucrose density gradient centrifugation and by gel filtration provided the strongest direct evidence for the presence of multiple elongation systems (Lessire *et al.*, 1985a). In the mid-1960s it was concluded that the epidermal elongating enzyme is specific for saturated acids on the basis that tissue slices readily incorporated exogenous labeled saturated but not unsaturated acids into very long-chain fatty acids, alkanes, and derivatives (Kolattukudy, 1966). This conclusion was confirmed by the recent results obtained with solubilized elongating enzyme(s) from leek epidermis (Agrawal and Stumpf, 1985). That CoA esters and not acyl carrier protein (ACP) derivatives are involved in elongation has been demonstrated (Agrawal and Stumpf, 1985; Lessire *et al.*, 1985b,c). Fatty acyl-CoA elongation using malonyl-CoA and NADPH as cosubstrates involves the same series of basic reactions as those catalyzed by fatty acid synthase, including condensation, ketoreduction, dehydration, and enoyl reduction. The elongating activities in the detergent-solubilized preparations were associated with relatively high molecular weight fractions (Agrawal and Stumpf, 1985; Lessire *et al.*, 1985a) and it has been speculated that all of the component activities involved in elongation might be catalyzed by large multifunctional proteins rather than by a multicomponent system similar to the plant fatty acid synthase. Until the elongating enzymes are solubilized, resolved, and purified, the nature of the enzymes involved cannot be elucidated.

The only chain-elongating enzyme that has been purified from any organism is the mycocerosic acid synthase from *Mycobacterium tuberculosis*. In this organism

2,4,6,8-tetramethyloctacosanoic acid and related polymethyl homologs are generated by elongation of n-fatty acids with methylmalonyl-CoA as demonstrated with cell-free preparations (Rainwater and Kolattukudy, 1983) (Fig. 1A). Fatty acid synthase of this organism generates very long chain acids with C_{26} as the dominant component (Fig. 1B). The purified soluble mycocerosic acid synthase, which was distinctly different from the fatty acid synthase (Fig. 1C), elongated C_6–C_{20} n-fatty acyl-CoA specifically with methylmalonyl-CoA as the elongating agent using NADPH as the preferred reductant (Rainwater and Kolattukudy, 1985). With all primers a preference for generating tetramethyl-branched (four elongation) products was found (Table I). A similar degree of preference for the number of elongation steps catalyzed by three or so elongation systems could account for the diversity of chain length classes found in plant waxes. The mycobacterial enzyme was found to be a dimer of ~240-kDa protomers, each containing a covalently attached phosphopantetheine (Rainwater and Kolattukudy, 1985). Thus this elongating enzyme is a multifunctional enzyme somewhat similar to the multifunctional fatty acid synthase. Whether an analogous situation exists in higher plants is yet to be determined.

B.   Discovery of Enzymatic Decarbonylation of Aldehydes

How the elongated fatty acid loses the carboxyl carbon to yield a hydrocarbon was not known until recently. It was known for a decade that in cell-free

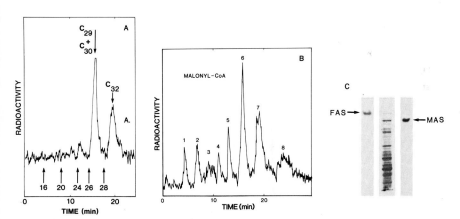

Fig. 1.   Synthesis of long-chain fatty acids by a cell-free extract of *Mycobacterium tuberculosis*. (A) Radio-GLC of methyl esters synthesized from [*methyl*-^{14}C]methylmalonyl-CoA using stearoyl-CoA and arachidoyl-CoA as primers. Locations of coinjected n-fatty acid methyl esters are indicated by their carbon numbers. (B) Radio-GLC of methyl esters synthesized from [2-^{14}C]malonyl-CoA. Peak 1, n-C_{16}; 2, n-C_{18}; 3, n-C_{20}; 4, n-C_{22}; 5, n-C_{24}; 6, n-C_{26}; 7, n-C_{28}; 8, n-C_{30}. (C) Dodecyl sulfate–gel electrophoresis of cell-free extract (center) flanked by Western blot analyses done using antibody to fatty acid synthase (FAS) or mycocerosic acid synthase (MAS). [From Rainwater and Kolattukudy (1983, 1985) and S. Kikuchi and P. E. Kolattukudy (1986 unpublished observations).]

TABLE I

Effect of Primer Chain Length on the Number of Elongations by Methylmalonyl-CoA Catalyzed by Mycocerosic Acid Synthase from *Mycobacterium tuberculosis*[a,b]

Number of Elongations	Primer chain length			
	8	12	16	20
1	—	13	9	6
2	2	10	11	2
3	22	26	14	9
4	33	40	26	26
5	3	12	—	—

[a]From Rainwater and Kolattukudy (1985).
[b]Given as picomoles product molecules.

preparations exogenous C_{32} fatty acid was converted to C_{30} alkane (Khan and Kolattukudy, 1974). More recently it was shown that cell-free preparations from pea leaves catalyzed the conversion of exogenous n-C_{18}, n-C_{22}, and n-C_{24} fatty acids to the corresponding alkanes containing two carbon atoms less than the precursor acid, with O_2 and ascorbate as required cofactors (Bognar *et al.*, 1984). Since the cell-free preparation also catalyzed α-oxidation, it was suspected that some α-oxidation intermediate, with one less carbon atom than the substrate acid, might lose another carbon to generate the alkane. A careful chromatographic examination of the products generated from [U-^{14}C]octadecanoic acid by a particulate fraction from pea leaves showed that at all time periods, including very short incubation periods, α-hydroxy-C_{18} acid, C_{17} aldehyde and C_{17} acid were the major products. Thus the only intermediate generated during the incubation which might preferentially lose another carbon to yield alkane was suspected to be the aldehyde. This conclusion was consistent with an earlier observation that inhibition of alkane synthesis by thiols in tissue slices resulted in accumulation of aldehydes containing one carbon more than the alkanes (Buckner and Kolattukudy, 1973). A possible mechanism by which aldehydes might serve as the precursors of alkanes was suggested by the finding that triphenylphosphine complex of Ru and Rh decarbonylated aldehydes to alkanes at 100–200°C (Doughty *et al.*, 1979). That such a reaction is plausible in a biological system was suggested by the finding that a porphyrin complex could catalyze decarbonylation at near ambient temperatures (Domazetis *et al.*, 1981). In fact, exogenous C_{18} and C_{24} aldehydes were converted to alkanes by the cell-free preparations from *Pisum sativum* (Bognar *et al.*, 1984). Although the alkane-synthesizing activity could be solubilized from the particulate preparation, α-oxidation activity could not be resolved from the alkane-synthesizing activity and thus the solubilized preparation also produced alkanes with two carbon atoms less than the parent acid.

A two-step density gradient procedure was developed to remove the α-oxidation activity and the resulting particulate preparation (sp. gr. 1.3 g/cm^3) catalyzed alkane synthesis from aldehydes by the novel biochemical reaction which involved loss of CO (Cheesbrough and Kolattukudy, 1984). The particulate preparation catalyzed the conversion of [1-^{3}H,1-^{14}C]octadecanal to [^{3}H]heptadecane (Fig. 2) and ^{14}CO. Thus the aldehydic hydrogen was retained during the enzymatic decarbonylation. This enzyme showed an optimal pH of 7.0 and an apparent K_m of 35 μM for octadecanal. Although the mechanistic details of this novel biochemical reaction remain unknown, the retention of the aldehydic hydrogen and the involvement of a metal ion suggested by the severe inhibition of the enzyme by chelators are consistent with a mechanism suggested for the nonenzymatic decarbonylation catalyzed by coordination complexes of Ru and Rh. According to this mechanism, the metal participates in holding the aldehydic hydrogen, and after removal of CO the alkyl moiety recombines with the aldehydic hydrogen to yield the alkane. Such a mechanism predicts that the aldehydic hydrogen would be retained in the alkane as observed with the enzymatic decarbonylation of [1-^{3}H,1-^{14}C]octadecanal. A similar decarbonylation is also responsible for the synthesis of alkanes in animals. A cell-free preparation from the uropygial gland of grebe (*Podiceps nigricollis*) catalyzed conversion of octadecanal to heptadecane and CO (T. M. Cheesbrough and P. E. Kolattukudy, unpublished observations). Thus decarbonylation might be a widely occurring reaction in biological systems.

The decarbonylase activity in *P. sativum* was found to be associated with a heavy particulate fraction containing cell wall/cuticle fragments. Electron microscopy revealed only amorphous particles; no vesicles were observed (Fig. 3). Chemical examination of the depolymerization products showed the presence of dihydroxy

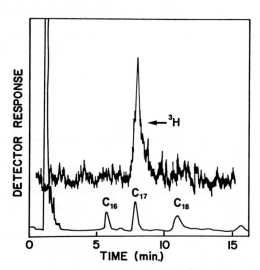

Fig. 2.   Radio-GLC of the alkane produced from [1-^{14}C,1-^{3}H]octadecanal by a particulate fraction from pea leaves. The lower tracing shows the location of coinjected *n*-alkanes. [From Cheesbrough and Kolattukudy (1984).]

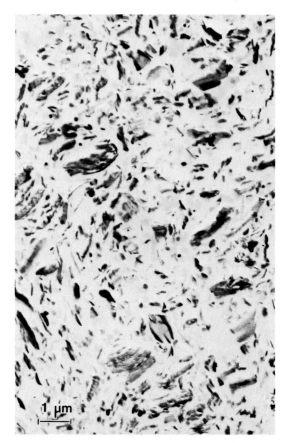

Fig. 3. Electron micrograph of decarbonylase-containing particulate fraction from pea leaves. [From T. M. Cheesbrough and P. E. Kolattukudy (unpublished observations).]

C_{16} acid, the characteristic component of cutin. This fraction showed little acyl-CoA reductase activity or α-oxidation activity, and free fatty acids were not converted to alkanes by this fraction (Table II). The particulate fraction from which the decarbonylase preparation was obtained contained vesicles and catalyzed acyl-CoA reduction to the aldehyde as well as conversion of acyl-CoA and free fatty acids to alkanes (Table II). Crude microsomal preparations from *P. sativum* catalyzed chain elongation of fatty acids. From these results it is concluded that alkane biosynthesis involves elongation of long-chain CoA esters and reduction of these elongated products to the corresponding aldehydes by the acyl-CoA reductase, both by enzymes localized in the endoplasmic reticulum, followed by excretion of the aldehydes to the extracellular matrix where the decarbonylase catalyzes decarbonylation to generate the alkane and CO. Thus, the overall process involved in the production of alkanes and derivatives can be revised as schematically shown in Fig.

TABLE II

Relative Efficiency of Conversions of Fatty Acids and Derivatives into Alkanes and α-Oxidation Activity of the Particulate Preparations from Pea Leaves[a,b]

Particulate fraction	Hydrocarbon formation nmol/min per mg			α-Oxidation nmol/min per mg
	Fatty acid	Acyl-CoA	Aldehyde	
P-2	0.02	0.06	0.0002	1.56 (18.7)
G-1	5.0	20.3	5.1	0.73 (0.023)
G-2	0.0	0.60	24.0	0

[a]Data from Cheesbrough and Kolattukudy (1984).

[b]P-2, Crude microsomes; G-1, particulate fraction from the density gradient step 1; G-2, the final decarbonylase preparation. Assays with P-2 and G-1 used [n-^{3}H] tetracosanoic acid or tetracosanoyl-CoA prepared from it, whereas for assays with G-2, [1-^{14}C]octadecanoic acid or the CoA ester prepared from it was used. [1-^{3}H,1-^{14}C]Octadecanal was used where aldehyde is indicated. α-Oxidation was measured using [1-^{14}C]palmitic acid. Values in parentheses indicate total activity in nmol/min.

4. Such a hypothesis is consistent with the observed apparent lack of chain length specificity of the decarbonylase preparation. Most probably the chain length composition of the alkanes found in each plant is determined by the enzyme(s) that provides the substrate for the decarbonylase. Since acyl-CoA reductase from *P. sativum* appeared to reduce acyl-CoA from C_{18}–C_{32}, the specificity that really determines the composition of the alkanes resides in the elongation process. Since multiple elongation systems are probably present in the microsomal preparations, the specificity of the elongating enzymes cannot be determined until the different elongating enzymes are solubilized and resolved.

III. SUBERIN

A. Structure and Composition of Suberin

Chemical analyses of suberized layers from a variety of anatomical regions have provided a general picture of the composition of this layer. According to these results, suberized layers consist of an insoluble polymeric structural component which has soluble waxes associated with it. The polymer itself consists of a phenolic domain attached to the cell wall and an aliphatic domain which is probably attached to the aromatic domain. The soluble waxes associated with this layer provide the major barrier to diffusion (Soliday *et al.*, 1979).

Since the aromatic components are held together by nonhydrolyzable bonds similar to those found in lignin, a glimpse at the chemical composition is provided

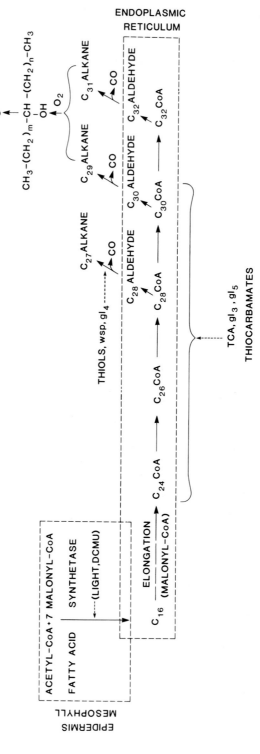

Fig. 4. Proposed pathway for synthesis of alkanes and derivatives. DCMU, N'-(3,4-Dichlorophenyl)-N,N'-dimethylurea; TCA, trichloroacetate; gl_3, gl_4, and gl_5, mutants of *Brassica oleracea*; *wsp*, mutant of *Pisum sativum*.

by oxidative cleavage and analysis of the aromatic aldehydes generated. Nitrobenzene oxidation of potato periderm suberin generated mainly p-hydroxybenzaldehyde and vanillin (Cottle and Kolattukudy, 1982a). Similar examination of suberized layers from a variety of anatomical regions confirmed the hypothesis that release of p-hydroxybenzaldehyde and vanillin is a characteristic feature of the aromatic domain of suberin (Table III). Lignin-containing polymeric fractions, on the other hand, yield syringaldehyde as a major component. More recently the above observations were confirmed with the use of CuO oxidation and thioglycolic acid derivatization of potato wound periderm suberin (Hammerschmidt, 1985).

   The aliphatic domains of suberin from a variety of anatomical regions have been analyzed by depolymerization by alkaline hydrolysis, transesterification, or reductive cleavage with $LiAlH_4$ followed by analyses of products by combined gas–liquid chromatography and mass spectrometry (Kolattukudy, 1980b; Kolattukudy et al., 1981; Kolattukudy and Espelie, 1985). Suberin samples characteristically contain high proportions of long-chain (C_{16}–C_{24}) ω-hydroxy acids and dicarboxylic acids with smaller amounts of very long chain (C_{20}–C_{30}) fatty acids and fatty alcohols. Often monounsaturated C_{18} monomers are the dominant components and saturated C_{16}, C_{20}, C_{22}, and C_{24} ω-hydroxy acids and diacids comprise a small proportion. In some cases, such as cotton fiber, saturated C_{22} ω-hydroxy acids and diacids constitute the dominant components (Ryser et al., 1983; Yatsu et al., 1983) and such long-chain components have been reported in bark samples (Kolattukudy

TABLE III

Aromatic Aldehydes Released by Nitrobenzene Oxidation of Cutin and Suberin-Enriched Samples[a,b]

Tissue	p-Hydroxybenzaldehyde	Vanillin	Syringaldehyde
Potato periderm	3.28	16.51	0.53
Potato periderm LAH residue	1.77	6.66	0.20
Chalazal region of grapefruit seed coat	1.19	15.63	1.80
Grapefruit inner seed coat	3.45	1.84	0.16
Grapefruit outer seed coat	0.40	38.72	1.09
Cottonseed coat	0.37	6.44	1.07
Pecky rot lignin	—	34.90	—
Agave idioblasts	0.02	0.26	0.12
Agave vascular leaf tissue	0.36	1.95	10.91
Wound-healed rutabaga	0.93	1.59	—
Parsnip periderm	—	0.25	—
Carrot periderm	—	0.51	—
Quercus suber bark	3.00	1.32	1.17
Zea mays root	2.49	2.13	0.13
Z. mays root (low Mg^{2+})	7.38	5.51	4.98
Tomato fruit cutin	5.07	—	—

[a]From Kolattukudy and Espelie (1985).

[b]Values are given as μg/mg dry tissue.

and Espelie, 1985). The polyfunctional acids that are the major components of cutin (Kolattukudy, 1980a) are usually present only in small quantities in suberin, although exceptions might be found.

Suberin-associated waxes contain the same type of components as those found in cuticular waxes. However, several features of the waxes found in suberized layers make them distinct from cuticular waxes. Free fatty acids, which are only minor components in cuticular waxes, are major components of suberin-associated waxes. For example, free acids constituted about one-half of the suberin waxes in the storage organs of *Daucus carota* and *Brassica napobrassica* (Espelie *et al.*, 1980) and the barks of *Pinus contorta* (Rowe and Scroggins, 1964) and *Ailanthus glandulosa* (Chiarlo and Tacchino, 1965). C_{20}, C_{22} and C_{24} saturated acids are the major components of the free fatty acid fraction. Fatty alcohols are also major constituents of suberin-associated waxes. The chain lengths of the major alcohols of suberin waxes are shorter than those found in cuticular waxes; in the former, C_{22} and C_{24} are often the major components whereas C_{26} and C_{28} are often the major alcohols found in the latter (Kolattukudy, 1980a). Wax esters are only minor components of suberin-associated waxes, constituting <10%. Hydrocarbons of suberin waxes are characterized by the presence of higher proportions of even-chain length components and shorter average chain length when compared to those found in cuticular waxes (Dean and Kolattukudy, 1977; Espelie *et al.*, 1980). In suberin waxes at least several alkanes dominate, whereas in cuticular waxes one or a few usually dominate.

B. Biosynthesis of Suberin

Although oxidation of the polymer by nitrobenzene or CuO gives only low yields of the aldehydes, such methods can be used to follow the time course of deposition of the aromatic components of suberin (Cottle and Kolattukudy, 1982a). The deposition of such phenolic components showed a lag period of nearly 3 days after wounding followed by several days of active deposition; the process ceased in about 6 or 7 days after wounding. A similar time course was observed when a thioglycolic acid method was recently used for measuring the aromatic components of suberin (Hammerschmidt, 1985). The deposition of the aliphatic components of suberin and the soluble waxes also followed a similar time course (Fig. 5).

The biosynthesis of the monomers from which the aromatic domains of the polymer are synthesized probably proceeds as described for lignin synthesis (Grisebach, 1981; Higuchi, 1985), and much information is available on phenolic biosynthesis in plants (see Vol. 7 of this treatise). Biosynthetic studies with wound-healing potato slices support the hypothesis that the pathway from pheny-lalanine to C_6–C_3 precursors of suberin is operative during suberization. Thus, exogenous labeled phenylalanine and cinnamic acid were incorporated into suberin during the wound-healing process, and nitrobenzene oxidation of the labeled insoluble material derived from the labeled precursors released labeled *p*-hydroxy-benzaldehyde and vanillin (Cottle and Kolattukudy, 1982a). Aminoxyphenyl-

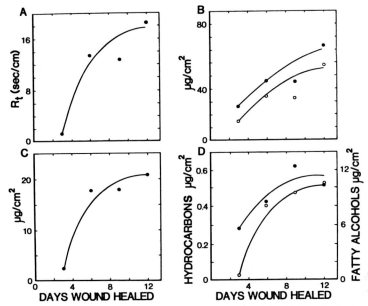

Fig. 5. Time course of development of diffusion resistance and deposition of suberin-associated components during the wound healing of potato tuber tissue slices. (A) Diffusion resistance. (B) Phenolics measured as aromatic aldehydes generated by alkaline nitrobenzene oxidation: p-hydroxybenzaldehyde (●); vanillin (○). (C) Polymeric aliphatics measured as diols generated by LiAlH₄ reduction. (D) Waxes recovered by solvent extraction: hydrocarbon (●); alcohols (○). [From W. Cottle and P. E. Kolattukudy (unpublished observations).]

propionic acid, a well-known inhibitor of phenylalanine ammonia lyase, inhibited suberization as measured by the deposition of the aromatic components and by the development of diffusion resistance in wound-healing potato slices (W. Cottle and P. E. Kolattukudy, unpublished observations).

Synthesis of the aromatic polymeric domain from the monomers probably involves a unique isoperoxidase just as in the case of lignification. The following evidence supports this conclusion: A specific anionic peroxidase was induced during the wound healing of *Solanum tuberosum* tuber slices, and this isoperoxidase was found to be localized specifically in the suberizing cells (Borchert, 1974). Both the time course of appearance and the tissue distribution of this wound-induced peroxidase activity correlated with that of the suberization process (Borchert, 1978; K. Chung and P. E. Kolattukudy, unpublished observations). Immunocytochemical examination of wound-healing tissue showed that the anionic isoperoxidase appeared only in the walls of suberizing cells and only during suberization (Espelie *et al.*, 1986). Abscisic acid induction of suberization in potato callus tissue culture was accompanied by induction of the specific isoperoxidase by the hormone (Cottle and Kolattukudy, 1982b). Plant roots appear to have peroxidase activity localized in epidermal and endodermal cell walls (Mueller and Beckman, 1978; Smith and

O'Brien, 1979), both of which are areas known to undergo suberization (Kolattukudy, 1984a). Parenchyma tissue of potato tubers treated with *p*-coumaric and ferulic acids generated polymeric material which appeared to be as highly condensed as suberin phenolics (Bland and Logan, 1965; Cottle and Kolattukudy, 1982a). An increased production of peroxidase is associated with the wound-healing lesions in the fruit of *Lycopersicon esculentum* (Fleuriet and Deloire, 1982; Fleuriet and Macheix, 1984), and wounding–healing of this tissue is known to involve suberization (Dean and Kolattukudy, 1976). The suppression of suberization caused by iron deficiency in bean roots was not because of a lack of hydroxylated fatty acids but due to a decrease in the level of anionic isoperoxidase (Sijmons *et al.*, 1985). This peroxidase may be induced as a result of fungal infection, and in such case the deposition of the phenolic polymers catalyzed by the enzyme probably restricts the ingress of the invading pathogen.

C. Regulation of Suberization

Plants resort to suberization whenever physiological or developmental changes or stress factors require the erecting of a diffusion barrier. For example, wounding causes the formation of suberized periderm on plant organs irrespective of the nature of the natural protective layer (Dean and Kolattukudy, 1976). Deficiency in Mg^{2+} caused increased suberization in corn roots (Pozuelo *et al.*, 1984), whereas iron deficiency caused suppression of suberization in bean roots (Sijmons *et al.*, 1985), suggesting that modulation of suberization might be used as a mechanism in adapting to mineral deficiency conditions. Cold adaptation of winter rye was found to involve increased deposition of the aliphatic components in the polymers in the epidermal and mestome sheath (Griffith *et al.*, 1985). The molecular mechanism by which suberization is regulated remains unclear.

Abscisic acid was found to promote suberization in potato tissue slices in which suberization was suppressed by washing (Soliday *et al.*, 1978). Induction of suberization by abscisic acid was confirmed also in callus cultures from potato tuber. To study the mechanism of this induction, the effect of this hormone on the level of ω-hydroxy fatty acid dehydrogenase, an enzyme uniquely involved in suberization (Agrawal and Kolattukudy, 1977, 1978), was determined in the tissue culture. This enzyme was found to be not a good model, because tissue cultures, unlike intact potato tubers, already had a high level of this enzyme and the hormone treatment resulted only in modest increases of the activity level (Cottle and Kolattukudy, 1982b). However, the suberization-associated highly anionic isoperoxidase was highly induced by abscisic acid (Fig. 6). This peroxidase was purified and characterized (Espelie and Kolattukudy, 1985). The cDNA for this enzyme has been cloned in λgt11 and the recombinants were identified using antibodies prepared against the enzyme (E. Roberts and P. E. Kolattukudy, unpublished observations). With this cDNA it should be possible to study the molecular mechanism of regulation of suberization.

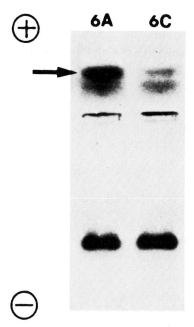

Fig. 6. Induction of the anionic isoperoxidase (arrow) involved in suberization in potato tissue culture by the addition of 10^{-4} M abscisic acid to the agar medium. Lane 6A, tissue on abscisic acid agar for 6 days; lane 6C, control. [From Cottle and Kolattukudy (1982b).]

D.   Biodegradation of Suberin

Many phytopathogenic fungi were screened for their ability to grow on suberin as the sole source of carbon, and several of them, including *Fusarium solani* f.sp. *pisi, Streptomyces scabies, Helminthosporium sativum, H. solani, Leptosphaeria conio-thyrium, Ascodichaena rugosa, Rhizoctonia crocorum, Phytophthora infestans,* and *Cyathus stercoreus,* grew well. From the extracellular fluid of suberin-grown cultures of *F. solani pisi* an enzyme that catalyzes hydrolysis of polyesters including cutin and suberin was purified to homogeneity (Fernando *et al.*, 1984). This enzyme was found to be identical to the cutinase generated by the same organism when grown on cutin in all of the physicochemical, catalytic, and immunological properties. Another protein fraction obtained from the extracellular fluid released radioactive aromatic monomers from potato periderm suberin which was biosyn-thetically labeled specifically in the aromatic domains with exogenous labeled cinnamic acid (Kolattukudy, 1984b). This protein fraction could release only ~10% of the label from the labeled suberin even after extensive treatment with the enzyme, and HPLC analyses of the products released by this treatment showed that the major components were *p*-coumaric acid and ferulic acid. Thus this enzyme probably released only esterified phenolics but not the nonhydrolyzable aromatic core of the polymer. If and how this core is biodegraded are yet to be determined.

IV. CUTINASES

Considerable progress has been made in our understanding of the structure of cutinase, the gene that codes for this enzyme, and the possible function of this enzyme in fungal interaction with plants. A unique bacterial cutinase was also recently discovered and characterized. This enzyme might play a role in providing carbon source for phyllospheric nitrogen-fixing bacteria. Recent progress in our understanding of both the fungal and bacterial enzyme are summarized below.

A. Fungal Cutinases

Cutinase has been purified to apparent homogeneity from a number of phytopathogenic fungi (Kolattukudy, 1980a, 1985). All of them show remarkably similar amino acid composition and molecular weight. Some key features, such as the catalytic mechanism, the presence of one or two histidines, one tryptophan, and four cysteines, which are all involved in disulfide bonds, are conserved in the fungal cutinases thus far examined.

1. Catalytic Triad

The first fungal cutinase to be purified was severely inhibited by organic phosphates that are known to react with active serine, and on this basis it was tentatively concluded that this fungal enzyme was a serine polyesterase (Purdy and Kolattukudy, 1975a,b). All of the ~15 fungal cutinases so far examined are inhibited by the organic phosphates. Chemical modification of the three amino acid residues characteristic of such enzymes provided the strongest evidence that cutinase contains the catalytic triad (Köller and Kolattukudy, 1982). Thus, modification of the serine residue by treatment of the enzyme with diisopropyl-fluorophosphate caused inactivation of the enzyme, and the extent of modification was correlated with the loss in enzyme activity so that it was clear that a unique serine was involved in catalysis. Modification of the carboxyl residues by treatment of the enzyme with water-soluble carbodiimide resulted in the modification of three carboxyl residues without causing any loss in enzymatic activity. In the presence of sodium dodecyl sulfate (SDS) an additional carboxyl group was modified by carbodiimide and this modification resulted in the loss of the enzyme activity, strongly suggesting that this carboxyl group is the member of the catalytic triad. All attempts to modify the imidazol group of the histidine residue by treatment with diethylpyrocarbonate failed except when the treatment was done in the presence of critical micellar concentrations of SDS. Initially it was thought that the critical histidine residue was not accessible to the reagent and that the detergent exposed this residue to the modifying reagent (Köller and Kolattukudy, 1982). However, it was found that cutinase was resistant to diethylpyrocarbonate because it catalyzed rapid hydrolysis of this reagent (Foster and Kolattukudy, 1986). Inactivation of the enzyme by the detergent allowed modification of the histidine residue. In any case, it was clear that catalysis by cutinase involved the catalytic triad characteristic of serine hydrolases (Fig. 7). The involvement of the predicted acyl enzyme intermediate was also demonstrated by the isolation of labeled acetyl enzyme after treatment of the carbethoxylated enzyme with *p*-nitrophenyl [1-^{14}C]acetate (Köller and Kolattukudy, 1982). The members of the catalytic triad were located within the primary structure of the enzyme deduced from the nucleotide sequence of the cloned cDNA (Soliday *et al.*, 1984). Modification of the three peripheral carboxyl groups by carbodiimide and glycine ethyl ester followed by similar treatment in the presence of SDS and labeled glycine ester yielded a protein that was selectively labeled at the critical carboxyl group. After proteolysis the labeled active-site peptide was isolated and the amino acid composition itself allowed the identification of the critical carboxyl group in the primary structure of the enzyme (W. Ettinger and P. E. Kolattukudy, unpublished observations). Active serine was modified by treatment of the enzyme with labeled diisopropylfluorophosphate, and the labeled active-site peptide was isolated after proteolysis of the labeled protein, and the amino acid sequence was determined (Soliday and Kolattukudy, 1983). This sequence matched with a segment of the primary structure of the enzyme (Soliday *et al.*, 1984) (Fig. 8). Since there was only one histidine in this enzyme, identification of this third member of the catalytic triad in the primary structure was simple.

The members of the catalytic triad in the cutinase from *F. solani pisi*, Asp^{99}, Ser^{136} and His^{204}, are located far apart in the primary structure (Fig. 8). Obviously

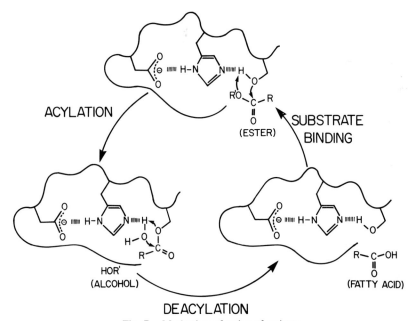

Fig. 7. Mechanism of action of cutinase.

AACCACAACTACCTTCACTTCATCAACATTCACTTCAACTTCTTCGCCTCTTCCTTTTCACTCTTTATCATCCTCACC

```
                                        10
MET LYS PHE PHE ALA LEU THR THR LEU LEU ALA ALA THR ALA SER ALA LEU PRO THR SER
ATG AAA TTC TTC GCT CTC ACC ACA CTT CTC GCC GCC ACG GCT TCG GCT CTG CCT ACT TCT
                          ┣━C-57
                                        30      32
ASN PRO ALA GLN GLU LEU GLU ALA ARG GLN LEU GLY ARG THR THR ARG ASP ASP LEU ILE
AAC CCT GCC CAG GAG CTT GAG GCG CGC CAG CTT GGT AGA ACA ACT CGC GAC GAT CTG ATC
                      ┣━C-4
                  47          50
ASN GLY ASN SER ALA SER CYS ARG ASP VAL ILE PHE ILE TYR ALA ARG ┃GLY SER THR GLU
AAC GGC AAT AGC GCT TCC TGC CGC GAT GTC ATC TTC ATT TAT GCC CGA GGT TCA ACA GAG
                                        70
THR GLY ASN LEU GLY THR LEU GLY PRO SER ILE ALA SER ASN LEU GLU SER ALA PHE GLY
ACG GGC AAC TTG GGA ACT CTC GGT CCT AGC ATT GCC TCC AAC CTT GAG TCC GCC TTC GGC
                                        90                              99
LYS ASP GLY VAL TRP ILE GLN GLY VAL GLY GLY ALA TYR ARG┃ALA THR LEU GLY ASP ASN
AAG GAC GGT GTC TGG ATT CAG GGC GTT GGC GGT GCC TAC CGA GCC ACT CTT GGA GAC AAT
                                        110
ALA LEU PRO ARG GLY THR SER SER ALA ALA ILE ARG┃GLU MET LEU GLY LEU PHE GLN GLN
GCT CTC CCT CGC GGA ACC TCT AGC GCC GCA ATC AGG GAG ATG CTC GGT CTC TTC CAG CAG
            125                 130                         136
ALA ASN THR LYS CYS PRO ASP ALA THR LEU ILE ALA GLY GLY TYR SER GLN GLY ALA ALA
GCC AAC ACC AAG TGC CCT GAC GCG ACT TTG ATC GCC GGT GGC TAC AGC CAG GGT GCT GCA
                                        150
LEU ALA┃ALA ALA SER ILE GLU ASP LEU ASP SER ALA ILE ARG ASP LYS ILE ALA GLY THR
CTT GCA GCC GCC TCC ATC GAG GAC CTC GAC TCG GCC ATT CGT GAC AAG ATC GCC GGA ACT
                                        170
VAL LEU PHE GLY TYR THR LYS ASN LEU GLN ASN ARG GLY ARG ILE PRO ASN TYR PRO ALA
GTT CTG TTC GGC TAC ACC AAG AAC CTA CAG AAC CGT GGC CGA ATC CCC AAC TAC CCT GCC
                        187         190             194
ASP ARG┃THR LYS VAL PHE CYS ASN THR GLY ASP LEU┃VAL CYS THR GLY SER LEU ILE VAL
GAC AGG ACC AAG GTC TTC TGC AAT ACA GGG GAT CTC GTT TGT ACT GGT AGC TTG ATC GTT
        204                         210
ALA ALA PRO HIS LEU ALA TYR GLY PRO ASP ALA ARG GLY PRO ALA PRO GLU PHE LEU ILE
GCT GCA CCT CAC TTG GCT TAT GGT CCT GAT GCT CGT GGC CCT GCC CCT GAG TTC CTC ATC
                        230
GLU LYS VAL ARG ALA VAL ARG GLY SER ALA ●●● GGAGGATGAGAATTTTAGCAGGCGGGCCTGTTAAT
GAG AAG GTT CGG GCT GTC CGT GGT TCT GCT TGA
```

TATTGCGAGGTTTCAAGTTTTTCTTTTGGTGAATAGCCATGATAGATTGGTTCAACACTCAATGTACTACAATGCCC

Fig. 8. Nucleotide sequence of cloned cDNA for cutinase from *Fusarium solani pisi* and the amino acid sequence deduced from it. C-57 and C-4 indicate two additional clones which were sequenced. Solid lines represent regions which were subjected to direct amino acid sequencing. [From Soliday *et al.* (1984).]

they must be held together in the proper juxtaposition by the secondary and tertiary structure. One essential structural feature involved in holding the enzyme in the active conformation is the disulfide bridges. Reduction of the disulfides renders the fungal cutinase inactive and the enzyme becomes insoluble in the aqueous medium (W. Köller and P. E. Kolattukudy, unpublished observations). The identity of the cysteine residues involved in the two disulfide bridges present in the enzyme from *F. solani pisi* has been determined by isolation of the disulfide-containing fragment generated by proteolysis followed by reductive cleavage, isolation of the two cysteine-containing peptides, and determination of their composition. From their composition the origin of the peptides could be identified in the primary structure of the proteins (W. Ettinger and P. E. Kolattukudy, unpublished observations). Thus Cys^{125} was found to be bound to Cys^{47} and Cys^{187} to Cys^{194} by disulfide bonds. The position of these cysteine residues were found to be conserved in *Colletotrichum capsici* enzyme, which has been completely sequenced via its cDNA (Kolattukudy, 1985; W. Ettinger and P. E. Kolattukudy, unpublished observations).

2. Cloning of cDNA for Cutinase

The finding that cutin monomers induce cutinase synthesis in glucose-grown *F. solani pisi* (Lin and Kolattukudy, 1978) gave a system with which cutinase synthesis could be studied. Cell-free translation of the $poly(A)^+$ mRNA isolated from the induced cultures, but not that from uninduced cultures, produced a protein that immunologically cross-reacted with anticutinase IgG. This primary translation product was slightly (2.1 kDa) larger than the mature protein, strongly suggesting that a leader sequence is involved in the secretion of the enzyme (Flurkey and Kolattukudy, 1981).

Clones containing cDNA, synthesized using $poly(A)^+$ RNA isolated from fungal cultures induced to synthesize cutinase, were screened with ^{32}P-labeled cDNA prepared with mRNA from the induced and uninduced cultures (Soliday *et al.*, 1984). Hybrid-selected translation and examination of the products with anticutinase IgG were used to detect the presence of the coding region for cutinase. When the inserts from the 15 cutinase-positive clones were examined by Southern blots, their size ranged from 279 to 950 nucleotides. Northern blot analysis of the $poly(A)^+$ RNA showed that cutinase mRNA contained 1050 nucleotides, strongly suggesting that the clone containing 950 nucleotides represented nearly the entire mRNA. This cDNA and its subcloned restriction fragments were sequenced by a combination of Maxam-Gilbert and the dideoxy chain termination techniques. The open reading frame translated into a protein with a molecular weight of 23,951, as shown in Fig. 8. Two other clones containing the bulk of the coding region for cutinase were also completely sequenced to confirm the sequence. To verify the amino acid sequence deduced from the nucleotide sequence, three portions of the protein were subjected to direct amino acid sequencing. The peptide containing the only tryptophan of the protein was isolated from a tryptic digest of the enzyme, and this peptide was sequenced yielding 37 amino acid residues. The labeled active-site peptide obtained by tryptic digestion of the enzyme labeled by treatment with [^{3}H]diisopropylfluorophosphate revealed the sequence of 30 amino acid residues. The sequence of another 10 residues around a cystine was also determined (W. Ettinger and P. E. Kolattukudy, unpublished observations). The nucleotide sequence and the amino acid sequences were in complete agreement, showing that the cloned cDNA did in fact represent cutinase.

The N-terminal sequence in the open reading frame showed characteristics quite typical of signal sequences. In cutinase lysine was the basic residue, preceding the hydrophobic core, which was composed of 15 hydrophobic amino acid residues in *F. solani pisi* (Soliday *et al.*, 1984). Since the N-terminal residue of the mature protein is known to be *N*-glucuronyl glycine (Lin and Kolattukudy, 1977), the first glycine from the N terminus of the proenzyme is presumed to be the processing site (residue 32). This would leave a post-hydrophobic core of 14 residues, which is larger than the usual 4–8 residues found in other proteins. However, in view of the unusual nature of the N terminus of the mature cutinase, it is probable that an unusual processing is involved in the synthesis of this protein, and the nature of this processing has not been elucidated.

3. Cutinase Gene

Genomic DNA fragments from *F. solani pisi* were cloned in Charon 35, and the recombinants were screened with ^{32}P-labeled cDNA. Subclones containing cutinase reading frame were prepared and the nucleotide sequence was determined (Fig. 9). The open reading frame for cutinase matched precisely with the structure of the cloned cDNA, except for the presence of one 51-bp intron which had splicing signals homologous to those found in other fungal introns (Kolattukudy *et al.*, 1985). The usual polyadenylation site was present in the expected region (C. L. Soliday and P. E. Kolattukudy, unpublished observations). In the 5′ end a TATA boxlike sequence was found, but there was no CAT box. In the 5′ region the nucleotide sequence of ∼ 1 kb has been determined.

Southern blot analyses of the genomic DNA from many strains of *F. solani pisi*, which varied greatly in their ability to produce cutinase, revealed two DNA fragments that cross-hybridized with the cDNA in the high producers of the enzyme, whereas only one of these was found in the case of some of the low producers (Kolattukudy and Soliday, 1985). Similar results were obtained when various fragments of the gene were used as labeled probes, suggesting that both fragments represented complete genes. The functional significance of the presence of the two copies of the gene is unknown. It is possible that one copy is responsible for the high level of induction and the other for the low constitutive level of the enzyme. cDNA from *Colletotrichum capsici* was cloned in the expression vector λ gt11, and cutinase-positive clones were identified using antibodies (W. Ettinger and P. E. Kolattukudy, unpublished observations). Southern blot analysis of the genomic DNA from *C. capsici* with a variety of restriction enzymes revealed two fragments that hybridized with the cDNA, indicating the presence of two copies of cutinase gene. Southern blot analyses of the genomic DNA from *C. orbiculare*, *C. coccodes*, *C. graminicola*, and *C. gloeosporioides* showed that cDNA for the enzyme from *C. capsici* hybridized with the genomic DNA from *C. graminicola* and *C. gloeosporioides*. In both cases, Southern blots of digests prepared with several restriction enzymes suggested that there was only one copy of the gene in these fungi.

```
960                -930                  -900                  -870                    -840
GTGGAAGGGGAGCCGTGTGGRGGCACARGARGCTGARARARACCGAGGTCARARGACCTGCTGARTAGTTTGTCACTGAGATGAGATGGGARCGGCATGARATCTTGGGCCTTATTCTGACACCATCGCCTCCTTCCT
     -810                 -780                  -750                   -720                    -690
GTACGCCTTGCATCARARGAGAGGGGTTCTACCCCCTARARGCACAGCTCGARTGARATTCCCATTTGAGTAGATCCARCGATGCTCCGTGTTTCTCTAGGTTGGGTAGGCAGGTTGACCCACCCATGATGAGCGGCA
     -660                  -630                  -600                    -570
GGGTCAGGTAGGTCGGTARATATCCGAGCCGCTGATTTGCGGATGARCGAGTCCARGCTTGGCATGATCTGTTGATCTCARARCCCTCCARTCACCGTARAGAGGCACTCGTARARGTCCTCCGATGCCTCTCCACCA
     -540                  -510                  -480                    -450                    -420
TCAGGTAGGTGGTGTCTATGCGCGCTCTGACACTCTTCGCRAGGTGTARCAGARTAGGCARAGCGGCCTTCCCGAGGTTTCATCTCTARARCCHHRCTCACGCTTGTCARAGTGACAGCTGARCAGCCGATATTCGCT
     -390                  -360                  -330                    -300
GATTGGGCTCTTTCATGTTTGCGGGARCGTTCCATCTACCGGTTTAGCGATCGGCACGGGTTGCARCTAGARGCGGARGAGCTTGCGGGGAGGGGCACGGGGTGGTTTCCTGARGCGACTAGGTTGCCTGARACTATC
-270                 -240                  -210                    -180                    -150
ACGACTCATAGCTGCGAGCGCATGGTCCAGTATCARGAGTTTTGACGTCCTTTGATGARARATGCCCCTCTCTTGACGCTAGRARCCGAGGARATGGATCGCGAGCCGAGGCTCGATTTCGRAGCTTGGACGATGATA
-120                 -90                   -60                    -30                     1
GTTTCATCTGTTCARGCTTARATATCGTTGTTCCAGACCACTCGGGARCGGARCCAGACARCCACACATACCTTCACTTCATCARCATTCACTTCARCTCTTCGCCTCTTCCTTTTCRCTCTTTATCATCATCACC ATG
                                                                                                      MET
ARA TTC TTC GCT CTC ACC ACA CTT CTC GCC GCC ACG GCT TCG GCT CTG CCT ACT TCT ARC CCT GCC CAG GAG CTT GAG GCG CGC CAG CTT GGT AGA ACA ACT CGC
LYS PHE PHE ALA LEU THR THR LEU LEU ALA ALA THR ALA SER ALA LEU PRO THR SER ASN PRO ALA GLN GLU LEU GLU ALA ARG GLN LEU GLY ARG THR THR ARG
              10                           20                          30
GAC GAT CTG ACC AAC GGC ART AGC GCT TCC TGC GCC GAT GTC ATC TTC ATT TAT GCC CGA GGT TCA ACA GAG ACG GGC AAC TTG GTTCGTAGARTTTCTTCTCTCATGACAR
ASP ASP LEU THR ASN GLY ASN SER ALA SER CYS ALA ASP VAL ILE PHE ILE TYR ALA ARG GLY SER THR GLU THR GLY ASN LEU
              40                          50                          60
CATCACTTTTCTTACACATCCCATTAG  GGA ACT CTC GGT CCT AGC ATT GCC TCC AAC CTT GAG TCC GCC TTC GGC AAG GAC GGT GTC TGG ATT CAG GGC GTT GGC GGT GCC
                             GLY THR LEU GLY PRO SER ILE ALA SER ASN LEU GLU SER ALA PHE GLY LYS ASP GLY VAL TRP ILE GLN GLY VAL GLY GLY ALA
                                                  70                          80                          90
TAC GCA GCC ACT CTT GGA GAC ART GCT CTC CCT CGC GGA ACC TCT AGC GCC GCA ATC AGG GAG ATG CTC GGT CTC TTC CAG CAG GCC ARC ACC ARG TGC CCT GAC
TYR ALA ALA THR LEU GLY ASP ASN ALA LEU PRO ARG GLY THR SER SER ALA ALA ILE ARG GLU MET LEU GLY LEU PHE GLN GLN ALA ASN THR LYS CYS PRO ASP
              100                          110                          120
GCG ACT TTG ATC GCC GGT GGC TAC AGC CAG GGT GCT GCA CTT GCA GCC GCC TCC ATC GAG GAC CTC GAC TCG GCC ATT CGT GAC AAG ATC GCC GGA ACT GTT CTG
ALA THR LEU ILE ALA GLY GLY TYR SER GLN GLY ALA ALA LEU ALA ALA ALA SER ILE GLU ASP LEU ASP SER ALA ILE ARG ASP LYS ILE ALA GLY THR VAL LEU
              130                          140                          150                         160
TTC GGC TAC ACC AAG AAC CTA CAG AAC CGT GGC CGA ATC CCC AAC TAC CCT GCC GAC AGG ACC AAG GTC TTC TGC AAT ACA GGG GAT CTC GTT TGT ACT GGT AGC
PHE GLY TYR THR LYS ASN LEU GLN ASN ARG GLY ARG ILE PRO ASN TYR PRO ALA ASP ARG THR LYS VAL PHE CYS ASN THR GLY ASP LEU VAL CYS THR GLY SER
              170                          180                          190
TTG ATC GTT GCT GCA CCT CAC TTG GCT TAT GGT CCT GAT GCT CGT GGC CCT GCC CCT GAG TTC CTC ATC GAG GTT CGG GCT GTC CGT GGT TCT GCT TGA GGA
LEU ILE VAL ALA ALA PRO HIS LEU ALA TYR GLY PRO ASP ALA ARG GLY PRO ALA PRO GLU PHE LEU ILE GLU LYS VAL ARG ALA VAL ARG GLY SER ALA ***
              200                          210                          220                         230
GGATGAGARTTTTAGCAGGCGGGCCTGTTARTTATTGCGAGGTTTCARGTTTTTCTTTTGGTGRRTAGCCRTGATRGATTGG TTCARCACTCARTGTRCTRCARTGGCCCATAGTTTCARRTTARRGARGCRRTGART
900                 930                  960                    990                     1020
GGTGATCTACATATCGCTTTGCCCAGRARTCCCRRCCAGGCTTCCATACCCTGAGCCAGTTGAGCACARATTTCGTGCCLT CTGCTGAGCTTGCCAGGRRRGGTCGATACATRRACCGGCCTTGACAGACAGGGCGC
     1050                1080                 1110                   1140
TRCCTGCACGARTTGGTCCCGCCAGGTGTGCGCTCARGGCGRAGTTCGCCGRTTTATRGACCRCCTCTCRTTCCCRTCRTGCRCRTCTGTCCCTGRCTCGCCTTCTCCATCRRTARCACCGRGRTTGGTTACRRTCCR
     1170                1200                 1230                   1260                   1290
GGRTAGCTCGCGATCCCTCTTTGCTTGATCTCCGTGRRTACTCCTGCCRATCRTGCRCTAGCTTCRTCRAGCCRACARTGTTG11TTTCRGGCCGGCGTTCRACCTTTCCTCGATATCCCCRCGGGRGRCCTTGATGCG
               1320                 1350                   1380                    1410
GRCCATATCTCCCTCTCRRGATCRCGGRCAGGTTGGTTTTCCCRGTTGTTGGCCCGGGCTGTGGCTCGRRTATCCGCRRCTRGGTCGGRGTCRRACGTATGGTGGRTRGTCGRCRCGCRGTTCTGCRCCTTCCGTTGG
1440                1470                 1500                   1530                   1560
GTCTCRGCTGCRTTGCCTTTTTCGGGGTACRTGRRTCTCCGCTGCCRTTGCRGTRGRGGCGGTGRRRGCGCGGGCCTTCTTTTCRGGGRCGTAGCRRGCCTRRRCRTGCTAGCCTGRTGCCGTGRRGRRGRCCRGT
     1590                1620                 1650                   1680                   1710
TRGRGTGGTGGTRCCRTGCTGRCGRCRGGCRCCRRGRRTGCGRCRRRGRGCTGCRTTTGGRTGCTRRRRGRRGTTGTCTGGGRRGCRTRTGRCCCGRGTTGRRGRGGRGCCCRCGTGGCCTTTGCCGRCTTGGRGGRGRGT
     1740                1770                 1800
RRCGRTGGRCCGRRGGTRTGCCRTRCTTGTGRRRRRGCRRRCCCGRGRGTTRTGGGGTGTTTGGCCRRCTTCTCCTGRGGRRGRGGGRGRTC
```

Fig. 9. Nucleotide sequence of genomic DNA for cutinase from *Fusarium solani pisi*. [From C. L. Soliday and P. E. Kolattukudy (unpublished observations).]

B. Hydroxy Fatty Acids as Signals of Contact of Fungal Spores with Plant Surface

If the germinating fungal spores use cutinase to gain access into the plant, the spores need to sense the contact with plant surface so that expression of cutinase gene can be triggered. In fact, presence of cutin was found to be necessary for the induction of cutinase in the spores of *F. solani pisi* (Woloshuk and Kolattukudy, 1986). It was suspected that the small amount of cutinase carried by the spore would produce a small amount of cutin hydrolysate, which would be used to sense the presence of cutin. In accordance with this postulate, chemical hydrolysate prepared from cutin or cutin monomers isolated from the hydrolysate induced cutinase synthesis in the spores (Fig. 10). The most unique cutin monomers, 10,16-dihydroxy-C_{16} acid and 9,10,18-trihydroxy-C_{18} acid, were found to be the most potent inducers of cutinase synthesis.

Cutinase induction by the monomers was shown to involve a pretranslational level control. The appearance of extracellular cutinase activity triggered by the presence of cutin correlated with the appearance of immunologically measured cutinase, which began 30–45 min after the spores contacted cutin (Woloshuk and Kolatukudy, 1986). Dot–blot analysis using ^{32}P-labeled cDNA as the probe revealed that cutinase mRNA became measurable within 15 min after the spore contacted cutin and the level increased for at least 5 h, whereas no mRNA could be detected during such periods without cutin in the medium (Fig. 11). When the unique monomers of cutin were used instead of cutin, the level of mRNA reached a maximum within 1 h and then the level began to decrease. That this decline was caused by metabolic removal of the monomers was demonstrated with the use of exogenous labeled monomers. In any case, it is clear that the fungal spores use the di- and trihydroxy acids, that are found nowhere in nature except as constituents of plant cuticle, as the specific markers to sense the contact with the plant so that cutinase gene can be turned on only when needed.

The ability of the fungal isolates to produce cutinase was correlated with their ability to cause disease on intact plant organs (Kolattukudy et al., 1985). Spores from isolates which varied greatly in their ability to produce disease were tested for

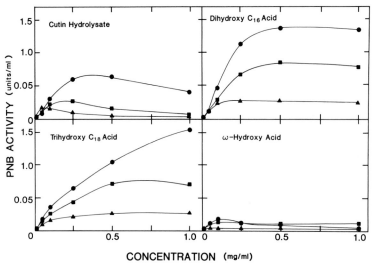

Fig. 10. Induction of cutinase in spore suspensions of *Fusarium solani pisi* by cutin hydrolysate and cutin monomers. Incubation times: 1 h (▲); 3 h (■); 6 h (●). PNB activity, hydrolysis of *p*-nitrophenyl butyrate. [From Woloshuk and Kolattukudy (1986).]

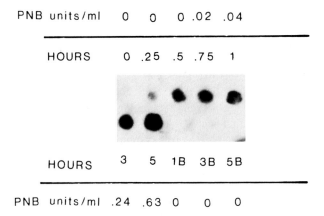

Fig. 11. Dot blots of RNA from the spores of *Fusarium solani* f. sp. *pisi* incubated with cutin for 0.25, 0.5, 0.75, and 1 h or without cutin for 1, 3, and 5 h. ^{32}P-labeled cutinase cDNA was used for the dot–blot assays. [From Kolattukudy *et al.* (1985).]

the inducibility of cutinase by cutin (C. P. Woloshuk and P. E. Kolattukudy, unpublished observations). Only the highly virulent ones showed cutinase induction (Fig. 12). Dot-blot analyses of cutinase transcripts also supported this conclusion. Thus cutinase inducibility in fungal spores might be a significant parameter that determines virulence on intact plant organs for the pathogens that use cuticular penetration as a mode of entry into plants.

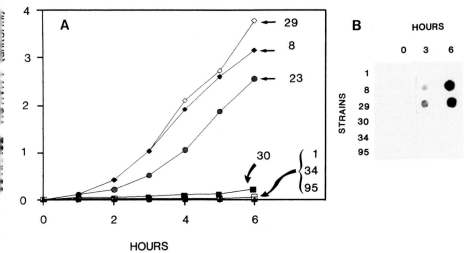

Fig. 12. Induction of cutinase in the spores of various isolates of *Fusarium solani pisi* by addition of cutin to the media. (A) Cutinase activity, measured spectrophotometrically; (B) dot blots of RNA isolated after various times of incubation of spores with cutin. ^{32}P-Labeled cDNA for cutinase was used as the probe. [(From C. P. Woloshuk and P. E. Kolattukudy, unpublished observations).]

C. Cutinase-Directed Antipenetrants for Plant Protection

Following the finding that *F. solani pisi* penetrating into its host secreted cutinase (Shaykh *et al.*, 1977), similar observations were made with *C. gloeosporioides* penetrating into papaya fruits (M. Dickman, C. L. Soliday, P. E. Kolattukudy, and S. Patil, unpublished observations). If cutinase is necessary for penetration, specific inhibition of cutinase might prevent penetration and therefore fungal infection of intact plant organs. The first such evidence wasobtained when it was demonstrated that inhibition of cutinase either with chemicals or with antibodies prevented infection of pea epicotyls (*P. sativum*) by *F. solani pisi* (Maiti and Kolattukudy, 1979). Similar experiments were subsequently performed to show that cutinase inhibition could also protect papaya fruits against *C. gloeosporioides* (Dickman *et al.*, 1982), apple seedlings against *Venturia inequalis*, and corn seedlings against *C. graminicola* (Kolattukudy, 1984b). Whether the protection observed under the controlled conditions of the bioassay can be achieved also under field conditions is not known. The only field test conducted so far was done on papaya fields in Hawaii. In this experiment biweekly spraying of a cutinase inhibitor for 11 weeks gave protection of papaya fruits against lesion formation by *C. gloeosporioides* (M. B. Dickman, W. Nishijima, S. S. Patil, and P. E. Kolattukudy, unpublished observations). Thus cutinase-directed antipenetrants can be used to protect plants against fungal attack.

D. Bacterial Cutinase

A culture of phyllospheric bacteria, which when sprayed on plants substituted for nitrogen fertilizers and increased yield (Patti and Chandra, 1981), was found to contain a cutinase-producing organism (Sebastian, *et al.,* 1987). This organism was identified as a fluorescent *Pseudomonas putida* and was found to be cohabiting with a nitrogen-fixing *Corynebacterium*. The two organisms together could grow on cutin as the sole carbon source without the addition of reduced N_2. The fluorescent *P. putida* was induced to generate an extracellular cutinase by the presence of cutin in the medium. This enzyme was purified to homogeneity and characterized (J. Sebastian and P. E. Kolattukudy, unpublished observations). It is a single peptide of 30 kDa and catalyzes cutin hydrolysis optimally at pH 8.5–10. Studies with inhibitors suggest that a catalytic triad involving active serine is involved in catalysis. This bacterial cutinase gene has been cloned and characterized.

ACKNOWLEDGMENTS

I thank Linda Rogers for assistance in preparing the manuscript. The work from the author's laboratory was supported by a U.S. Public Health Service Grant (GM 18278) and two National Science Foundation Grants (PCM-8007908 and PCM-8306835).

REFERENCES

Agrawal, V. P., and Kolattukudy, P. E. (1977). *Plant Physiol.* **59**, 667–672.
Agrawal, V. P., and Kolattukudy, P. E. (1978). *Arch. Biochem. Biophys.* **191**, 452–465.
Agrawal, V. P., and Stumpf, P. K. (1985). *Arch. Biochem. Biophys.* **240**, 154–165.
Agrawal, V. P., Lessire, R., and Stumpf, P. K. (1984). *Arch. Biochem. Biophys.* **230**, 580–589.
Bland, D. E., and Logan, A. F. (1965). *Biochem. J.* **95**, 515–520.
Bognar, A. L., Paliyath, G., Rogers, L., and Kolattukudy, P. E. (1984). *Arch. Biochem. Biophys.* **235**, 8–17.
Borchert, R. (1974). *Dev. Biol.* **36**, 391–399.
Borchert, R. (1978). *Plant Physiol.* **62**, 789–793.
Buckner, J. S., and Kolattukudy, P. E. (1972). *Biochem. Biophys. Res. Commun.* **46**, 801–807.
Buckner, J. S., and Kolattukudy, P. E. (1973). *Arch. Biochem. Biophys.* **156**, 34–45.
Cheesbrough, T. M., and Kolattukudy, P. E. (1984). *Proc. Natl. Acad. Sci. U.S.A.* **81**, 6613–6617.
Chiarlo, B., and Tacchino, E. (1965). *Riv. Ital. Sostanze. Grasse* **42**, 122–124.
Cottle, W., and Kolattukudy, P. E. (1982a). *Plant Physiol.* **69**, 393–399.
Cottle, W., and Kolattukudy, P. E. (1982b). *Plant Physiol.* **70**, 775–780.
Dean, B. B., and Kolattukudy, P. E. (1976). *Plant Physiol.* **58**, 411–416.
Dean, B. B., and Kolattukudy, P. E. (1977). *Plant Physiol.* **59**, 48–54.
Dickman, M. B., Patil, S. S., and Kolattukudy, P. E. (1982). *Physiol. Plant Pathol.* **20**, 333–347.
Domazetis, G., James, B. R., Tarpey, B., and Dolphin, D. (1981). *In* "Catalytic Activation of Carbon Monoxide" (P. C. Ford, ed.), ACS Symp. Ser., pp. 243–252. Am. Chem. Soc., Washington, D.C.
Doughty, D. H., McGuiggan, M. F., Wang, H., and Pignalet, L. H. (1979). *In* "Fundamental Research in Homogeneous Catalysis," (M. Tsutsui, ed.), Vol. 3, pp. 909–919. Plenum, New York.
Espelie, K. E., and Kolattukudy, P. E. (1985). *Arch. Biochem. Biophys.* **240**, 539–545.
Espelie, K. E., Sadek, N. Z., and Kolattukudy, P. E. (1980). *Planta* **148**, 468–476.
Espelie, K. E., Franceschi, V. R., and Kolattukudy, P. E. (1986). *Plant Physiol* **81**, 487–492.
Fernando, G., Zimmermann, W., and Kolattukudy, P. E. (1984). *Physiol. Plant Pathol.* **24**, 143–155.
Fleuriet, A., and Deloire, A. (1982). *Z. Pflanzenphysiol.* **107**, 259–268.
Fleuriet, A., and Macheix, J.-J. (1984). *Physiol. Plant.* **61**, 64–68.
Flurkey, W. H., and Kolattukudy, P. E. (1981). *Arch. Biochem. Biophys.* **212**, 154–161.
Foster, R. J., and Kolattukudy, P. E. (1987). *Int. J. Biochem.* (in press).
Griffith, M., Huner, N. P. A., Espelie, K. E., and Kolattukudy, P. E. (1985). *Protoplasma* **125**, 53–64.
Grisebach, H. (1981). *In* "The Biochemistry of Plants" (E. E. Conn, ed.), Vol. 7, pp. 301–316. Academic Press, New York.
Hammerschmidt, R. (1985). *Potato Res.* **28**, 123–127.
Higuchi, T. (1985). *In* "Biosynthesis and Biodegradation of Wood Components" (T. Higuchi, ed.), pp. 141–160. Academic Press, New York.
Khan, A. A., and Kolattukudy, P. E. (1974). *Biochem. Biophys. Res. Commun.* **61**, 1379–1386.
Kolattukudy, P. E. (1966). *Biochemistry* **5**, 2265–2275.
Kolattukudy, P. E. (1980a). *In* "The Biochemistry of Plants, Vol. 4, Lipids: Structure and Function" (P. K. Stumpf and E. E. Conn, eds.), pp. 571–645. Academic Press, New York.
Kolattukudy, P. E. (1980b). *Science* **208**, 990–1000.
Kolattukudy, P. E. (1984a). *Can. J. Bot.* **62**, 2918–2933.
Kolattukudy, P. E. (1984b). *In* "Structure, Function, and Biosynthesis of Plant Cell Walls" (W. M. Dugger and S. Bartnicki-Garcia, eds.), pp. 302–343. Waverly Press, Baltimore, Maryland.
Kolattukudy, P. E. (1985). *Annu. Rev. Phytopathol.* **23**, 223–250.
Kolattukudy, P. E., and Espelie, K. E. (1985). *In* "Biosynthesis and Biodegradation of Wood Components" (T. Higuchi, ed.), pp. 161–207. Academic Press, New York.
Kolattukudy, P. E., and Soliday, C. L. (1985). *In* "Cellular and Molecular Biology of Plant Stress," UCLA Symposia on Molecular and Cellular Biology, New Series, Vol. 22, pp. 381–400. Alan R. Liss, New York.

Kolattukudy, P. E., Espelie, K. E., and Soliday, C. L. (1981). *In* "Encyclopedia of Plant Physiology, New Series, Plant Carbohydrates" (F. A. Loewus and W. Tanner, eds.), Vol. 13B, pp. 225–254. Springer-Verlag, Berlin and New York.

Kolattukudy, P. E., Soliday, C. L., Woloshuk, C. P., and Crawford, M. (1985). *In* "Molecular Genetics of Filamentous Fungi," UCLA Sumposia on Molecular and Cellular Biology, New Series, Vol. 34, pp. 421–438. Alan R. Liss, New York.

Köller, W., and Kolattukudy, P. E. (1982). *Biochemistry* **21**, 3083–3090.

Lessire, R., Bessoule, J.-J., and Cassagne, C. (1985a). *FEBS Lett.* **187**, 314–320.

Lessire, R., Juguelin, H., Moreau, P., and Cassagne, C. (1985b). *Arch. Biochem. Biophys.* **239**, 260–269.

Lessire, R., Juguelin, H., Moreau, P., and Cassagne, C. (1985c). *Phytochemistry* **24**, 1187–1192.

Lin, T. S., and Kolattukudy, P. E. (1977). *Biochem. Biophys. Res. Commun.* **75**, 87–93.

Lin, T. S., and Kolattukudy, P. E. (1978). *J. Bacteriol.* **133**, 942–951.

Maiti, I. B., and Kolattukudy, P. E. (1979). *Science* **205**, 507–508.

Mueller, W. C., and Beckman, C. H. (1978). *Can. J. Bot.* **56**, 1579–1587.

Patti, B. R., and Chandra, A. K. (1981). *Plant Soil* **61**, 419–427.

Pozuelo, J. M., Espelie, K. E., and Kolattukudy, P. E. (1984). *Plant Physiol.* **74**, 256–260.

Purdy, R. E., and Kolattukudy, P. E. (1975a). *Biochemistry* **14**, 2824–2831.

Purdy, R. E., and Kolattukudy, P. E. (1975b). *Biochemistry* **14**, 2832–2840.

Rainwater, D. L., and Kolattukudy, P. E. (1983). *J. Biol. Chem.* **258**, 2979–2985.

Rainwater, D. L., and Kolattukudy, P. E. (1985). *J. Biol. Chem.* **260**, 616–623.

Rowe, J. W., and Scroggins, J. H. (1964). *J. Org. Chem.* **29**, 1554–1562.

Ryser, U., Meier, H., and Halloway, P. J. (1983). *Protoplasma* **117**, 196–205.

Sebastian, J., Chandra, A. K., Kolattukudy, P. E. (1987). *J. Bacteriol.* **169**, 131–136.

Shaykh, M., Soliday, C., and Kolattukudy, P. E. (1977). *Plant Physiol.* **60**, 170–172.

Sijmons, P. C., Kolattukudy, P. E., and Bienfait, H. F. (1985). *Plant Physiol.* **78**, 115–120.

Smith, M. M., and O'Brien, T. P. (1979). *Aust. J. Plant Physiol.* **6**, 201–219.

Soliday C. L., and Kolattukudy, P. E. (1983). *Biochem. Biophys. Res. Commun.* **114**, 1017–1022.

Soliday, C. L., Dean, B. B., and Kolattukudy, P. E. (1978). *Plant Physiol.* **61**, 170–174.

Soliday, C. L., Kolattukudy, P. E., and Davis, R. W. (1979). *Planta* **146**, 607–614.

Soliday, C. L., Flurkey, W. H., Okita, T. W., and Kolattukudy, P. E. (1984). *Proc. Natl. Acad. Sci. U.S.A.* **81**, 3939–3943.

Woloshuk, C. P., and Kolattukudy, P. E. (1986). *Proc. Natl. Acad. Sci., U.S.A.* **83**, 1704–1708.

Yatsu, L. Y., Espelie, K. E., and Kolattukudy, P. E. (1983). *Plant Physiol.* **73**, 521–524.

Lipids of Blue-Green Algae (Cyanobacteria)

12

N. MURATA
I. NISHIDA

I. INTRODUCTION

The blue-green algae (or cyanobacteria) constitute a large group within the prokaryotic kingdom (Stanier and Cohen-Bazire, 1977). They are, however, similar to eukaryotic plants and algae in performing oxygenic photosynthesis with two types of photochemical reactions, and having chlorophyll *a* as the major photosynthetic pigment. The blue-green algae are classified as Gram-negative bacteria, and their cell envelope is composed of the outer membrane and plasma (inner) membrane separated by a peptidoglycan layer. In addition, they have intracellular photosynthetic membranes, that is, the thylakoid membranes. Their membrane structure is similar to that of eukaryotic plant chloroplast, which contains outer and inner envelope membranes, surrounding the thylakoid membranes.

The thylakoid membranes of the blue-green algae contain chlorophyll a and β-carotene (Omata and Murata, 1983, 1984a), and contain phycobilisomes attached to their outer surface (Gantt and Conti, 1969). They are the site of both photosynthetic and respiratory electron transport (Omata and Murata, 1984b, 1985). The plasma membrane and outer membrane, on the other hand, contain xanthophylls, but little or no chlorophyll a or β-carotene (Murata et al., 1981; Omata and Murata, 1983, 1984a; Resch and Gibson, 1983; Jürgens and Weckesser, 1985), and phycobilisomes are never seen attached to these membranes (Stanier and Cohen-Bazire, 1977).

The lipids are found only in the membranes within blue-green algae. The thylakoid and plasma membranes contain glyceroglycolipids and phosphatidylglycerol (Section II), whereas the outer membrane contains lipopolysaccharides and hydrocarbons in addition to the glycerolipids (Section V). The biosynthetic pathway of glyceroglycolipids in the blue-green algae differs from that in higher plants and green algae (Section III).

When some nitrogen-fixing species of filamentous blue-green algae are grown in the absence of combined nitrogen sources, certain vegetative cells undergo characteristic morphological changes to give rise to heterocysts (Haselkorn, 1978). Prominent among these morphological changes is the development of a refractile multilayered heterocyst envelope. This specially differentiated heterocyst envelope contains unique glycolipids (Section V).

Changes in the lipid composition, membrane fluidity, and physiological activities induced by ambient temperature changes have been extensively studied in the blue-green algae. Thus, these algae can be regarded as model systems for temperature-induced alterations in eukaryotic plants.

Previously published reviews related to this chapter include those by Nichols (1973), and Quinn and Williams (1978, 1983).

II. GLYCEROLIPIDS: DISTRIBUTION

A.   Lipid Classes

The blue-green algae contain four major glycerolipids: monogalactosyldiacylglycerol (MGDG), digalactosyldiacylglycerol (DGDG), sulfoquinovosyl diacylglycerol (SQDG), and phosphatidylglycerol (PG) (Nichols et al., 1965; Hirayama, 1967; Nichols and Wood, 1968a; Fogg et al., 1973; Wolk, 1973; Stanier and Cohen-Bazire, 1977), and a minor component, monoglucosyldiacylglycerol

(GlcDG) (Feige *et al.*, 1980; Sato and Murata, 1982a). The molecular structures of these lipids are shown in Fig. 1. Although Zepke *et al.* (1978) reported that *Tolypothrix tenuis* and *Oscillatoria chalybea* contain another glycolipid, trigalactosyldiacylglycerol, this lipid may have been artificially produced, as is the case in envelope membranes of higher plant chloroplasts (Block *et al.*, 1983). The blue-green algae do not contain phosphatidylcholine (PC), phosphatidylinositol (PI), phosphatidylethanolamine, diphosphatidylglycerol, or phosphatidylserine (Nichols *et al.*, 1965; Hirayama, 1967; Nichols and Wood, 1968a; Appleby *et al.*, 1971). By comparison, the chloroplasts of higher plants contain MGDG, DGDG, SQDG, and PG as major components, and PC and PI as minor components (Douce *et al.*, 1973; Mackender and Leech, 1974; Block *et al.*, 1983). The glycerolipid composition of *Anabaena variabilis* is listed in Table I. MGDG represents ~ 50% of the total glycerolipid content (see also Zepke *et al.*, 1978). The GlcDG content apparently does not exceed 1% (Feige *et al.*, 1980).

Feige (1978) first discovered a glycerolipid (termed X-MGD and later identified as GlcDG), which migrated a little faster than MGDG in thin-layer chromatography on silica gel, when he labeled lichens containing blue-green algae with $H^{14}CO_3^-$. Feige *et al.* (1980) surveyed 30 species of blue-green algae, and

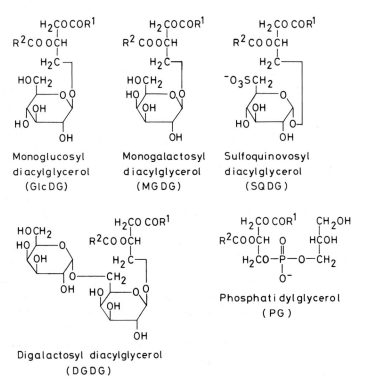

Fig. 1. Glycerolipids of blue-green algae.

TABLE I
Glycerolipid Composition of Blue-Green Algae and *Prochloron* sp.

Organism[a]	Lipid				
	GlcDG	MGDG	DGDG	SQDG	PG
Anabaena variabilis (1)	1	54	17	11	17
Prochloron sp. (2)	3	55	11	26	5
Anacystis nidulans					
Intact cells (3)	nd[b]	57	11	11	21
Thylakoid membranes (4)	nd	55	12	11	22
Plasma membranes (4)	nd	63	12	7	18
Outer membranes (5)	nd	53	24	5	18

[a]References: 1. Sato and Murata (1982a); 2. Murata and Sato (1983); 3. Murata *et al.* (1979); 4. Omata and Murata (1983); 5. Murata *et al.* (1981).
[b]nd, Not determined.

concluded that this lipid is ubiquitously present in these algae but not in eukaryotic algae. The lipid was identified as β-glucosyl diacylglycerol by mass spectrometry of its acetylated derivative, enzymatic hydrolysis with β-glucosidase of the deacylated derivative, and peroxide oxidation of the lipid (Feige *et al.*, 1980). This structure was substantiated by mass-spectrometric analysis of its trimethylsilylated derivative and stoichiometric determination of the glycerol, acyl groups, and glucose unit (Sato and Murata, 1982a). As will be described in Section III, the GlcDG is a precursor in the biosynthesis of MGDG in the blue-green algae. The GlcDG is found also in the prokaryotic green alga, *Prochloron* (Murata and Sato, 1983).

Table I shows a comparison of the polar lipid composition of the thylakoid membranes, plasma membrane, and outer membrane of *Anacystis nidulans*, although the GlcDG content has not been determined. No great differences in lipid composition exist among the three types of membranes from this alga. The lipid content relative to the total dry weight is 19%, 57%, and 3% in the thylakoid, plasma, and outer membranes, respectively (Murata *et al.*, 1981; Omata and Murata, 1983).

The lipid composition of specially differentiated cells of the blue-green algae is unique. Heterocysts isolated from *Anabaena cylindrica* contain the same acyl glycerolipids in about the same proportion as occurs in vegetative cells (Winkenbach *et al.*, 1972), and in addition, unique non-glycero glycolipids (Walsby and Nichols, 1969; Bryce *et al.*, 1972; Lambein and Wolk, 1973). Details of the heterocyst glycolipids will be described in Section V. Reddy (1983) has reported that the akinetes (Nichols and Adams, 1982) from *Anabaena fertilissima* and *Anabaenopsis arnoldii* contained MGDG and SQDG as well as the heterocyst glycolipids, but neither DGDG nor PG.

B. Fatty Acids

The fatty acid composition of glycerolipids from blue-green algae was studied first by Holton *et al*. (1964) in *Anacystis nidulans*, and by Levin *et al*. (1964) in *Anabaena variabilis*. Subsequent analyses have characterized the fatty acids from a number of blue-green algae. The fatty acids thus far known to be present in blue-green algae are hexadecanoic acid or palmitic acid (16:0), Δ^9-*cis*-hexadecenoic acid or palmitoleic acid (16:1), hexadecadienoic acid (double-bond positions undetermined) (16:2), octadecanoic acid or stearic acid (18:0), Δ^9-*cis*-octadecenoic acid or oleic acid (18:1), $\Delta^{9,12}$-*cis*-octadecadienoic acid or linoleic acid (18:2), $\Delta^{9,12,15}$-*cis*-octadecatrienoic acid or α-linolenic acid (α-18:3), $\Delta^{6,9,12}$-*cis*-octadecatrienoic acid or γ-linolenic acid (γ-18:3), and octadecatetraenoic acid (double-bond positions undetermined) (18:4) (see Table II). Some species, such as *A. nidulans* (Allen *et al*., 1966; Sato *et al*., 1979), *Nostoc muscorum* (Han *et al*., 1969), and *Synechococcus* sp. of marine origin (Goodloe and Light, 1982), contain Δ^{11}-*cis*-octadecenoic acid (*cis*-vaccenic acid). There are no branched-chain fatty acids (Parker *et al*. 1967) or Δ^3-*trans*-hexadecenoic acid (Δ^3-*trans*-16:1) (Nichols *et al*., 1965; Allen *et al*., 1966; Nichols and Wood, 1968a). The latter is a ubiquitous fatty acid in PG from higher plant chloroplasts (Weenink and Shorland, 1964).

According to Kenyon (1972) and Kenyon *et al*. (1972), the blue-green algae can be classified into four groups with respect to the composition and metabolism of fatty acids. Strains in the first group contain only saturated and monounsaturated fatty acids, whereas those in the other groups contain, in addition, 18:2, as well as polyunsaturated fatty acids characteristic of each group. The second group is characterized by the presence of α-18:3, the third group by γ-18:3, and the fourth group by 18:4. The fatty acid composition of some strains from each of the four groups is listed in Table II. Unicellular strains of blue-green algae belong to the first and third groups (Kenyon, 1972). Filamentous blue-green algae are distributed among all four groups (Kenyon *et al*., 1972; Oren *et al*., 1985). The prokaryotic green alga *Prochloron* contains only saturated and monounsaturated fatty acids, as is the case in blue-green algae of the first group (Perry *et al*., 1978; Johns *et al*., 1981; Murata and Sato, 1983; Kenrick *et al*., 1984). On the other hand, fatty acids in chloroplasts of most higher plants correspond to those of the second group of blue-green algae. Chloroplasts of some higher plants also contain 18:4 (Sastry, 1974), and in this respect would correspond to the fourth group.

The fatty acid composition of individual lipids from *Anabaena variabilis* and *Anacystis nidulans* was first studied by Nichols *et al*. (1965). These and many subsequent studies have indicated that the fatty acid composition of MGDG is similar to that of DGDG, and the fatty acid composition of SQDG is similar to that of PG in each alga. In *A. nidulans* (Nichols *et al*., 1965; Hirayama, 1967; Sato *et al*., 1979) and *Synechococcus lividus* (Fork *et al*., 1979), both first-group species, the ratio of the monounsaturated to saturated acids is higher in MGDG and DGDG than in SQDG and PG. Among filamentous strains, MGDG and DGDG contain

TABLE II
Fatty Acid Composition of the Total Lipids of Blue-Green Algae

Group	Strain[a]	Growth temp. (°C)	Fatty acid										
			14:0	16:0	18:0	14:1	16:1	18:1	16:2	18:2	α-18:3	γ-18:3	18:4
1	*Anacystis nidulans* (1)	28	1	46	1	3	46	3[c]	0	0	0	0	0
	Synechococcus lividus (2)	55	0	54	22	0	10	14	0	0	0	0	0
	Prochloron sp. (3)	—	12	38	1	17	24	3[c]	0	0	0	0	0
2	*Plectonema terebrans* (4)	30	1	35	3	1	13	20	5	11	6	0	0
	Nostoc muscorum (5)	26–30	2	32	2	2	15	7[c]	0	10	21	0	0
	Anabaena variabilis (1)	28	0	32	1	0	22	11	1	17	16	0	0
3	*Synechocystis 6714* (6)	33	Trace	28	1	1	4	5	0	17	0	31	0
	Spirulina platensis (7)	—	0	44	3	0	10	5	Trace	13	Trace	22	0
4	*Tolypothrix tenuis*[b] (8)	25	1	22	3	2	3	16	0	15	6	13	11
	Phormidium sp. (8)	25	1	18	3	Trace	9	21	0	32	1	0	2

[a] References: 1. Sato *et al.* (1979); 2. Fork *et al.* (1979); 3. Johns *et al.* (1981); 4. Parker *et al.* (1967); 5. Holton *et al.* (1964); 6. Kenyon (1972); 7. Nichols and Wood (1968a); 8. Kenyon *et al.* (1972).
[b] A different result was obtained by Zepke *et al.* (1978).
[c] Mixture of Δ^9 and Δ^{11}.

greater proportions of polyunsaturated acids than SQDG and PG. This was demonstrated in *A. variabilis* (Nichols *et al.*, 1965; Sato *et al.*, 1979), *Anabaena cylindrica* (Zepke *et al.*, 1978), *Nostoc calcicola* (Zepke *et al.*, 1978), and *0. chalybea* (Zepke *et al.*, 1978) in the second group, *Spirulina platensis* (Nichols and Wood, 1968a) in the third group, and *T. tenuis* (Zepke *et al.*, 1978) in the fourth group. Table III shows the fatty acid composition of five individual lipids from *A. variabilis*. The C_{16} and C_{18} acids each represent ∼ 50% of the total fatty acids. MGDG and DGDG contain both 16:0 and 16:1 and a small proportion of 16:2, whereas in GlcDG, SQDG and PG, 16:0 accounts for more than 90% of the total C_{16} content. In contrast, the composition of the C_{18} acids is similar in the four major lipids. As will be described in Section IV, the fatty acid composition depends markedly on growth temperature.

In the chloroplasts of higher plants, in contrast to blue-green algae, the fatty acid composition of each individual lipid is unique (Douce *et al.*, 1973; Mackender and Leech, 1974). More than 95% of the fatty acid content of MGDG is α-18:3 plus $\Delta^{7,10,13}$-*cis*-hexadecatrienoic acid (16:3), while the major fatty acids in DGDG are 16:0 and α-18:3. SQDG contains 16:0 and α-18:3, whereas PG contains Δ^3-*trans*-16:1 in addition to 16:0 and α-18:3. PC and PI contain 16:0, 18:1, 18:2 and α-18:3.

Distribution of specific fatty acids between the C-1 and C-2 positions of the *sn*-glycerol moiety can be determined by selective enzymatic hydrolysis of the ester linkage at the C-1 position with *Rhizopus* lipase (Fischer *et al.*, 1973). Two types of the positional distribution are known. In *Anacystis nidulans*, most of the monounsaturated fatty acids, 14:1, 16:1 and 18:1, are esterified at the C-1 position and most of 16:0 is at the C-2 position in all lipid classes (Sato *et al.*, 1979). In

TABLE III
Fatty Acid Composition of Glycerolipids from *Anabaena variabilis*[a]

Growth temp. (°C)	Lipid	Fatty acid						
		16:0	16:1	16:2	18:0	18:1	18:2	α-18:3
38	GlcDG	51	2	0	12	30	5	0
	MGDG	26	25	1	0	19	28	0
	DGDG	22	29	1	0	14	33	1
	SQDG	56	2	0	5	24	13	0
	PG	51	3	0	2	25	19	0
22	GlcDG	51	3	0	11	20	12	3
	MGDG	26	20	6	0	2	6	40
	DGDG	21	24	7	0	1	7	40
	SQDG	53	4	0	1	5	8	29
	PG	54	2	0	0	3	8	32

[a]Sato and Murata (1982b).

the filamentous blue-green algae such as *Anabaena variabilis* (Sato *et al.*, 1979), *Anabaena cylindrica*, *O. chalybea*, *N. calcicola*, and *T. tenuis* (Zepke *et al.*, 1978), which contain approximately the same amounts of C_{16} and C_{18} acids, the C_{18} acids are esterified to the C-1 position, and the C_{16} acids to the C-2 position in all lipid classes (see also Table V). In the chloroplasts of higher plants (Auling *et al.*, 1971; Tulloch *et al.*, 1973) and in green algae (Safford and Nichols, 1970), the positional distribution of fatty acids in MGDG, SQDG, and PG is similar to that of the filamentous blue-green algae. However, chloroplast fatty acids are not as strictly distributed with respect to chain length as in the blue-green algae.

The fatty acid composition of individual lipids in the thylakoid membranes, plasma membrane, and outer membrane prepared from *Anacystis nidulans* is compared in Table IV. A characteristic distribution is seen in MGDG, DGDG, SQDG, and PG, regardless of the types of membrane from which they are derived (Murata *et al.*, 1981; Omata and Murata, 1983).

C. Molecular Species

Molecular species of glycerolipids are characterized by the combination of the acyl groups at the C-1 and C-2 positions, and by the polar substituent group at the C-3 position of *sn*-glycerol. Molecular species have been characterized in those

TABLE IV

Fatty Acid Composition of Lipids from the Thylakoid, Plasma, and Outer Membranes of *Anacystis nidulans* Grown at 28°C[a]

Membrane[b]	Lipid	Fatty acid						
		14:0	14:1	16:0	16:1	17:0	18:0	18:1[c]
TM	MGDG	2	3	45	46	0	1	3
	DGDG	2	2	48	41	1	2	4
	SQDG	2	1	56	38	1	0	2
	PG	2	0	53	40	0	0	5
PM	MGDG	1	3	48	44	0	1	3
	DGDG	2	2	51	41	0	1	3
	SGDG	2	1	58	35	0	1	3
	PG	0	0	54	41	0	0	5
OM[d]	MGDG	2	5	45	46	0	1	1
	DGDG	4	10	42	42	0	1	1
	SQDG	2	7	52	36	1	2	1
	PG	3	3	49	40	1	2	3

[a]Omata and Murata (1983) and Murata *et al.* (1981).

[b]TM, Thylakoid membranes; PM, plasma membranes; OM, outer membranes.

[c]Mixture of Δ^9 and Δ^{11}.

[d]The cell envelope fraction in Murata *et al.* (1981) is regarded as the outer membranes.

species which contain an *sn*-1-C_{18}/2-C_{16}-glycerol moiety (these are represented as C_{18}/C_{16} species) (Zepke *et al.*, 1978; Sato and Murata, 1980a).

The most abundant molecular species in all the lipid classes of *Anabaena cylindrica*, *O. chalybea*, and *N. calcicola* is α-18:3/16:0, and that in *T. tenuis* is 18:4/16:0 (Zepke *et al.*, 1978). The molecular species of the five glycerolipids from *Anabaena variabilis* are shown in Table V. GlcDG is unique in containing a high proportion of 18:0/16:0 and 18:1/16:0. MGDG and DGDG contain highly unsaturated species such as 18:2/16:1, α-18:3/16:1 and α-18:3/16:2, whereas the other lipids are devoid of these molecular species. As will be described in Section IV, the molecular species present depends on the growth temperature.

In chloroplasts of spinach (Nishihara *et al.*, 1980), in contrast, the major molecular species are α-18:3/α-18:3 and α-18:3/16:3 in MGDG, while they are α-18:3/α-18:3 and 16:0/α-18:3 in DGDG. The most abundant molecular species in SQDG and PG are 16:0/α-18:3 and α-18:3/Δ^3-*trans*-16:1, respectively.

III. BIOSYNTHESIS OF GLYCEROLIPIDS

A. Lipid Classes

As mentioned in the previous section, the blue-green algae contain MGDG, DGDG, SQDG, and PG as major components and GlcDG as a minor component. Nichols (1968) labeled *Anabaena cylindrica* and *Anacystis nidulans* with

TABLE V
Molecular Species Composition of Glycerolipids in *Anabaena variabilis*[a]

Growth temp. (°C)	Lipid	C-1 C-2	Molecular species							
			18:0 16:0	18:1 16:0	18:2 16:0	α-18:3 16:0	18:1 16:1	18:2 16:1	α-18:3 16:1	α-18:3 16:2
38	GlcDG		24	60	10	0	0	0	0	0
	MGDG		1	25	23	1	11	35	0	0
	DGDG		1	16	24	0	16	38	0	0
	PG		1	56	41	0	0	0	0	0
	SQDG[b]		10	48	26	0	0	0	0	0
22	GlcDG		22	40	24	6	0	0	0	0
	MGDG		2	2	12	34	0	3	32	12
	DGDG		1	4	20	19	1	9	37	4
	PG		0	10	26	61	0	0	0	0
	SQDG[b]		2	10	16	58	0	0	0	0

[a]Sato and Murata (1982b).
[b]Estimated from the positional distribution of fatty acids.

[^{14}C]acetate and noted that the radioactivity was incorporated into all the major lipid classes; the ^{14}C distribution remained essentially constant during incubation periods for 45 min to 24 h. Feige *et al.* (1980) conducted $H^{14}CO_3$ pulse-labeling experiments in 30 species of blue-green algae, and observed that ^{14}C is first incorporated into GlcDG, then radioactivity appears in MGDG. They proposed that GlcDG is a precursor of MGDG in the biosynthesis of glycolipids in the blue-green algae.

Sato and Murata (1982a) showed by pulse-labeling experiments that in *Anabaena variabilis* the conversion from GlcDG to MGDG results from epimerization of glucose to galactose—that is, stereochemical isomerization at the C-4 atom of the glucose unit—but not by replacement of glucose by galactose. They further demonstrated that DGDG is produced by transfer of newly synthesized galactose to MGDG (Sato and Murata, 1982a). SQDG and PG are also rapidly labeled, suggesting that these lipids are directly synthesized, but not via GlcDG (Sato and Murata, 1982a).

Sato and Murata (1982c) also observed in *Anabaena variabilis* a membrane-associated activity of UDP-glucose:diacylglycerol glucosyltransferase which synthesizes GlcDG by transfer of glucose from UDP-glucose to diacylglycerol. The GlcDG synthesis activity seems to be located in both the thylakoid and plasma membranes of *Anacystis nidulans* (Omata and Murata, 1986). Interestingly, in higher plant chloroplasts the activity of galactolipid synthesis is confined to the envelope membranes (Douce and Joyard, 1979). However, in etioplasts of wheat seedlings the galactosyltransferase activity is detected in both envelope and prothylakoid membranes (Sandelius and Selstam, 1984).

The activities of glycerol-3-phosphate acyltransferase and 1-acylglycerol-3-phosphate acyltransferase, the enzymes involved in the first steps of glycerolipid synthesis, have been characterized in crude extracts of *A. variabilis* (Lem and Stumpf, 1984b). These enzymes from *Anabaena variabilis* use acyl-[acyl-carrier protein] (acyl-ACP), but not acyl-coenzyme A (acyl-CoA) as the substrate. These observations indicate that the precursor of the glycerolipids is phosphatidic acid, which is converted to glyco-, sulfo-, and phospholipids in this blue-green alga. Interestingly, the crude extract of *A. variabilis* is active in transferring 18:1 from 18:1-ACP to glycerol-3-phosphate, although, in contrast to the case of higher plants, the 18:1 is not synthesized in the ACP-bound form in this alga (Lem and Stumpf, 1984b).

A tentative scheme for glycerolipid biosynthesis, based on limited information, is proposed in Fig. 2.

B. Fatty Acids

Fatty acid biosynthesis has been extensively studied in *Anabaena variabilis*, which contains polyunsaturated fatty acids. After a 6-h labeling of the algal cells with [2-^{14}C]acetate, Appleby *et al.* (1971) observed that radioactivity was incorporated into 18:0 (5%), 18:1 (52%), 18:2 (13%), 16:0 (25%), and 16:1 (6%). Externally added [1-^{14}C]18:1 was also incorporated into the glycerolipids, and a

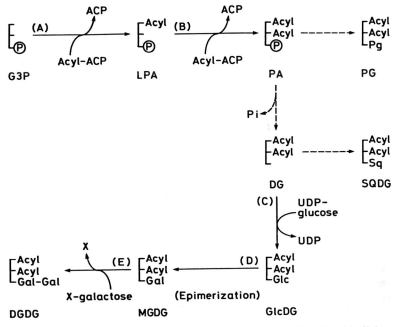

Fig. 2. A scheme for glycerolipid biosynthesis in blue-green algae. Reactions (A–C) have been demonstrated *in vitro* in *Anabaena variabilis* (Stapleton and Jaworski, 1984a,b; Lem and Stumpf, 1984b; Sato and Murata, 1982c), reaction (C) *in vitro* in *Anacystis nidulans* (Omata and Murata, 1986), and reactions (D) and (E) *in vivo* in *A. variabilis* (Sato and Murata, 1982a). The reactions indicated by broken arrows have not yet been experimentally demonstrated. G3P, Glycerol 3-phosphate; LPA, 1-acyl-glycerol 3-phosphate or lysophosphatidic acid; PA, phosphatidic acid; DG, **1,2**-diacylglycerol; ACP, acyl-carrier protein; Acyl, fatty acid; Pg, phosphoglycerol; Sq, sulfoquinovose; Glc, glucose; Gal, galactose; X, unidentified galactose carrier.

small proportion of the 18:1 was also converted to 18:2 (Appleby *et al.*, 1971). Sato and Murata (1982b) showed, by pulse-labeling for 6 min with $H^{14}CO_3^-$, that radioactivity is incorporated into 16:0, 18:0 and 18:1 of GlcDG, SQDG, and PG. An extension of labeling time up to 1 h resulted in an increase in radioactivity of 18:1 relative to that of the total fatty acids and corresponding decrease in that of 18:0. These changes in radioactivity of the 18:0 and 18:1 proceeded further during a subsequent chase for 10 h. It is suggested that, in *A. variabilis*, only saturated fatty acids (i.e., 16:0 and 18:0) are first esterified to the lipids and then desaturated while they are bound to the glycerol moiety. The lipid-linked desaturation of 16:0 to 16:1 in MGDG was confirmed by a combination of $H^{13}CO_3^-$ feeding and mass spectrometric analysis of the 2-acylglycerol moiety of ^{13}C-enriched MGDG (Sato *et al.*, 1986).

Fatty acid biosynthesis was studied *in vivo* in *Anacystis nidulans*, which contains 16:0 and 16:1 as major components, but no polyunsaturated fatty acids (Nichols, 1968). Upon labeling with [^{14}C]acetate for 45 min, radioactivity in MGDG and

SQDG was detected only in 14:0 and 16:0 but little or none in 14:1 and 16:1. After a 6-h labeling, a considerable proportion of radioactivity was distributed in 14:1 and 16:1, suggesting that in *A. nidulans* also saturated acids are esterified into glycerolipids and then desaturated.

The lack of fatty acid desaturation in the form of acyl-ACP or acyl-CoA was also suggested by *in vitro* experiments of Lem and Stumpf (1984a), and Stapleton and Jaworski (1984b), who studied fatty acid synthesis in crude extracts from *Anabaena variabilis*. Under optimal conditions for fatty acid synthesis, only saturated fatty acids, 14:0, 16:0 and 18:0, were produced from [^{14}C]acetate. Neither added 18:0-ACP nor 18:0-CoA was desaturated. Similar results were obtained in crude extracts of *Anabaena cylindrica* (Al'Araji and Walton, 1980). All of these observations suggest that saturated acyl thioesters are the final products in fatty acid synthesis, and that all the desaturation reactions take place after the fatty acids are esterified to the glycerol moiety of the lipids.

The biochemical characteristics of the enzymes involved in fatty acid synthesis have been studied in crude extracts of *Anabaena variabilis* by Lem and Stumpf (1984a), and Stapleton and Jaworski (1984a,b). The fatty acid synthetase in the crude extracts was found to be of a non-associated and ACP-dependent type, and therefore similar to the prokaryotic (or plant) type found in most bacteria and the chloroplasts of eukaryotic plants. Stapleton and Jaworski (1984a) purified malonyl-CoA:ACP transacylase from *A. variabilis* and found that it is similar to that from spinach in molecular mass, isoelectric point, thermal stability, pH optimum, and various kinetic parameters. Fatty acid elongation from 14:0 to 16:0 and from 16:0 to 18:0 which has been demonstrated in the crude cell extracts requires the ACP and NADPH, but not CoA or NADH (Lem and Stumpf, 1984a; Stapleton and Jaworski, 1984b).

Some species, such as *Anacystis nidulans*, contain high proportions of Δ^{11}-*cis*-octadecenoic acid. This fatty acid is known to be produced by an anaerobic pathway in bacteria, resulting in only saturated and monounsaturated fatty acids like those of the first group of blue-green algae (see Section II,B). It is uncertain if these species of blue-green algae synthesize the Δ^{11}-*cis*-octadecenoic acid as the acyl-ACP form and then esterify it to the glycerol of glycerolipids. However, the glycerolipids primarily synthesized by *A. nidulans* seem to contain only saturated fatty acids, and their desaturation to produce Δ^{11}-*cis*-octadecenoic acid seems to take place after they are esterified to the glycerol moiety of the lipids (Nichols, 1968).

C. Molecular Species

The biosynthesis of lipid molecular species was studied in *Anabaena variabilis* by pulse-labeling with $H^{14}CO_3^-$ and a subsequent chase (Sato and Murata, 1982b; Murata and Sato, 1982). The primary products of lipid synthesis were 18:0/16:0-GlcDG, 18:0/16:0-PG and 18:0/16:0-SQDG. They were converted to unsaturated molecular species by desaturation of the fatty acids. In GlcDG, the desaturation of fatty acids and the epimerization of glucose to galactose seem to take place independently. The pathways of biosynthesis of molecular species of GlcDG, MGDG, SQDG, and PG in *A. variabilis* are shown in Fig. 3. DGDG is produced by transfer of galactose from an unidentified galactose carrier to MGDG, and apparently no desaturation of fatty acids takes place in DGDG (Sato and Murata, 1982b).

IV. CHANGES IN GLYCEROLIPIDS IN RESPONSE TO ENVIRONMENT

A. Temperature

1. Isothermal Growth

Holton *et al.* (1964) first studied the effect of growth temperature on the fatty acid composition of the total lipids in *Anacystis nidulans*. They found that with a lowering of growth temperature, 16:0 decreases and 16:1 increases, and that the average chain length of fatty acids decreases. Similar changes in fatty acid composition with growth temperature were observed in *Synechococcus cedrorum* (Sherman, 1978). Sato *et al.* (1979) and Murata *et al.* (1979) found that in *A.*

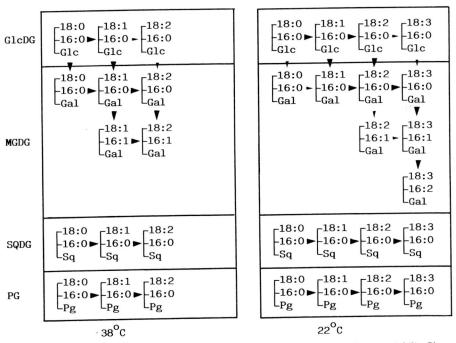

Fig. 3. A scheme for the biosynthesis of glycerolipid molecular species in *Anabaena variabilis*. Glc, Glucose; Gal, galactose; Pg, phosphoglycerol; Sq, sulfoquinovose; 18:3, α-linolenic acid (α-18:3).

nidulans the mode of changes with growth temperature in the fatty acid composition of all the major lipid classes is similar to that of the total lipids. They also demonstrated that following a decrease in growth temperature, the average chain length of monounsaturated acids is reduced at the C-1 positions of all the major lipid classes, and the amount of 16:1 increased and that of 16:0 concomitantly decreased at the C-2 positions of MGDG and DGDG. A decrease in growth temperature of a thermophilic unicellular blue-green alga, *Synechococcus lividus*, induced a decrease of 16:0 and 18:0 and an increase in 16:1 in MGDG and DGDG, whereas 18:0 decreased and 16:1 and 18:1 increased in PG and SQDG (Fork *et al.*, 1979).

In contrast to the unicellular blue-green algae in the first group (see Section II,B), *Anabaena variabilis* in the second group contains polyunsaturated fatty acids and responds to changes in growth temperature differently (see Table III). When the growth temperature is lowered, 18:1 and 18:2 decrease and α-18:3 increases at the C-1 position of all the major lipid classes, while the levels of C_{16} acids remain nearly constant except for a minor decrease in 16:1 and increase in 16:2 at the C-2 positions of MGDG and DGDG (Sato *et al.*, 1979; Sato and Murata, 1980b). The temperature dependence of the molecular species composition in *A. variabilis* (Table V) indicates that a decrease in growth temperature from 38° to 22°C induces decreases in 18:1/16:0 and 18:2/16:0 species and increases in α-18:3/ 16:0 in all the major lipid classes, and decreases in 18:1/16:1 and 18:2/16:0 and increases in α-18:3/16:1 and α-18:3/16:2 in MGDG and DGDG. Fatty acid analysis of the total lipid fraction of *Spirulina platensis* from the third group (Baasch *et al.*, 1984) demonstrates that the amounts of 16:0 and 18:2 decrease while 16:1 and γ-18:3 increase when the temperature is lowered from 32° to 22°C.

An increase in unsaturation and decrease in the chain length of esterified fatty acids are known to lower the phase transition temperature of membrane lipids (Chapman *et al.*, 1974). The changes in fatty acid composition with growth temperature can be regarded as an adaptive response to changes in the ambient temperature (Ono and Murata, 1982; Murata *et al.*, 1984).

2. Temperature Shift

When the growth temperature is suddenly increased or decreased, the fatty acid composition of *Anabaena variabilis* is rapidly altered (Sato and Murata, 1980a,b). Figure 4 shows the temperature shift-induced changes in the fatty acid composition of MGDG and DGDG. For 10 h after the temperature shift from 38° to 22°C, the total amount of lipids stays at a constant level (Sato and Murata, 1980a), but a decrease in 16:0 and a concomitant increase in 16:1 take place at the C-2 position of MGDG. Then, as fatty acid and lipid synthesis resumes, the ratio of 16:0 to 16:1 is slowly restored to that seen prior to the temperature shift. This type of transient desaturation of 16:0 to 16:1 is not observed in the other major lipid classes such as DGDG (see Fig. 4). The rapid and transient introduction of a cis double bond into 16:0 which takes place at the C-2 position in MGDG is regarded as an emergency acclimation to compensate for the decrease in membrane fluidity due to the decrease in temperature (Sato and Murata, 1980a,b). The rapid change in levels of

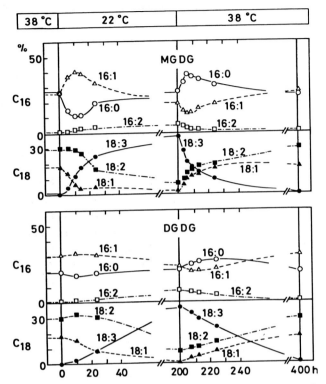

Fig. 4. Temperature shift-induced changes in fatty acid composition of MGDG and DGDG in *Anabaena variabilis*. Fatty acids: 16:0 (○); 16:1 (△); 16:2 (□); 18:1 (▲); 18:2 (■); α-18:3 (●) [Reconstructed from the data in Sato and Murata (1980a).]

unsaturation of the C_{16} acids after the temperature shift does not require *de novo* synthesis of fatty acids (Sato and Murata, 1981). The mass-spectrometric analysis of ^{13}C-enriched MGDG indiates that during the change in unsaturation 16:0 is desaturated in the lipid-bound form and not replaced by previously synthesized 16:1 (Sato *et al.*, 1986). Molecular oxygen, but not light, is necessary for the desaturation (Sato and Murata, 1981). The findings that desaturation is greatly suppressed by chloramphenicol, a protein synthesis inhibitor, and rifampicin, an RNA-synthesis inhibitor, suggest that a fatty acid desaturase, specific to 16:0 at the C-2 position of MGDG, is transiently synthesized after the downward shift in temperature (Sato and Murata, 1981). Decreases in the contents of 18:1 and 18:2 and an increase in α-18:3 take place in MGDG, SQDG, and PG, but these occur more slowly than the changes in C_{16} acids of MGDG. DGDG most slowly responds to the temperature shift (Fig. 4). The desaturation of 18:1 and 18:2 after a temperature shift is also abolished by inhibitors of protein synthesis or RNA synthesis (Sato and Murata, 1981). The amount of 18:1/16:0 molecular species of

MGDG decreases within 10 h after the temperature shift while those of SQDG and PG do so more slowly (Sato and Murata, 1980a).

An increase in growth temperature in *Anabaena variabilis* transiently stimulates *de novo* synthesis of fatty acids but suppresses the desaturation of existing lipids to produce 16:1, 16:2 and α-18:3 (Sato and Murata, 1980a). As a result, the relative content of 16:0 rapidly increases and that of 16:1 rapidly decreases, but these changes are slowly reversed after *de novo* fatty acid synthesis and desaturation restore the original steady-state levels (Fig. 4). The rapid and transient changes in unsaturation level of the C_{16} acids occur exclusively in MGDG. Slower changes in the unsaturation level of the C_{18} acids take place in all the major lipid classes (e.g., as in the case of DGDG shown in Fig. 4.)

In *Anacystis nidulans*, a downward temperature shift from 38° to 28°C produces both rapid and slow changes in the fatty acid composition (Murata *et al.*, 1979). However, in contrast to *Anabaena variabilis*, almost all of the 16:0 at the C-1 positions of MGDG, SQDG, and PG is converted to 16:1 within 10 h after the temperature shift. Decreases occur in chain lengths of the monounsaturated fatty acids at the C-1 positions of all the lipid classes. Also desaturation of 16:0 to 16:1 at the C-2 positions of MGDG and DGDG occurs, but only slowly. It is reasonable to assume that the rapid desaturation of 16:0 to 16:1 at the C-1 position of the lipids plays an important role in the acclimation of *A. nidulans* to the lower temperature.

B.   Other Environmental Factors

Piorreck *et al.* (1984) studied the effect of various concentrations of NO_3^- on the fatty acid composition of four species of blue-green algae. In *Anacystis nidulans* and *Spirulina platensis*, the NO_3^- concentration affected the fatty acid composition; an increase in NO_3^- concentration decreases 16:0 and increases 16:1 in the former alga, and increases 18:2 and decreases γ-18:3 in the latter. On the other hand, in *Microcystis aeruginosa* and *Oscillatoria rubescens*, the NO_3^- concentration had no effect on the fatty acid composition. Kalacheva and Trubachev (1981) studied the effect of different nitrogen sources on the fatty acid composition of total lipids from *Synechococcus elongatus*. The fatty acid compositions were almost the same in cultures grown with KNO_3, $(NH_4)_2HPO_4$, or urea.

Some species of blue-green algae can grow heterotrophically in darkness. The fatty acid composition of heterotrophically grown cells of species in the second group (see Section II,B) is similar to that of the autotrophically grown cells of the same species (Kenyon *et al.*, 1972).

Olson and Ingram (1975) compared the fatty acid composition of exponential and stationary-phase cultures of *Agmenellum quadruplicatum*. At growth temperature of 39°C, the stationary-phase cells contained more amounts of 16:1 and 18:1 and less amounts of 18:2 and 18:3 than the exponential-phase cells. At 30°C, in contrast, the degree of unsaturation of fatty acids was higher in the stationary-phase cells than the exponential-phase cells. Gusev *et al.* (1980) studied the fatty acid composition of *Anabaena variabilis* as a function of culture age. With increasing age the levels of 16:0 and 18:2 dramatically decreased and that of 18:1 markedly increased, whereas the content of 16:1 remained essentially constant. In contrast, Piorreck and Pohl (1984) observed that the fatty acid composition of *M. aeruginosa* remained almost constant throughout growth phase of a batch culture.

The effect of light intensity and quality on the lipid and fatty acid composition was studied by Döhler and Datz (1980) and by Datz and Döhler (1981) by exposing *Anacystis nidulans* to low-intensity white light, high-intensity white light, and high-intensity red light. High-intensity white light causes a decrease in MGDG and an increase in DGDG with respect to the amounts seen in low-intensity white light. In contrast, high-intensity red light induces an increase in amount of MGDG and decrease in DGDG with respect to the low-intensity white light. In all four major lipids, the amounts of 16:0 and 18:1 increase while 16:1 decreases with increase in intensity of white light.

V. NONGLYCEROLIPIDS

A. Heterocyst Glycolipids

Nichols and Wood (1968b) first reported unsaponifiable glycolipids in the nitrogen-fixing blue-green algae, *Anabaena cylindrica*, *Anabaena flos-aquae*, and *Chlorogloea fritschii*. These glycolipids appeared in both filamentous heterocystous strains (Walsby and Nichols, 1969; Wolk and Simon, 1969; Reddy, 1983), and in nitrogen-fixing unicellular strains (Lorch and Wolk, 1974), when they were grown under nitrogen-fixing conditions. In *A. cylindrica* the glycolipids account for 4.4% of the total lipids and 0.5% of the total dry cell weight (Bryce *et al.*, 1972). They are localized in the laminated layer of the heterocyst envelope (Winkenbach *et al.*, 1972), and are termed heterocyst glycolipids.

The chemical structures of the heterocyst glycolipids studied by Bryce *et al.* (1972) and Lambein and Wolk (1973) are presented in Fig. 5. Major components are α-glycosides of very long-chain tri- or tetrahydroxy fatty alcohols. In addition to these glycosidic glycolipids, Lambein and Wolk (1973) discovered α-glycosyl esters of very long-chain mono- and dihydroxy fatty acids as minor components.

Abreu-Grobois *et al.* (1977) studied the biosynthesis of the heterocyst glycolipids in *Anabaena variabilis* during heterocyst formation in a nitrate-free medium. Under these conditions, [^{14}C]acetate is incorporated into both sugar and hydrocarbon moieties of the glycolipids. About 90% of the total radioactivity incorporated into the heterocyst glycolipids is found in the glycosidic glycolipids during the early stages of heterocyst formation. Later during the development process 70% of the radioactivity is in the glycosidic glycolipids. These results suggest that the activity for the glycosidic and glycosyl ester glycolipid formation is a function of the maturity of the heterocysts. Abreu-Grobois *et al.* (1977) also observed that radioactivity is incorporated into 1-hexacosanol as well as 1,25- and 1,3-hexacosanediol, even though 1,3,25-hexacosanetriol is the only major aglycone of the

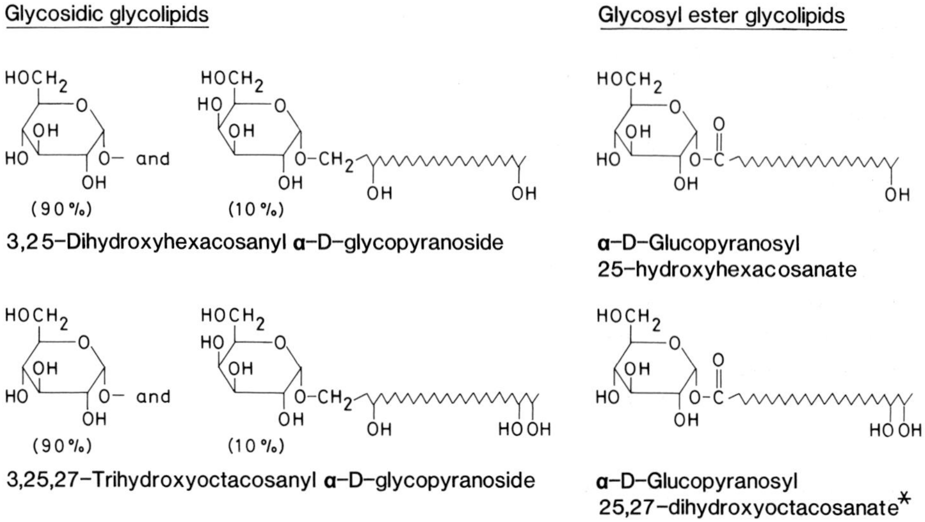

Glycosidic glycolipids   Glycosyl ester glycolipids

3,25-Dihydroxyhexacosanyl α-D-glycopyranoside

α-D-Glucopyranosyl
25-hydroxyhexacosanate

3,25,27-Trihydroxyoctacosanyl α-D-glycopyranoside

α-D-Glucopyranosyl
25,27-dihydroxyoctacosanate[*]

Fig. 5. Structures of heterocyst glycolipids (Bryce *et al.*, 1972; Lambein and Wolk, 1973). The glycosyl ester glycolipids also contain lesser amounts of α-D-galactopyranosyl derivatives. *The positions of hydroxy groups on the hydrocarbon chain are tentatively determined (Lambein and Wolk, 1973).

glycosidic glycolipids in the heterocysts. This suggests that hydroxylation of the aliphatic chain at the C-3 and C-25 positions takes place after the fatty alcohol is linked to glycosides. Krespki and Walton (1983) compared the biosynthesis of heterocyst glycolipids and glycerolipids during heterocyst formation. After cell cultures were transferred to nitrogen-fixing conditions heterocyst glycolipid synthesis increased for the first 6 h, reached a maximum at 15–28 h, then declined. In contrast, the activity of glycerolipid synthesis remained almost constant during heterocyst formation. Therefore, the biosynthesis of the hydrocarbon moieties of heterocyst glycolipids is regulated independently of the biosynthesis of the fatty acids. Krepski and Walton (1983) also suggested that an enzyme system for the hydroxylation of aliphatic hydrocarbon chains of heterocyst glycolipids is transiently activated during heterocyst formation.

Heterocyst formation and heterocyst glycolipid synthesis are both accelerated by 7-azatriptophan (Krepski and Walton, 1979). Mohy-Ud-Dhin *et al.* (1982) suggested that 7-azatriptophan accelerates one or several steps in the biosynthetic pathway of the very long-chain primary alkanol. Insertion of a hydroxyl group at the C-25 position of the fatty alcohol is likely to limit the biosynthesis of the glycoside containing 1,3,25-hexacosanetriol in the presence of 7-azatriptophan (Mohy-Ud-Dhin *et al.*, 1982).

B. Sterols

Several earlier reviews have discussed the sterols of blue-green algae (Nichols, 1973; Nes and McKean, 1977; Nes and Nes, 1980).

Reitz and Hamilton (1968) detected cholesterol and 24-ethylcholesterol in the unsaponifiable lipid fractions of *Anacystis nidulans* and *Fremyella diplosiphon*, and de Souza and Nes (1968) isolated various derivatives of 24-ethylsterols from acetone extracts of *Phormidium luridum*. Since then the occurrence of sterols has been reported in various strains of blue-green algae (Martinez-Nadal, 1971; Forin *et al.*, 1972; Paoletti *et al.*, 1976b; Duperon *et al.*, 1980). However, sterols represent only 0.001–0.03% of the dry weight of blue-green algal cells (R. Durepon, personal communication). Table VI summarizes the sterol composition reported in several strains of blue-green algae. Little is known about the chemical form of sterols in blue-green algae. Duperon *et al.* (1980) reported that *Nostoc commune* contains, in addition to free sterols, conjugated sterols, such as esterified sterols, steryl glycosides, and acylated steryl glycosides. In *A. nidulans* 80% of the sterol content is cholesterol; apparently none of the sterol is conjugated (R. Dureron, personal communication). In most species of blue-green algae examined, cholesterol is a minor component, whereas 24-methylsterols and/or 24-ethylsterols, at different degrees of unsaturation, predominate (Table VI). Nichols (1973) reported that the sterol content of *Ch. fritschii* varied quantitatively and qualitatively with the nitrogen source (Table VI).

C.   Hydrocarbons

Hydrocarbons are found in blue-green algae (Han *et al.*, 1968a,b; Oró *et al.*, 1967; Winters *et al.*, 1969; Jones, 1969; Gelpi *et al.*, 1970; Paoletti *et al.*, 1976a). The most abundant are straight-chain compounds with a chain length varying from C_{15} to C_{19}, among which *n*-heptadecane is predominant (Table VII). Branched hydrocarbons are also found, in which 7- and 8-methylheptadecanes are the major components. Some marine strains of blue-green algae contain an unsaturated hydrocarbon, nonadecene, as a principal hydrocarbon (Winters *et al.*, 1969). *Anacystis montana* also contains unsaturated hydrocarbons of C_{21}–C_{27} chain length.

Hydrocarbons are produced via decarboxylation of fatty acids in blue-green algae. Han *et al.* (1969) showed that, in *N. muscorum*, stearate is enzymatically decarboxylated to produce *n*-heptadecane. McInnes *et al.* (1980) confirmed this pathway of *n*-heptadecane synthesis in *Anacystis nidulans*. Han *et al.* (1969) suggested that the methyl group of methylheptadecane in *N. commune* originates from a methionine methyl group, which might be added to the double bond of Δ^{11}-*cis*-octadecenoic acid. Fehler and Light (1970) showed that in *A. nidulans* no cyclopropane intermediate such as 11,12-methyleneoctadecanoic acid, as observed in *Lactobacillus arabinosus* (Liu and Hofmann, 1962), is involved in the addition of a methyl group to the double bond.

D.   Lipopolysaccharides

The outer membrane of blue-green algae contains lipid-containing heteropolymers, that is, lipopolysaccharides (Weise *et al.*, 1970; Weckesser *et al.*, 1974; Buttke

TABLE VI

Sterol Composition of Blue-Green Algae

	Sterol									
	24-H[a]	24-Methyl[a]			24-Ethyl[a]				24-Ethylidene[a]	
Organism	Δ5	Δ5	Δ5,22	Δ7	Δ5	Δ5,22	Δ7	Δ7,22	Δ5	Unidentified
Spirulina platensis[b]	7	1	3	tr[c]	51	7	2	1	22	6
Calothrix sp.[b]	10	7	20	7	8	10	3	24	11	—
Nostoc commune[b]	10	3	1	19	4	4	13	42	4	—
Chlorogloea fritschii[d]	15	—	—	—	—	75	—	—	—	10
Chlorogloea fritschii[e]	70	—	—	—	—	—	—	—	—	30

[a]The substituent at the C-24 of cholestan-3-ol backbone.
[b]Paoletti *et al.* (1976b).
[c]Trace.
[d]Nitrogen supplied as NH_4^+ (Nichols, 1973).
[e]Nitrogen supplied as NO_3^- (Nichols, 1973).

TABLE VII
Hydrocarbon Composition of Blue-Green Algae

Organism[a]	Hydrocarbon							
	$C_{15:0}$	$C_{16:0}$	$C_{17:0}$	$C_{17:1}$	$CH_3-C_{17:0}$[b]	$C_{18:0}$	$C_{19:0}$	Unidentified
Oscillatoria williamsii (1)	9	tr[c]	91	tr	0	tr	0	0
Anacystis nidulans (2)	23	8	44	20	0	2	0	3
Spirulina platensis (2)	10	20	70	tr	0	0	0	0
Lyngbya aestuarii (2)	2	6	35	0	38	1	0	17
Nostoc muscorum (3)	tr	tr	83	0	16	tr	0	1
Calothrix sp. (4)	tr	tr	47	tr	1	31	10	1

[a]References: 1. Winters *et al.* (1969); 2. Gelpi *et al.* (1970); 3. Han *et al.* (1968b); 4. Paoletti *et al.* (1976a).
[b]Mixture of 7- and 8-methylheptadecane.
[c]Trace.

and Ingram, 1975; Katz *et al.*, 1977; Mikheyskaya *et al.*, 1977; Keleti *et al.*, 1979; Jones and Yopp, 1979; Schmidt *et al.*, 1980a,b; Raziuddin *et al.*, 1983; Jürgens *et al.*, 1985; Tornabene *et al.*, 1985; see also reviews in Weckesser *et al.*, 1979; Drews and Weckesser, 1982). *Anabaena flos-aquae*, however, lacks lipopolysaccharides (Wang and Hill, 1977).

When dispersed in water, lipopolysaccharides form colloidal particles. The isopycnic density of the lipopolysaccharide particles from *M. aeruginosa* is 1.49 g/cm^3 in CsCl solution (Raziuddin *et al.*, 1983).

The fatty acid composition of the lipopolysaccharides from several blue-green algae is presented in Table VIII. These lipopolysaccharides contain various 3-hydroxy straight- and branched-chain fatty acids (Schmidt *et al.*, 1980a,b), in contrast to the lipopolysaccharides from *Escherichia coli* and related bacteria, in which 3-hydroxymyristic acid is the prominent 3-hydroxy fatty acid (Galanos *et al.*, 1977). The 3-hydroxy fatty acids are presumed to form amide linkages to the lipopolysaccharides (Weckesser *et al.*, 1974; Katz *et al.*, 1977; Mayer and Weckesser, 1984). However, no 3-hydroxy fatty acids are detected in *Phormidium* strains (Mikheyskaya *et al.*, 1977) or in *Aphanothece halophytica* (Jones and Yopp, 1979). Straight-chain fatty acids are also detected in the lipopolysaccharides from blue-green algae, and are assumed to bind to the lipopolysaccharides by an ester linkage (Weckesser *et al.*, 1974; Katz *et al.*, 1977; Mayer and Weckesser, 1984).

Mild acid hydrolysis of the lipopolysaccharides from Gram-negative bacteria liberates a lipid moiety (termed lipid A), which is composed of fatty acids, glucosamine, and phosphorus (Galanos *et al.*, 1977). Lipid A is proposed to bind to a polysaccharide residue via an acid-labile sugar, 2-keto-3-deoxyoctonate (Galanos *et al.*, 1977). Similar lipid moieties (also termed lipid A) are separated by mild acid hydrolysis of the lipopolysaccharides of *Anacystis nidulans* (Mayer and Weckesser, 1984). However, this lipid A is not released from all blue-green algal lipopolysaccharides (Weckesser *et al.*, 1974; Mikheyskaya *et al.*, 1977; Schmidt *et*

TABLE VIII
Fatty Acid Composition of Lipopolysaccharides of Blue-Green Algae

Organism[a]	FA[b] (%)	Normal fatty acid					3-Hydroxy fatty acid						Unidentified fatty acid
		14:0	16:0	18:0	16:1	18:1	14:0	16:0	18:0	Anteiso 15:0	Iso 17:0	Anteiso 17:0	
Anabaena variabilis (1)	11	—	14	—	—	—	29	16	16	—	—	—	15
Anacystis nidulans (2)	12	—	2	5	—	—	5	64	—	—	—	—	24
Synechococcus													
PCC6307 (3)	21	21	46	2	5	—	7	13	—	2	0	3	1
PCC6603 (3)	8	tr[c]	5	4	tr	—	0	73	—	0	17	0	1
Synechocystis													
PCC6807 (4)	9	6	5	—	tr	6	21	48	—	5	—	—	9
Aphanothece halophytica (5)[d]	—	8	21	11	44	—	—	—	—	—	—	—	16

[a] References: 1. Weckesser et al. (1974); 2. Katz et al. (1977); 3. Schmidt et al. (1980a); 4. Schmidt et al. (1980b); 5. Jones and Yopp (1979).
[b] Fatty acid content relative to the total lipopolysaccharides.
[c] Trace
[d] 10:0 (3%) and 12:0 (4%) were also detected.

al., 1980b). Presumably, the 2-keto-3-deoxyoctonate content in the lipopolysaccharides is very low (Schmidt *et al.*, 1980b; Weckesser *et al.*, 1974). The phosphorus content of the algal lipid A is much lower than that of Gram-negative bacterial lipid A. Schmidt *et al.* (1980a,b) suggested that the lipid moieties from lipopolysaccharides of *Synechococcus* and *Synechocystis* strains differ from those of most Gram-negative bacteria.

VI. LIPID PHASE OF MEMBRANES

A. Physical Phase of Membrane Lipids

1. Anacystis nidulans

Phase transitions in the membrane lipids has long been proposed as a primary event in chilling injury of higher plants of tropical origin (Lyons, 1973; Raison, 1973). However, such transition has not been clearly demonstrated, probably because of the complicated membrane systems in higher plant cells. Blue-green algae are similar to chloroplasts of eukaryotic plants in their membrane structure, and therefore can be regarded as model systems for the study of molecular mechanisms of low-temperature stress in the plants.

Temperature-dependent changes in the physical phases of membrane lipids in blue-green algae have been studied by several techniques such as differential scanning calorimetry, X-ray diffraction analysis, spin probe electron-paramagnetic resonance, freeze–fracture electron microscopy, carotenoid absorption, chlorophyll absorption, chlorophyll fluorescence, and carotenoid photooxidation. Some of these techniques can be applied to the plasma membranes, and others to the thylakoid membranes.

Freeze–fracture electron microscopy was used to observe the phase separation state of the plasma and thylakoid membranes of *Anacystis nidulans* by Verwer *et al.* (1978), Armond and Staehelin (1979), Brand *et al.* (1979), and Furtado *et al.* (1979). Temperatures for the onset of phase separation in the plasma membranes determined by this method (Ono and Murata, 1982) are presented in Table IX. However, the method cannot be applied to determine the temperature for the onset of phase separation in the thylakoid membrane, since this membrane is fractured only when a large proportion of the membrane area is in the gel state (Ono and Murata, 1982).

Yamamoto and Bangham (1978) observed that a xanthophyll, zeaxanthin, incorporated into phosphatidylcholine liposomes reveals a characteristic absorption change when the liposome membrane goes from a liquid crystalline to a phase separation state. A very similar absorption change can be measured in intact cells, membrane fragments, and extracted lipids of *Anacystis nidulans* when they are exposed to low temperature (Brand, 1977, 1979; Ono and Murata, 1981a). The spectral change is observed predominantly in the plasma membrane preparation in

TABLE IX

Temperatures for the Onset of Phase Separation of Plasma and Thylakoid Membranes in *Anacystis nidulans* Grown at 28° and 38°C and *Anabaena variabilis* Grown at 22° and 38°C

Method	Sample[a]	A. nidulans 28°C PM[b]	28°C TM[b]	38°C PM	38°C TM	A. variabilis 22°C PM	22°C TM	38°C PM	38°C TM	References
Freeze–fracture electron microscopy	C	5	0	16	0	<0	—	<0	—	Ono and Murata (1982)
Carotenoid absorption	C	5	0	15	0	—	—	—	—	Ono and Murata (1981a); Vigh et al. (1985)
Chlorophyll *a* fluorescence	C	0	13	0	25	—	—	—	—	Murata and Fork (1975); Murata and Ono (1981); Vigh and Joó (1983)
Chlorophyll *a* absorption	C	0	15[c]	0	24[d]	—	—	—	—	Hoshina et al. (1984)
Carotenoid photooxidation	C	0	0	0	25	—	—	—	—	Szalontai and Csatorday (1980)
Spin probe	M	5	14	13	23	—	7	—	17	Wada et al. (1984); Murata et al. (1975)
X-Ray diffraction	M	0	16	0	26	—	—	—	—	Tsukamoto et al. (1980)
Differential scanning	L	0	13	0	25	—	—	—	—	Ono et al. (1983)
calorimetry		0	19	0	26	—	—	—	—	Mannock et al. (1985b)

[a]C, Intact cells; M, isolated membranes; L, extracted lipids.
[b]PM, Plasma membrane; TM, thylakoid membranes.
[c]Cells grown at 25°C.
[d]Cells grown at 39°C.

which zeaxanthin amounts to 70% of the total carotenoids, but little change is observed in the thylakoid membrane preparations in which β-carotene is the major carotenoid component (Omata and Murata, 1983). These findings suggest that the carotenoid absorption change is an indicator for the lipid phase of the plasma membrane but not that of the thylakoid membranes. The temperatures for the onset of phase separation of the plasma membrane determined by carotenoid absorption are presented in Table IX. Gombos and Vigh (1986) reached the same conclusion by using the nitrate-starved cells of *A. nidulans* which retain a normal plasma membrane, but contain highly degraded thylakoid membranes.

In model membrane systems composed of synthetic lipids and chlorophyll *a*, the chlorophyll *a* absorption in the red region (Hoshina, 1981, 1983) and the chlorophyll *a* fluorescence (Lee, 1975; Knoll *et al.*, 1980) show characteristic changes at phase transition temperature of the membranes. Chlorophyll *a* is localized in the thylakoid membranes, but not in the plasma or outer membranes (Omata and Murata, 1983, 1984a), and therefore can be regarded as a probe for lipid phase of the thylakoid membranes. The temperature dependence of fluorescence yield in intact cells shows a characteristic peak at temperatures (Table IX), which correspond to those for the onset of phase separation (Murata and Fork, 1975; Murata and Ono, 1981; Vigh and Joó, 1983). The temperature dependence of chlorophyll *a* absorption in the thylakoid membranes from *Anacystis nidulans* reveal shoulders or peaks (Hoshina *et al.*, 1984) which may correspond to the onset of phase separation in the membranes.

The spin probe method can be applied to the plasma membrane and thylakoid membranes prepared from *Anacystis nidulans* (Wada *et al.*, 1984; Murata *et al.*, 1975). Clear breaks in the Arrhenius plot of the rotational correlation time of a fatty acid spin probe are observed (Wada *et al.*, 1984), indicating an abrupt change occurring in the fluidity of the membrane lipids (i.e., the phase transition; Table IX).

X-Ray diffraction of lipid bilayers gives a sharp reflection line corresponding to a Bragg spacing of 4.2 Å when the membrane is in the gel state and a diffuse line corresponding to a spacing of 4.5 Å when they are in the liquid crystalline state (Rivas and Luzzati, 1969). Tsukamoto *et al.* (1980) studied the temperature dependence of the 4.2- and 4.5-Å reflections in thylakoid membranes, and suggested that the membranes from *Anacystis nidulans* grown at 28°C are in the phase separation state between 16° and −5°C, while those from cells grown at 38°C are in the phase separation state between 26° and 0°C (Table IX).

Physical phase transitions can also be studied by differential scanning calorimetry (Melchior and Steim, 1976). This method is not sensitive enough to be applied to the thylakoid or plasma membranes of intact cells but is applicable to liposomes made of lipids extracted from the algal cells (Ono *et al.*, 1983). Temperatures for the onset of thermotropic phase transition of the liposomes are presented in Table IX. Since almost all the lipids from the algal cells originate from the thylakoid membranes, the phase of the extracted lipids is presumed to correspond to the phase transition of thylakoid membranes. However, the lipid phase of liposomes made from extracted lipids may not directly correspond to the membrane phase, because the membrane proteins, which modify the lipid phase transition temperature of the biological membranes, are excluded.

The temperatures critical for the onset of phase separation of the plasma membrane and thylakoid membranes studied by various methods are summarized in Table IX. The methods of freeze–fracture electron microscopy, carotenoid absorption, and spin labeling provide nearly the same values for the temperature of the onset of phase separation in the plasma membranes from cells grown at 28° and 38°C. The six methods, chlorophyll fluorescence, chlorophyll absorption, carotenoid photooxidation, spin labeling, X-ray diffraction, and differential scanning calorimetry, all provide about the same values for the temperature of the onset of phase separation in the thylakoid membranes from cells grown at 28°C or 38°C, respectively. According to the data summarized in Table IX, the following facts are clearly suggested with respect to the lipid phase of the membranes in *Anacystis nidulans*: (1) At growth temperatures both types of membranes are in the liquid crystalline state. (2) With decrease in temperature to below those which permit growth, the thylakoid membranes first enter the phase separation state. (3) About 10°C below the onset of phase separation in the thylakoid membranes, the plasma membrane enters the phase separation state. (4) The temperatures for the onset of phase separation of both types of membranes depend on the growth temperature.

Irrespective of different features of the phase transition of the plasma and thylakoid membranes, there is no discernible difference between the plasma membranes and thylakoid membranes with respect to their polar lipid and fatty acid composition (Omata and Murata, 1983). The higher phase transition temperature of the thylakoid membrane with respect to the plasma membrane may be due to the high protein content of the thylakoid membranes, 70% of the dry weight, compared with the corresponding value of 40% found in plasma membranes (Omata and Murata, 1983). The shift in phase transition temperature due to the growth temperature corresponds to the growth temperature-dependent alteration in the fatty acid composition (Holton *et al.*, 1964; Sato *et al.*, 1979).

Mannock *et al.* (1985a,b) studied the phase behavior of MGDG, DGDG, SQDG, and PG isolated from *Anacystis nidulans* by differential scanning calorimetry and X-ray diffraction. In a cooling scan of aqueous dispersions of MGDG, DGDG, SQDG, and PG from cells grown at 38°C, the phase separation begins to appear at 17°C, 11°C, 15°C and 13°C, respectively.

2.   Other Blue-Green Algae

Anabaena variabilis is tolerant of low temperatures. Freeze–fracture electron microscopy reveals that the plasma membrane of cells grown at 22° and 38°C are both in the liquid crystalline state at 0°C (Ono and Murata, 1982). The electron-paramagnetic resonance signal of a spin probe suggests that the thylakoid membranes from cells grown at 22°C and 38°C enter the phase separation state at ~5°C and ~15°C, respectively (Wada *et al.*, 1984). Again in this alga, the temperature-dependent phase transition depends on the growth temperature. This is related to the growth temperature-dependent decrease in saturation level of their component fatty acids (Sato *et al.*, 1979).

The phase transition temperature of the thylakoid membranes of *Synechococcus lividus* was studied by chlorophyll *a* fluorescence (Fork *et al.*, 1979). It entered the phase separation state at 40°–43°C and 22°–25°C in cells grown at 55° and 38°C, respectively. The lipid phase of plasma membrane of *S. lividus* and *Synechococcus* strain 6910 was examined by freeze–fracture electron microscopy (Golecki, 1979). This membrane of *S. lividus* grown at 52°C was in the phase separation state below 30°C, and that of *Synechococcus* strain 6910 was below 15°C.

B. Physiological Effect of Membrane Lipid Phases

A number of studies have demonstrated that *Anacystis nidulans* is susceptible to low temperature, since Forrest *et al.* (1957) first observed the release of pteridines and glutamic acid from intact cells, and a corresponding loss of photosynthetic activity, at 4°C. When the algal cells are exposed to temperatures near 0°C, viability declines (Rao *et al.*, 1977), and activities of overall photosynthesis, photosynthetic electron transport, and phosphorylation all diminish (Forrest *et al.*, 1957; Jansz and MacLean, 1973; Rao *et al.*, 1977; Ono and Murata, 1981a; Vigh and Joó, 1983; Vigh *et al.*, 1985). Also, potassium ions, glutamate, and pteridines are released from the cells (Forrest *et al.*, 1957; Jansz and MacLean, 1973; Ono and Murata, 1981b; Vigh *et al.*, 1985), and morphological alterations of the plasma and the thylakoid membranes are detectable (Brand *et al.*, 1979). These low temperature-dependent phenomena are irreversible, and even when the cells are again warmed to growth temperature, only partial recoveries or none at all are observed. The critical temperatures for irreversible damage of the cells depend on growth temperature (Rao *et al.*, 1977; Ono and Murata, 1981a,b; Vigh *et al.*, 1985), suggesting that the temperature-dependent alteration of membrane lipids (Holton *et al.*, 1964; Sato *et al.*, 1979) may be involved in the susceptibility to low temperature.

Temperatures critical for irreversible damage and ion leakage, and for breaks in Arrhenius plots are listed in Table X. The critical temperatures for the irreversible damage of photosynthesis and the Hill reaction are 5° and 15°C in cells grown at 28° and 38°C, respectively, and correspond to the onset of phase separation in the plasma membranes. At about the same temperatures, potassium ions, free amino acids (Ono and Murata, 1981b), and electrolytes (Murata *et al.*, 1984) begin to leak from the algal cells to the surrounding medium. In contrast, breaks in the Arrhenius plots of photosynthesis, phosphorylation, and state 1–state 2 transition, and the declines of delayed fluorescence appear at ~15° and 25°C in cells grown at 28° and 38°C, respectively (Murata *et al.*, 1975, 1983; Ono and Murata, 1977, 1979), which correspond to the onset of phase separation in the thylakoid membranes.

Based on these observations, a mechanism as shown in Fig. 6 can be proposed for low temperature-induced irreversible phenomena in *Anacystis nidulans*. The plasma membrane and thylakoid membranes are barriers for ions and small molecules, whereas the outer membrane of the cell envelope is permeable to them. At the

TABLE X

Temperatures for the Characteristic Changes in Physiological Phenomena in *Anacystis nidulans* Grown at 28° and 38°C and *Anabaena variabilis* Grown at 22° and 38°C

Phenomenon	Sample[a]	A. nidulans 28°C	A. nidulans 38°C	A. variabilis 22°C	A. variabilis 38°C	References
Critical for irreversible damage						
Photosynthesis	C	5	16	<0	<0	Ono and Murata (1981a); Murata *et al.* (1984)
Hill reaction	C	5	15	—	—	Ono and Murata (1981a)
Critical for leakage						
Electrolytes	C	8	14	<0	<0	Murata *et al.* (1984)
Potassium ion	C	7	17	—	—	Ono and Murata (1981b)
Amino acids	C	7	17	—	—	Ono and Murata (1981b)
Break in Arrhenius plot						
Photosynthesis	C	14	22	7	15	Murata *et al.* (1983, 1984)
Phosphorylation	M	—	24	—	—	Ono and Murata (1979)

[a]C, Intact cells; M, isolated thylakoid membranes.

growth temperature, the plasma membrane and thylakoid membranes are both in the liquid crystalline state and are impermeable to ions and small molecules. With decrease in temperature, the thylakoid membranes go into the phase separation state and become permeable to ions and small molecules. Under these conditions, physiological activities such as photosynthesis and photosynthetic ATP formation are reversibly diminished as revealed by a break in the Arrhenius plot (Murata *et al.*, 1975, 1983; Ono and Murata, 1979). With a further decrease in temperature, the plasma membrane enters the phase separation state and becomes permeable. Under these conditions, ions and small molecules in the cytoplasm leak out, and those in the surrounding medium leak in. This diminishes cellular metabolism leading to the death of the cell. Even when the temperature is raised to that which supports growth in unchilled cells, the concentrations of ions and small molecules are not recovered, thus resulting in irreversible damage of all physiological activities of the cells.

The suggestion that the irreversible damage at low temperature in *Anacystis nidulans* is induced by a phase transition in the plasma membrane, but not in the thylakoid membrane, has been verified by Vigh and his colleagues. Vigh *et al.* (1985) selectively hydrogenated most of the unsaturated fatty acids in the plasma membrane, but not those of the thylakoid membrane of *A. nidulans*. They observed parallel shifts in the degree of fatty acid saturation, the temperature for the onset of phase separation in the plasma membrane, and the temperature critical for K^+ leakage and the irreversible decline of photosynthesis. Gombos and Vigh (1986), using nitrate-starved cells of *A. nidulans* which contain an intact plasma membrane

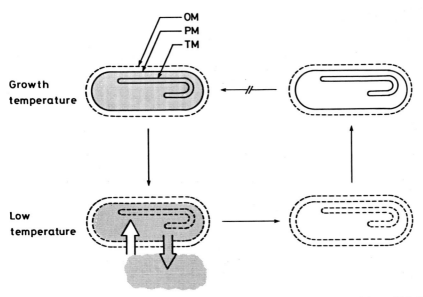

Fig. 6. A scheme for the cold-induced irreversible damage in *Anacystis nidulans*. OM, Outer membrane; PM, plasma membrane; TM, thylakoid membrane. Bold lines indicate membranes impermeable to ions and small molecules; dashed lines, membranes leaky. At growth temperature, the plasma membrane is in the liquid crystalline state and impermeable. With decrease in temperature, the plasma membrane enters the phase separation state and therefore becomes leaky. Under these conditions, ions and small molecules in the cytoplasm leak out and those in the outer medium leak in. Upon rewarming the cells to growth temperature, the plasma membrane resumes the liquid crystalline state and impermeability. Under these conditions, however, the ions and small molecules which have once been leaked out are not restored, leading to the irreversible damage of the cells.

but degraded thylakoid membranes, observed that K^+ leakage occurs at the same temperature as the temperature which induces the onset in phase separation in the plasma membrane.

In *Anabaena variabilis* which, however, is tolerant of low temperature, no irreversible damage to photosynthesis nor electrolyte leakage occurs at 0°C (Murata *et al.*, 1984). These results are related to the finding that the plasma membrane is in the liquid crystalline state above 0°C (Ono and Murata, 1982; Wada *et al.*, 1984). In contrast, the thylakoid membranes enter the phase separation state above 0°C, and a clear break in the Arrhenius plot of photosynthesis appears (Wada *et al.*, 1984; Murata *et al.*, 1984). Therefore, in *A. variabilis* as in *Anacystis nidulans*, phase transition of the thylakoid membranes does not induce irreversible damage in healthy cells, which is consistent with the mechanism for low temperature stress proposed for *A. nidulans*; that is, a phase transition in the plasma membrane is directly related to low-temperature damage.

ACKNOWLEDGMENTS

The authors are grateful to Dr. Margrit Frentzen, Universität Hamburg, Dr. Jerry Brand, University of Texas, and Dr. David G. Bishop, Macquarie University, for their critical reading of the manuscript and correction of English. This work was supported in part by a Grant-in-Aid for Special Research (61129005) from the Japanese Ministry of Education, Science and Culture, to N.M.

REFERENCES

Abreu-Grobois, F. A., Billyard, T. C., and Walton, T. J. (1977). *Phytochemistry* **16**, 351–354.
Al'Araji, Z. T., and Walton, T. J. (1980). *In* "Biogenesis and Function of Plant Lipids" (P. Mazliak, P. Benveniste, C. Costes, and R. Douce, eds.), pp. 259–262. Elsevier/North-Holland, Amsterdam.
Allen, C. F., Hirayama, O., and Good, P. (1966). *In* "Biochemistry of Chloroplasts" (T. W. Goodwin, ed.), Vol. 1, pp. 195–200. Academic Press, London.
Appleby, R. S., Safford, R., and Nichols, B. W. (1971). *Biochim. Biophys. Acta* **248**, 205–211.
Armond, P. A., and Staehelin, L. A. (1979). *Proc. Natl. Acad. Sci. U.S.A.* **76**, 1901–1905.
Auling, G., Heinz, E., and Tulloch, A. P. (1971). *Hoppe-Seyler's Z. Physiol. Chem.* **352**, 905–912.
Baasch, K.-H., Kohlhase, M., and Pohl, P. (1984). *In* "Structure, Function and Metabolism of Plant Lipids" (P.-A. Siegenthaler and W. Eichenberger, eds.), pp. 587–590. Elsevier, Amsterdam.
Block, M. A., Dorne, A.-J., Joyard, J., and Douce, R. (1983). *J. Biol. Chem.* **258**, 13281–13286.
Brand, J. J. (1977). *Plant Physiol.* **59**, 970–973.
Brand, J. J. (1979). *Arch. Biochem. Biophys.* **193**, 385–391.
Brand, J. J., Kirchanski, S. J., and Ramirez-Mitchell, R. (1979). *Planta* **145**, 63–68.
Bryce, T. A., Welti, D., Walsby, A. E., and Nichols, B. W. (1972). *Phytochemistry* **11**, 295–302.
Buttke, T. M., and Ingram, L. O. (1975). *J. Bacteriol.* **124**, 1566–1573.
Chapman, D., Urbina, J., and Keough, K. M. (1974). *J. Biol. Chem.* **249**, 2512–2521.
Datz, G., and Döhler, G. (1981). *Z. Naturforsch.* **C36**, 856–862.
de Souza, N. J., and Nes, W. R. (1968). *Science* **162**, 363.
Döhler, G., and Datz, G. (1980). *Z. Pflanzenphysiol.* **100**, 427–435.
Douce, R., and Joyard, J. (1979). *Adv. Bot. Res.* **7**, 1–116.
Douce, R., Holtz, R. B., and Benson, A. A. (1973). *J. Biol. Chem.* **248**, 7215–7222.
Drews, G., and Weckesser, J. (1982). *In* "The Biology of Cyanobacteria" (N. G. Carr and B. A. Whitton, eds.), pp. 333–358. Blackwell, Oxford.
Duperon, R., Doireau, P., Verger, A., and Duperon, P. (1980). *In* "Biogenesis and Function of Plant Lipids" (P. Mazliak, P. Benveniste, C. Costes, and R. Douce, eds.), pp. 445–447. Elsevier/North-Holland, Amsterdam.
Fehler, S. W. G., and Light, R. J. (1970). *Biochemistry* **9**, 418–422.
Feige, G. B. (1978). *Ber. Dtsch. Bot. Ges.* **91**, 595–602.
Feige, G. B., Heinz, E., Wrage, K., Cochems, N., and Ponzelar, E. (1980). *In* "Biogenesis and Function of Plant Lipids" (P. Mazliak, P. Benveniste, C. Costes, and R. Douce, eds.), pp. 135–140. Elsevier/North-Holland, Amsterdam.
Fischer, W., Heinz, E., and Zeus, M. (1973). *Hoppe-Seyler's Z. Physiol. Chem.* **354**, 1115–1123.
Fogg, G. E., Stewart, W. D. P., Fay, P., and Walsby, A. E. (1973). "The Blue-Green Algae." Academic Press, London.
Forin, M.-C., Maume, B., and Baron, C. (1972). *C. R. Hebd. Seances Acad. Sci., Ser. D* **274**, 133–136.
Fork, D. C., Murata, N., and Sato, N. (1979). *Plant Physiol.* **63**, 524–530.
Forrest, H. S., Van Baalen, C., and Myers, J. (1957). *Science* **125**, 699–700.
Furtado, D., Williams, W. P., Brain, A. P. R., and Quinn, P. J. (1979). *Biochim. Biophys. Acta* **555**, 352–357.
Galanos, C., Lüderitz, O., Rietschel, E. T., and Westphal, O. (1977). *Int. Rev. Biochem.* **14**, 239–335.

Gantt, E., and Conti, S. F. (1969). *J. Bacteriol.* **97**, 1486–1493.

Gelpi, E., Schneider, H., Mann, J., and Oró, J. (1970). *Phytochemistry* **9**, 603–612.

Golecki, J. R. (1979). *Arch. Microbiol.* **120**, 125–133.

Gombos, Z., and Vigh, L. (1986). *Plant Physiol.* **80**, 415–419.

Goodloe, R. S., and Light, R. J. (1982). *Biochim. Biophys. Acta* **710**, 485–492.

Gusev, M. V., Nikitina, K. A., Grot, A. V., and Vshivtsev, V. S. (1980). *Fiziol. Rast.* **27**, 1083–1087.

Han, J., McCarthy, E. D., Van Hoeven, W., Calvin, M., and Bradley, W. H. (1968a). *Proc. Natl. Acad. Sci. U.S.A.* **59**, 29–33.

Han, J., McCarthy, E. D., Calvin, M., and Benn, M. H. (1968b). *J. Chem. Soc. C*, 2785–2791.

Han, J., Chan, H. W.-S., and Calvin, M. (1969). *J. Am. Chem. Soc.* **91**, 5156–5159.

Haselkorn, R. (1978). *Annu. Rev. Plant Physiol.* **29**, 319–344.

Hirayama, O. (1967). *J. Biochem. (Tokyo)* **61**, 179–185.

Holton, R. W., Blecker, H. H., and Onore, M. (1964). *Phytochemistry* **3**, 595–602.

Hoshina, S. (1981). *Biochim. Biophys. Acta* **638**, 334–340.

Hoshina, S. (1983). *Plant Cell Physiol.* **24**, 937–940.

Hoshina, S., Mohanty, P., and Fork, D. C. (1984). *Photosyn. Res.* **5**, 347–360.

Jansz, E. R., and MacLean, F. I. (1973). *Can. J. Microbiol.* **19**, 381–387.

Johns, R. B., Nichols, P. D., Gillan, F. T., Perry, G. J., and Volkman, J. K. (1981). *Comp. Biochem. Physiol. B* **69**, 843–849.

Jones, J. G. (1969). *J. Gen. Microbiol.* **59**, 145–152.

Jones, J. H., and Yopp, J. H. (1979). *J. Phycol.* **15**, 62–66.

Jürgens, U. J., and Weckesser, J. (1985). *J. Bacteriol.* **164**, 384–389.

Jürgens, U. J., Golecki, J. R., and Weckesser, J. (1985). *Arch. Microbiol.* **142**, 168–174.

Kalacheva, G. S., and Trubachev, I. N. (1981). *Fiziol. Rast.* **28**, 519–525.

Katz, A., Weckesser, J., Drews, G., and Mayer, H. (1977). *Arch. Microbiol.* **113**, 247–256.

Keleti, G., Sykora, J. L., Lippy, E. C., and Shapiro, M. A. (1979). *Appl. Environ. Microbiol.* **38**, 471–477.

Kenrick, J. R., Deane, E. M., and Bishop, D. G. (1984). *Phycologia* **23**, 73–76.

Kenyon, C. N. (1972). *J. Bacteriol.* **109**, 827–834.

Kenyon, C. N., Rippka, R., and Stanier, R. Y. (1972). *Arch. Mikrobiol.* **83**, 216–236.

Knoll, W., Baumann, J., Korpiun, P., and Theilen, U. (1980). *Biochem. Biophys. Res. Commun.* **96**, 968–974.

Krepski, W. J., and Walton, T. J. (1979). *Biochem. Soc. Trans.* **7**, 1268–1269.

Krepski, W. J., and Walton, T. J. (1983). *J. Gen. Microbiol.* **129**, 105–110.

Lambein, F., and Wolk, C. P. (1973). *Biochemistry* **12**, 791–798.

Lee, A. G. (1975). *Biochemistry* **14**, 4397–4402.

Lem, N. W., and Stumpf, P. K. (1984a). *Plant Physiol.* **74**, 134–138.

Lem, N. W., and Stumpf, P. K. (1984b). *Plant Physiol.* **75**, 700–704.

Levin, E., Lennarz, W. J., and Bloch, K. (1964). *Biochim. Biophys. Acta* **84**, 471–474.

Liu, T. Y., and Hofmann, K. (1962). *Biochemistry* **1**, 189–191.

Lorch, S. K., and Wolk, C. P. (1974). *J. Phycol.* **10**, 352–355.

Lyons, J. M. (1973). *Annu. Rev. Plant Physiol.* **24**, 445–466.

McInnes, A. G., Walter, J. A., and Wright, J. L. C. (1980). *Lipids* **15**, 609–615.

Mackender, R. O., and Leech, R. M. (1974). *Plant Physiol.* **53**, 496–502.

Mannock, D. A., Brain, A. P. R., and Williams, W. P. (1985a). *Biochim. Biophys. Acta* **817**, 289–298.

Mannock, D. A., Brain, A. P. R., and Williams, W. P. (1985b). *Biochim. Biophys. Acta* **821**, 153–164.

Martinez Nadal, N. G. (1971). *Phytochemistry* **10**, 2537–2538.

Mayer, H., and Weckesser, J. (1984). *In* "Handbook of Endotoxin" (E. T. Rietschel, ed.), Vol. 1, pp. 221–247. Elsevier, Amsterdam.

Melchior, D. L., and Steim, J. M. (1976). *Annu. Rev. Biophys. Bioeng.* **5**, 205–238.

Mikheyskaya, L. V., Ovodova, R. G., and Ovodov, Y. S. (1977). *J. Bacteriol.* **130**, 1–3.

Mohy-Ud-Dhin, M. T., Krepski, W. J., and Walton, T. J. (1982). *In* "Biochemistry and Metabolism of Plant Lipids" (J. F. G. M. Wintermans and P. J. C. Kuiper, eds.), pp. 209–212. Elsevier, Amsterdam.

Murata, N., and Fork, D. C. (1975). *Plant Physiol.* **56,** 791–796.

Murata, N., and Ono, T. (1981). *In* "Photosynthesis" (G. Akoyunoglou, ed.), Vol. 6, pp. 473–481. Balaban Int. Sci. Serv., Philadelphia, Pennsylvania.

Murata, N., and Sato, N. (1982). *In* "Biochemistry and Metabolism of Plant Lipids" (J. F. G. M. Wintermans and P. J. C. Kuiper, eds.), pp. 165–168. Elsevier, Amsterdam.

Murata, N., and Sato, N. (1983). *Plant Cell Physiol.* **24,** 133–138.

Murata, N., Troughton, J. H., and Fork, D. C. (1975). *Plant Physiol.* **56,** 508–517.

Murata, N., Ono, T., and Sato, N. (1979). *In* "Low Temperature Stress in Crop Plants: The Role of the Membrane" (J. M. Lyons, D. Graham, and J. K. Raison, eds.), pp. 337–345. Academic Press, New York.

Murata, N., Sato, N., Omata, T., and Kuwabara, T. (1981). *Plant Cell Physiol.* **22,** 855–866.

Murata, N., Wada, H., Omata, T., and Ono, T. (1983). *In* "Effects of Stress on Photosynthesis" (R. Marcelle, H. Clijsters, and M. van Poucke, eds.), pp. 193–199. Nijhoff/Junk, The Hague.

Murata, N., Wada, H., and Hirasawa, R. (1984). *Plant Cell Physiol.* **25,** 1027–1032.

Nes, W. R., and McKean, M. L. (1977). "Biochemistry of Steroids and Other Isopentenoids," pp. 411–533. University Park Press, Baltimore, Maryland.

Nes, W. R., and Nes, W. D. (1980). "Lipids in Evolution," pp. 136–139. Plenum, New York.

Nichols, B. W. (1968). *Lipids* **3,** 354–360.

Nichols, B. W. (1973). *In* "The Biology of Blue-Green Algae" (N. G. Carr and B. A. Whitton, eds.), pp. 144–161. Univ. of California Press, Berkeley.

Nichols, J. M., and Adams, D. G. (1982). *In* "The Biology of Cyanobacteria" (N. G. Carr and B. A. Whitton, eds.), pp. 387–412. Blackwell Publications, Oxford.

Nichols, B. W., and Wood, B. J. B. (1968a). *Lipids* **3,** 46–50.

Nichols, B. W., and Wood, B. J. B. (1968b). *Nature (London)* **217,** 767–768.

Nichols, B. W., Harris, R. V., and James, A. T. (1965). *Biochem. Biophys. Res. Commun.* **20,** 256–262.

Nishihara, M., Yokota, K., and Kito, M. (1980). *Biochim. Biophys. Acta* **617,** 12–19.

Olson, G. J., and Ingram, L. O. (1975). *J. Bacteriol.* **124,** 373–379.

Omata, T., and Murata, N. (1983). *Plant Cell Physiol.* **24,** 1101–1112.

Omata, T., and Murata, N. (1984a). *Arch. Microbiol.* **139,** 113–116.

Omata, T., and Murata, N. (1984b). *Biochim. Biophys. Acta* **766,** 395–402.

Omata, T., and Murata, N. (1985). *Biochim. Biophys. Acta* **810,** 354–361.

Omata, T., and Murata, N. (1986). *Plant Cell Physiol.* **27,** 485–490.

Ono, T., and Murata, N. (1977). *Biochim. Biophys. Acta* **460,** 220–229.

Ono, T., and Murata, N. (1979). *Biochim. Biophys. Acta* **545,** 69–76.

Ono, T., and Murata, N. (1981a). *Plant Physiol.* **67,** 176–181.

Ono, T., and Murata, N. (1981b). *Plant Physiol.* **67,** 182–187.

Ono, T., and Murata, N. (1982). *Plant Physiol.* **69,** 125–129.

Ono, T., Murata, N., and Fujita, T. (1983). *Plant Cell Physiol.* **24,** 635–639.

Oren, A., Fattom, A., Padan, E., and Tietz, A. (1985). *Arch. Microbiol.* **141,** 138–142.

Oró, J., Tornabene, T. G., Nooner, D. W., and Gelpi, E. (1967). *J. Bacteriol.* **93,** 1811–1818.

Paoletti, C., Pushparaj, B., Florenzano, G., Capella, P., and Lercker, G. (1976a). *Lipids* **11,** 258–265.

Paoletti, C., Pushparaj, B., Florenzano, G., Capella, P., and Lercker, G. (1976b). *Lipids* **11,** 266–271.

Parker, P. L., Van Baalen, C., and Maurer, L. (1967). *Science* **155,** 707–708.

Perry, G. J., Gillan, F. T., and Johns, R. B. (1978). *J. Phycol.* **14,** 369–371.

Piorreck, M., and Pohl, P. (1984). *Phytochemistry* **23,** 217–223.

Piorreck, M., Baasch, K.-H., and Pohl, P. (1984). *Phytochemistry* **23,** 207–216.

Quinn, P. J., and Williams, W. P. (1978). *Prog. Biophys. Mol. Biol.* **34,** 109–173.

Quinn, P. J., and Williams, W. P. (1983). *Biochim. Biophys. Acta* **737,** 223–266.

Raison, J. K. (1973). *J. Bioenerg.* **4,** 285–309.

Rao, V. S. K., Brand, J. J., and Myers, J. (1977). *Plant Physiol.* **59,** 965–969.
Raziuddin, S., Siegelman, H. W., and Tornabene, T. G. (1983). *Eur. J. Biochem.* **137,** 333–336.
Reddy, P. M. (1983). *Biochem. Physiol. Pflanz.* **178,** 575–578.
Reitz, R. C., and Hamilton, J. G. (1968). *Comp. Biochem. Physiol.* **25,** 401–416.
Resch, C. M., and Gibson, J. (1983). *J. Bacteriol.* **155,** 345–350.
Rivas, E., and Luzzati, V. (1969). *J. Mol. Biol.* **41,** 261–275.
Safford, R., and Nichols, B. W. (1970). *Biochim. Biophys. Acta* **210,** 57–64.
Sandelius, A. S., and Selstam, E. (1984). *Plant Physiol.* **76,** 1041–1046.
Sastry, P. S. (1974). *Adv. Lipid Res.* **12,** 251–310.
Sato, N., and Murata, N. (1980a). *Biochim. Biophys. Acta* **619,** 353–366.
Sato, N., and Murata, N. (1980b). *In* "Biogenesis and Function of Plant Lipids" (P. Mazliak, P. Benveniste, C. Costes, and R. Douce, eds.), pp. 207–210. Elsevier/North-Holland, Amsterdam.
Sato, N., and Murata, N. (1981). *Plant Cell Physiol.* **22,** 1043–1050.
Sato, N., and Murata, N. (1982a). *Biochim. Biophys. Acta* **710,** 271–278.
Sato, N., and Murata, N. (1982b). *Biochim. Biophys. Acta* **710,** 279–289.
Sato, N., and Murata, N. (1982c). *Plant Cell Physiol.* **23,** 1115–1120.
Sato, N., Murata, N., Miura, Y., and Ueta, N. (1979). *Biochim. Biophys. Acta* **572,** 19–28.
Sato, N., Seyama, Y., and Murata, N. (1986). *Plant Cell Physiol.* **27,** 819–835.
Schmidt, W., Drews, G., Weckesser, J., Fromme, I., and Borowiak, D. (1980a). *Arch. Microbiol.* **127,** 209–215.
Schmidt, W., Drews, G., Weckesser, J., and Mayer, H. (1980b). *Arch. Microbiol.* **127,** 217–222.
Sherman, L. A. (1978). *J. Phycol.* **14,** 427–433.
Stanier, R. Y., and Cohen-Bazire, G. (1977). *Annu. Rev. Microbiol.* **31,** 225–274.
Stapleton, S. R., and Jaworski, J. G. (1984a). *Biochim. Biophys. Acta* **794,** 240–248.
Stapleton, S. R., and Jaworski, J. G. (1984b). *Biochim. Biophys. Acta* **794,** 249–255.
Szalontai, B., and Csatorday, K. (1980). *J. Mol. Struct.* **60,** 269–272.
Tornabene, T. G., Bourne, T. F., Raziuddin, S., and Ben-Amotz, A. (1985). *Mar. Ecol. Prog. Ser.* **22,** 121–125.
Tsukamoto, Y., Ueki, T., Mitsui, T., Ono, T., and Murata, N. (1980). *Biochim. Biophys. Acta* **602,** 673–675.
Tulloch, A. P., Heinz, E., and Fischer, W. (1973). *Hoppe-Seyler's Z. Physiol. Chem.* **354,** 879–889.
Verwer, W., Ververgaert, P. H. J. T., Leunissen-Bijvelt, J., and Verkleij, A. J. (1978). *Biochim. Biophys. Acta* **504,** 231–234.
Vigh, L., and Joó, F. (1983). *FEBS Lett.* **162,** 423–427.
Vigh, L., Gombos, Z., and Joó, F. (1985). *FEBS Lett.* **191,** 200–204.
Wada, H., Hirasawa, R., Omata, T., and Murata, N. (1984). *Plant Cell Physiol.* **25,** 907–911.
Walsby, A. E., and Nichols, B. W. (1969). *Nature (London)* **221,** 673–674.
Wang, A. W., and Hill, A. (1977). *J. Bacteriol.* **130,** 558–560.
Weckesser, J., Katz, A., Drews, G., Mayer, H., and Fromme, I. (1974). *J. Bacteriol.* **120,** 672–678.
Weckesser, J., Drews, G., and Mayer, H. (1979). *Annu. Rev. Microbiol.* **33,** 215–239.
Weenink, R. O., and Shorland, F. B. (1964). *Biochim. Biophys. Acta* **84,** 613–614.
Weise, G., Drews, G., Jann, B., and Jann, K. (1970). *Arch. Mikrobiol.***71,** 89–98.
Winkenbach, F., Wolk, C. P., and Jost, M. (1972). *Planta* **107,** 69–80.
Winters, K., Parker, P. L., and Van Baalen, C. (1969). *Science* **163,** 467–468.
Wolk, C. P. (1973). *Bacteriol. Rev.* **37,** 32–101.
Wolk, C. P., and Simon, R. D. (1969). *Planta* **86,** 92–97.
Yamamoto, H. Y., and Bangham, A. D. (1978). *Biochim. Biophys. Acta* **507,** 119–127.
Zepke, H. D., Heinz, E., Radunz, A., Linscheid, M., and Pesch, R. (1978). *Arch. Microbiol.* **119,** 157–162.

Index

Contents of Other Volumes

Volume 1—The Plant Cell

Volume 4—Lipids: Structure and Function

Volume 5—Amino Acids and Derivatives

Volume 8—Photosynthesis

Volume 10—Photosynthesis

Volume 11—Biochemistry of Metabolism

Volume 12—Physiology of Metabolism

Volume 13—Methodology